1 MONTH OF FREE READING

at

www.ForgottenBooks.com

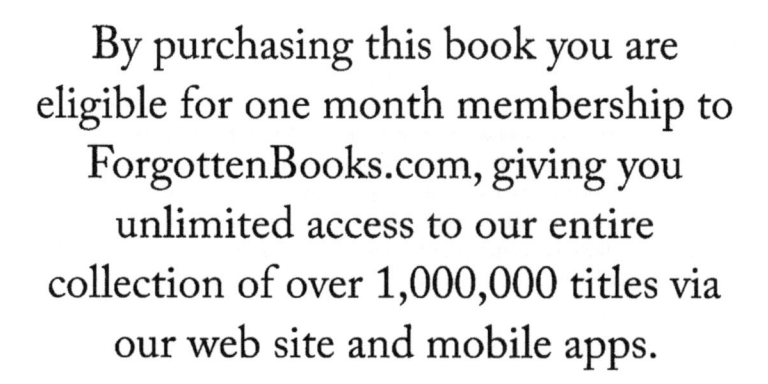

By purchasing this book you are eligible for one month membership to ForgottenBooks.com, giving you unlimited access to our entire collection of over 1,000,000 titles via our web site and mobile apps.

To claim your free month visit:

www.forgottenbooks.com/free712327

* Offer is valid for 45 days from date of purchase. Terms and conditions apply.

ISBN 978-0-267-77332-9
PIBN 10712327

This book is a reproduction of an important historical work. Forgotten Books uses
state-of-the-art technology to digitally reconstruct the work, preserving the original format
whilst repairing imperfections present in the aged copy. In rare cases, an imperfection in
the original, such as a blemish or missing page, may be replicated in our edition. We do,
however, repair the vast majority of imperfections successfully; any imperfections that
remain are intentionally left to preserve the state of such historical works.

Forgotten Books is a registered trademark of FB &c Ltd.
Copyright © 2018 FB &c Ltd.
FB &c Ltd, Dalton House, 60 Windsor Avenue, London, SW19 2RR.
Company number 08720141. Registered in England and Wales.

For support please visit www.forgottenbooks.com

THE JOURNAL

——OF THE——

FRANKLIN INSTITUTE

DEVOTED TO

SCIENCE AND THE MECHANIC ARTS.

EDITED BY

Dr. H. W. Jayne, Chairman; Mr. Edwin Swift Balch, Ph.D., Dr. Persifo; Frazer, Mr. Louis E. Levy, Mr. James Christie, Committee on Publicationsr with the Assistance of
Dr. Wm. H. Wahl, Secretary of the Institute.

VOL. CLXI—Nos. 961-966

(81ST YEAR)

JANUARY-JUNE, 1906

PHILADELPHIA:
Published by the Institute, at the Hall, 15 South Seventh Street
1906

1
F8
v.161

621352
24.10.55

JOURNAL OF THE FRANKLIN INSTITUTE

Vol. CLXI—January-June, 1906

INDEX

Alumino-silicides 318
Aluminum for patterns..172
Aluminum, production of, in Europe..................................318
Aluminum industry and the Hall patent..............................383
Aluminum-zinc alloy..100
A huge towage undertaking..450
An immense dock..428
American Foundrymen's Association. (Moldenke)....................145
American locomotives 319
Armor-plate, improvement in..178
Artificial graphite 212

Barytes mining in Virginia..320
Battleship "Dreadnaught" 318
Bearing metals, bronze, lead in......................................316
Bigelow, W. D. The administration of the imported food law..........213
Boiler furnaces, cause of failure of.................................113
BOOK NOTICES...142, 237, 473
Borax, production of, in 1904...69
Bromide, production of, in 1904.......................................70

Carter, Oscar C. S. The plateau country of the Southwest and the
    "Enchanted Mesa"..451
Canadian Rockies, exploration of the.................................428
Clay-working industries in 1905......................................235
Colles, Geo. Wetmore. Mica and the mica industry................43, 81
Color photography, improvements in the diffraction process of. (Ives)..439
Concrete blocks in building..316
Cowell, E. P. (See Perrott.)

Diamonds, South American, origin of..................................130
Dock, an immense..428
Dyestuffs, analysis of. (Matthews)...................................229

Electric furnace process..211
Electric reduction at Niagara Falls....................................42
Electrochemical calculations. (Richards)........................131, 162
Electrolysis of gas-pipes...227
"Enchanted Mesa," the plateau country of the Southwest, and. (Carter) 451
Etching by machinery. (Levy)..59
Exploration of the Canadian Rockies..................................428

Falkenau, Arthur. The selection of material for the construction of hydraulic machinery 173
Finances of engineering enterprises. (Marks)..........................197
Fire-escapes in Philadelphia..100
Fire-tests of window glass..130
Fire-protection, improved method of......................................450
Food law, administration of the imported. (Bigelow)...................213
Forced draft 114
FRANKLIN as a man of science and an inventor. (Houston)........241, 321
  influence of, abroad. (Straus-Frank)...................................429
  the social and domestic life of. (Irwin)..............................431
  "Trust Funds" to the cities of Boston and Philadelphia.............358
FRANKLIN INSTITUTE:
  Committee on Science and the Arts:
    REPORTS:
      New physical and diagnostic instruments (Wetherill), 80; Speed-jack (Reed), 239; Friction indicator (Weidlog), 239; Quartz-glass mercury lamp (Heraeus), 240; concrete pile (Chenoweth), 240; Automatic pistol (Browning), 398; System of electric distribution, &c. (Hallberg), 398; Moving platforms (Schmidt), 399; Historic collection of incandescent lamps (Hammer), 400.
  Proceedings of Stated Meetings:.....................79, 148, 238, 397, 398, 474
  Sections: Proceedings of stated meetings..............78, 146, 237, 395, 473

Geographic survey, topographic maps of...................................467
Glass, window, fire-tests of..130
Graphite, artificial...212
Gravitation, the problem of. (Morris)....................................115

Hall patents for aluminum, expiration of.................................383
Hardening steel 320
Haupt, Lewis M. Notes on great tunnels..................................401
Heating the Pennsylvania University Dormitories. (Spangler).........179
Hexamer, C. A. (See Perrott.))
High kite flying..472
Hone stones, Belgian...320
Hydraulic machinery, selection of material for. (Falkenau)...........173

Improved method of fire protection.......................................450
Iron ores, titaniferous...100
Irwin, Agnes. The social and domestic life of Franklin................431
Isherwood, B. F. (See Viallate.)
Ives, Herbert E. Improvements in the diffraction process of color photography 439

Japan, the economic future of. (Viallate-Isherwood)...................413

INDEX

Keller, Edward. Labor-saving appliances in the laboratory...........101
Kite flying, high...472

Labor-saving appliances in the laboratory. (Keller)...................101
Lead in bronze bearing-metals..316
Locomotives, American 319

March exports..428
Marks, Wm. D. The finances of engineering problems.................197
Matthews, J. Merritt. The analysis of dyestuffs......................229
MERRICK, JOHN VAUGHAN. Memorial of...............................469
Merritt, James S. (See Perrott.)
Metric system in the United States....................................394
Mica and the mica industry. (Colles).............................43, 81
Mining industries in 1905..178
Moldenke, Richard. (See Outerbridge.)................................145
Morris, Charles. The problem of gravitation..........................115

Niagara Falls, preservation of..228

Obituary. JOHN VAUGHAN MERRICK......................................469
Open-hearth steel process, Talbot's continuous........................317
Outerbridge, A. E., Jr. High-grade silicon for purifying cast-iron......144

Perrott, Emile G. Reinforced concrete in building construction. (Discussion) 1
Photography, color diffraction process of. (Ives)......................439
Photo-printing machine, Rondinella's, report on........................71
Plateau country of the Southwest and the "Enchanted Mesa." (Carter)....451
Platinum in 1904, production of......................................239

Radium Tubes, explosion of...317
Rails, unusually heavy..42, 172
Reinforced concrete in building construction. (Discussion)...............1
Richards, Jos. W. Electrochemical calculations..................131, 162
Rondinella's photo-printing machine. (Report on)......................71

Salt deposits, New Mexico..114
Silicon, high-grade, for purifying cast-iron. (Outerbridge)............144
Sky-scrapers, American, in England...................................196
Slate, Maine, improved variety of....................................196
Smoke suppression 428
Spangler, H. W. Some data relating to the heating of the dormitories of the University of Pennsylvania..................................179
Steam and turbine engines, comparative economy of....................319
Steam turbine, a 10,000 H. P..41
Steel, hardening...320
Stone, production of, in 1904..234
Straus-Frank, Victor. The influence of Franklin abroad...............429

Talbot's continuous open-hearth steel process............................317
Tantalum, physical properties of..113
"Thermit" repair, a..394
The World's merchant marine..468
The first turbine steamship..428
Tin in New South Wales...384
Titaniferous iron ores...100
Topographic maps of the Geographic Survey..............................467
Towage undertaking, a huge...450
Tunnels, great, notes on. (Haupt)......................................401
Turbine, and steam, engines, comparative economy of....................319
Turbine steamship, the first...428
Typewriter, an improved German...172

Viallate, Achille. The economic future of Japan........................451

Warships, increasing speed of...172
Webb, Walter Loring. (See Perrott.)
Windmill, proper site for..35
World's merchant marine, the...468

Zambesi power scheme..211
Zuni salt deposits, New Mexico...114

JOURNAL

OF THE

FRANKLIN INSTITUTE

OF THE STATE OF PENNSYLVANIA

FOR THE PROMOTION OF THE MECHANIC ARTS

Vol. CLXI, No. 1   81st YEAR   JANUARY, 1906

The Franklin Institute is not responsible for the statements and opinions advanced by contributors to the *Journal*.

Mechanical and Engineering Section.

(Stated Meeting, held Thursday, October 5th, 1905.)

Reinforced Concrete in Building Construction.

DISCUSSION: Mr. Emile G. Perrot, Mr. Walter Loring Webb, Mr. C. A. Hexamer, Mr. James S. Merritt, Mr. E. P. Cowell.

MR. E. G. PERROT:—In the preparation of the paper which I present to you to-night, it has been my aim to treat the subject from the standpoint of the designer and constructor, giving such data and information as has been found in the writer's experience to be applicable to everyday use; also the general practice of the art throughout the country.

With the constant increasing use of plain concrete in construction and the knowledge of its inherent weaknesses when applied to special forms of construction, it is but natural that the recent rapid development of concrete reinforced with bars of metal distributed throughout the material should have been employed to overcome these weaknesses. So successfully has this been attained, that the introduction of reinforced concrete marks a new epoch in the history of building, the possibilities of which seem unlimited.

Furthermore, I feel that we are approaching an era where the preference for steel construction fireproofed with terracotta or other fireproof material for general building work is likely to be superseded by this more recent construction. To quote from a current periodical, "We cannot measure to-morrow by the yard-stick of yesterday. It takes prophetic insight to lay the lines broad enough and wide enough for the needs of the future, and the world moves so rapidly that even the highest wisdom finds its best calculations none too ample for the requirements of this growing world."

Viewing the great buildings which stand before us to-day as the exponent of that high constructional skill possessed by the Gothic architects, pre-eminent in the list are the French Gothic Cathedrals. These buildings are of masonry, construction solely, and have stood the ravages of time.

But, on the other hand, knowing that the life of steel and iron in construction is a very indefinite quantity, may we not ask ourselves what is to become of the steel skyscraper in the ages to come? As the protection afforded steel or iron imbedded in cement is, for all practical purposes, absolute, is not the logic of reinforced concrete established, and may we not, with much more certainty, secure a permanent construction which will not only outlive steel or iron as generally used today, but possesses more truth of expression in art, which is one of the fundamental principles of the great works of the Middle Ages.

Generally speaking, what has been accomplished with the steel skeleton construction is possible in reinforced concrete, and furthermore, if we so will, we can make not only the structural elements of a building of reinforced concrete, but also the ornamental features as well. This, for certain classes of buildings, makes it more desirable as a method of construction, while for the more monumental class, it hardly seems a fitting material in which to carry out any artistic expression.

However, for business buildings and factories, there seems to be no good reason why the exterior walls and ornamentation, such as is used for these structures, should not be built of reinforced concrete, our main difficulty at the present being the lack of sufficient knowledge and experience to make a satisfactory finish; although the practice at present seems to tend to a very simple treatment of the concrete, some going so far as to advocate that no further labor be put upon the concrete after the moulds are removed, a proper texture being obtained by well-constructed forms and a proper spading of concrete.

The invention of reinforced concrete is generally attributed to J. Monier, a French gardener, who, in 1868, constructed flower pots with concrete strengthened with a metal network in order to reduce their thickness.

This modest beginning was the starting point of numerous other applications.

At the Paris Exposition of 1855, however, there was exhibited a boat constructed on this system by Mr. Lambot. In America, the first example of this construction was in the year 1874, when W. E. Ward constructed a house of concrete in which the floors were reinforced with bars; but it is to E. L. Ransome that we owe the use of reinforced concrete in this country.

The principle of this construction, as used for supporting members in buildings, is based on the following facts:

Every simple beam, loaded either uniformly over its length, or concentrated at any point thereof, is in compression at the top and in tension at the bottom. By using steel rods of the proper area and at the proper location to resist the tensile stresses, and arranging the concrete so as to resist the compression, the beam will be in equilibrium, as the resistance of concrete in compression is so much greater than it is in tension, the latter being of a low value, no dependence is placed on the concrete to resist the tensile forces, sufficient area of steel being introduced to resist these strains. The force of adhesion between the concrete and the steel is, however, sufficient to transmit the internal stresses across the section of the beam; so that the fundamental principle of the theory of the beam is applicable to this system of construction the same as to a beam composed of a homogeneous material.

I show on the screen a drawing giving ten different methods of reinforced concrete beams. (See Plate I.)

In Fig. 1, we have a simple beam with the reinforcement run-

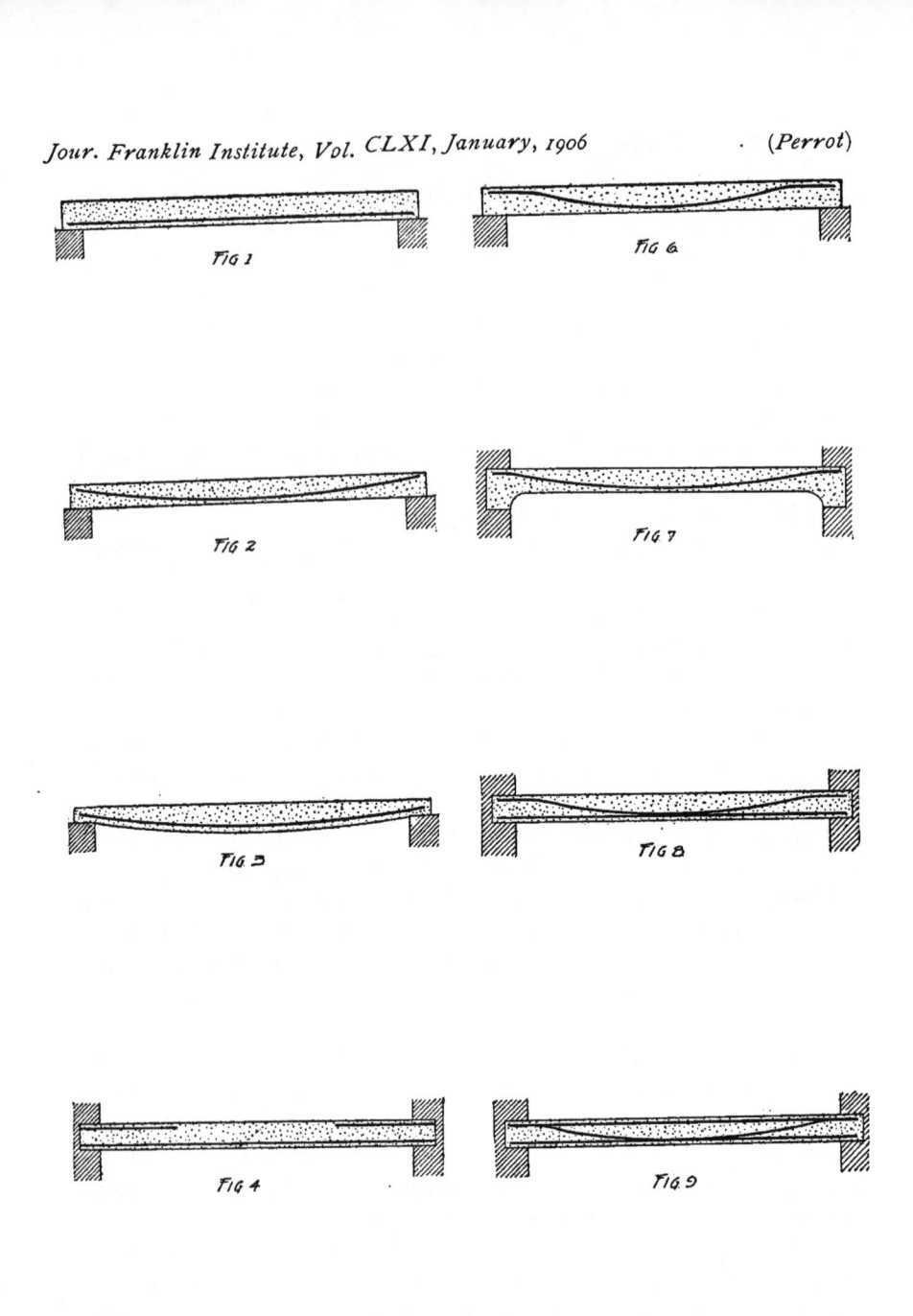

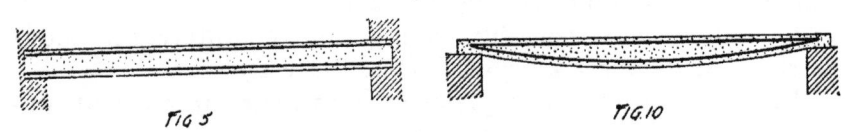

Plate I. Methods of reinforcing concrete beams.

ning in a straight line near the bottom side. This form of reinforcing concrete is more generally used for the slab connecting the beams than the beams themselves.

In Fig. 2, we have the bars curved, being nearer the bottom of the center of the span, where the bending moment is greater, and curving upwards towards the ends, where the bending moment finally reaches zero.

In Fig. 3, we have the steel rods placed as before, and the bottom of the beam curved, following the outline of the rod; this gives the beam a continuous strength, the resisting moment being just equal to the bending moment at every point along the beam.

Fig. 4 shows one method of reinforcing restrained beams; as the ends of the beams are held rigidly in the walls, there is a negative bending moment at the top of the beam near the supports, gradually changing into the positive bending moment at the bottom. The upper rods should be long enough to resist the tension produced on the top of the beam. The point of contraflexure, or where the negative bending moment changes, is about ,21 of the span from the support.

In Fig. 5, the top rods run the entire length of the span; they not only resist the tension near the support, but add to the compressive value of the concrete at the center of the span.

Fig. 6 shows a method by which one rod is made to perform the office of two rods, the rod being bent in such a manner as to take care of both the positive and negative bending moment.

In Fig. 7, the ends of the beams are made deeper and coved; this gives additional sectional area to the concrete where the vertical shear is greatest.

Fig. 8 is a combination of Figs. 1 and 6, and is an admirable disposition of the rods, since the positive and the negative moments are resisted by separate rods, and the straight rods permit of securing stirrups or hangers which run vertically at close intervals to resist the internal shear in the beam. This arrangement is used in the Hennebique System.

Fig. 9 is a fuller development of Fig. 8, but not as extensively used.

Fig. 10 is a development of Fig. 3, having a straight bar in the upper part of the beam; this not only reinforces the compression side of the beam, but also is justified by practical reasons such as those mentioned for the straight rods running the full length of the beam shown in Fig. 8.

For very long spans and special cases of loading, other methods of reinforcing the beam are used, the above illustrations applying to normal conditions only.

Among the numerous patented systems, those most familiar in this country are the Ransome, Kahn, Thatcher, Hennebique, De Valliere, Cummings, Johnson, Columbian, Expanded Metal, Roebling, International, Unit, Visintini, etc.

These may be divided into five classes: (1) Those using "deformed" bars; (2) Those using plain bars; (3) Those using a "webbing;" (4) those using a "frame;" and those in which the parts are moulded in advance and then set in position.

In the first class are the Ransome, which uses a square twisted bar; the Kahn, which uses a square bar with side wings bent upwards to form stirrups; the Thatcher bar, which is a round rolled bar with flattened sections; the Johnson bar, which is square in section and has ridges on the four sides.

In the second class are the Hennebique and De Valliere, in which half the number of bars are bent in the form shown in Fig. 8; the stirrups in the two systems being the points of difference. The Cummings combines the rod and stirrup in one bar. The Columbian uses a specially-rolled bar similar to a double cross in section, and supports it at the ends on steel beams or walls.

In the third class are Expanded Metal; Roebling and International; these employ a netting or sheeting in the concrete slab, in some instances strengthing with rods or cables.

In the fourth class may be mentioned the "Unit" Girder Frame, in which the entire reinforcement is made at a shop and delivered as one piece ready to set in the moulds.

The fifth class is represented by such systems as Visintini, which is moulded in advance, making a complete beam.

The following illustrations show the practical application of the foregoing principles as used in the different patented systems:

EXTRACTS FROM REPORT OF FIRE TEST ON UNIT SYSTEM.

"The superiority of concrete as a fireproof material was again demonstrated in a recent fire, water and load test made in

Philadelphia, June 16th, 1905, on the "Unit" system of reinforced concrete. The test was conducted by the engineering staffs of the Building Bureaus of the cities of New York and Philadelphia, directed by Professor Ira H. Woolson, E.M., of Columbia University, New York, assisted by two students of the University.

The test house, or furnace, was a brick building, 10 ft. 6 in. wide by 19 ft. 6 in. long inside. The clear height from the ground to the bottom of the beans was 10 ft. 6 in. The roof consisted of three concrete beams, 14 in. deep, spaced 4 ft. centers to centers, two beams resting at the ends on 12x12 in. reinforced concrete columns. The middle beams rested on 11x16 inch reinforced girders, which in turn were supported by the columns. The slab was $3\frac{1}{2}$ in. thick.

The columns were reinforced with four $\frac{7}{8}$ in. rods tied together every 18 inches with $\frac{1}{4}$ in. wire ties. The beams were reinforced with a girder frame made up in three different manners. The 11x14 in. beam was reinforced with a girder frame made of $\frac{3}{4}$ inch patented quadruple bar, with $\frac{1}{8}$x1 in. stirrups secured to the bar at intervals of about 12 inches, except at the ends, where they were closer, by means of the prongs struck out of the webs. The middle beam was 9 in. wide and 14 in. deep, reinforced with four $1\frac{3}{16}$ in. round steel rods clamped together with $\frac{1}{8}$x1 in. stirrups spaced as in the other beams. The remaining beam was the same size as the middle beam, reinforced with two patented Siamese bars punched their entire length with stirrups clinched thereto by means of the prongs thus formed and spaced the same as in the other beams. The slab was reinforced with $\frac{5}{16}$ in. plain, round rods run at right angles to the beams, spaced every 6 in., with $\frac{5}{16}$ in. round rods run at right angles thereto about every two feet. Additional short rods, $\frac{5}{16}$ in. in diameter were run near the top of the slab over each beam to reinforce this portion and passed through holes in the stirrups. The distance from center to center of the columns was 16 feet. The beams were continuous over the columns, and rested in the end walls. The concrete was hand-mixed: one part of Vulcanite cement, two parts of Jersey bank gravel, four parts of trap rock, $\frac{1}{2}$ and $\frac{3}{4}$ in. size. The steel had an ultimate strength of 60,000 to 70,000 pounds per square inch. The concrete was two months old.

The test slab and beams were to carry a total live load of 30,700 lbs. during the fire test and 122,800 lbs the following day. The average temperature of the fire during the test was 1708.6 degrees F. The maximum deflection of the middle beam under the action of the fire and water was $1\frac{27}{32}$ in. On June 17th, under the full load of 650 pounds per square foot, which by an error was 50 pounds more than the requirements, the deflection was only two inches, which was within one inch of the deflection allowed by the department.

When the total load was removed some two weeks afterwards, it was noticed that the beams had recovered $\frac{3}{4}$ inch of the deflection, showing that the elastic quality of the construction had not been destroyed. The effect of the water under the high velocity and pressure was to cause some of the concrete on the under side of the beams to spall off. The concrete on the girders, columns and slab was undisturbed, except for the small spalling of a corner of two of the columns. An interesting fact developed by the test was the expansion of the four columns from heat. They elongated about one inch and recovered their normal position in cooling."

Plate II shows the load test of 650 pounds per square foot, applied after the fire.

We have here some samples of steel which is embedded in concrete. One sample is taken from the pavement of the City Hall, and shows that the steel was not affected by the elements in any manner, the concrete forming an absolute protection, although this particular piece had been embedded in the pavement for at least nineteen years. The other specimen I show is taken from Binder's pavement, 13th street above Chestnut. This shows the steel in contact with the concrete and the thorough protection afforded by it. This specimen was at least twelve years in the pavement.

I now give you the result of load tests made under my directions last May on four full-size tee-beams having a clear span of twenty feet. These tests were made to determine the strength of this form of beam reinforced with Unit Girder Frames: Plate III shows the cross-section of the beams and Plate IV the method of reinforcing the beams.

Beam No. 1.—One $\frac{3}{4}''$ quadruple unit bar with solid web for 6′ 7″ in center—Area, 2.14 sq. in. See Fig 1, Plate V.

Jour. Franklin Institute, Vol. CLXI, January, 1906. (*Perrot*)

Plate II. Test of floor after fire.

Beam No. 2.—One $\frac{3}{4}''$ quadruple unit bar with punched web entire length—Area, 1.86 sq. in. See Fig. 2, Plate V.

Beam No. 3.—Four $\frac{13}{63}''$ plain round rods clamped together —Area, 2.07 sq. in. See Fig. 1, Plate VI.

Beam No. 4.—Two $\frac{3}{4}''$ siamese bars clamped together, web punched entire length—Area 1.86 sq. in. See Fig. 2, Plate VI.

I have scheduled the results of the tests in the following tables, which show that the Christophe's formula as we have been using it corresponds very closely with the actual conditions found in the tests. See Tables Nos. 1 and 2 of Plate VII.

Table No. 1 gives the necessary data concerning the general composition and physical properties of the materials. Table No. 2 gives the comparison of the actual bending moments to the theoretical bending moment.

By reference to table No. 1, it will be noticed that beam No. 1 failed by the breaking of the bar; beams Nos. 2 and 4 by crushing of the concrete at the top of the beam; and beam No. 3 by the slipping of the rods in the concrete. This slipping is clearly shown in the photograph of the beam, the concrete at the top of the beam being intact. From observations of the tests, it is evident that beams Nos. 1 and 3 possessed nearer the proper proportion of steel to concrete than beams Nos. 2 and 4, and by actual computation, the area of steel required for the Tee section used, is 2.14 square inches. This is based on safe working stresses on concrete and steel of 500 and 16,000 pounds respectively, and is obtained by using formula No. 3, to locate the neutral axis, and then substituting the value of "a" thus found in formula No. 4, which gives the area of the steel required.

The hypotheses assumed in our computations for all beams and girders, while not entirely correct, are sufficiently near the truth for all practical purposes, for, even though they were entirely correct, it would be very difficult to derive a formula suitable for practice which would take into account all the varying conditions of the concrete, such as produced by the age of the mixture, the proportion, the coefficient of elasticity, etc.

The most important hypothesis used in calculating pieces subject to flexure is the conservation of plane sections, since it furnishes the starting point, that two sections originally parallel will, after loading, rotate about the neutral plane and remain true planes.

In deducing a formula on the above assumption, we also neglect the effect of the concrete in tension. Some authorities, however, consider the effect of concrete in tension in deducing their formula, but it is better to err on the side of safety and not count the concrete in tension, depending solely on the metal reinforcement to resist tensile strains.

The formula for tee-sections which we have adopted, when the neutral axis falls below the slab, is the same as that employed by Paul Christophe, the celebrated French engineer, changing the ratio of the coefficient of elasticity of the steel and concrete from 10 to 20. If 10 was used, Beam No. 1 would show an excess of strength of 16 per cent. above the formula, whereas the variation when 20 is used, is only 5.6 per cent., showing this ratio to be nearly accurate.

$$M = \frac{p}{6\,a} \left[a^2 w\,(3h'-a) - (a-g)^2(w-e)\,(3h'-a-2g) \right] \quad (1)$$

p = the unit compressive strain on the concrete.
e = the width of the beam or rib.
a = the distance from the neutral axis to the most remote fibre in compression.
w = the width of the slab.
h' = the distance from the top of the slab to the center of action of the reinforcement.
g = the thickness of slab.
t = safe unit stress on steel.
S = area of steel.

The position of the neutral axis, (a) is found by the following formula, after which it is inserted in the formula for bending moment:

$$a = -\frac{1}{e}[g(w-e)+m\,t] + \sqrt{\frac{1}{e^2}[g(w-e)+m\,t]^2 + \frac{2}{e}\left[\tfrac{1}{2}g^2\,(w-e) + m\,t\,h'\right]} \quad (2)$$

This method is applicable when the area of the steel is determined upon in advance. Where the area of the steel is to be found, having the exterior dimensions of the concrete given, the formula for the position of the neutral axis may be simplified considerably when the two materials are considered to be

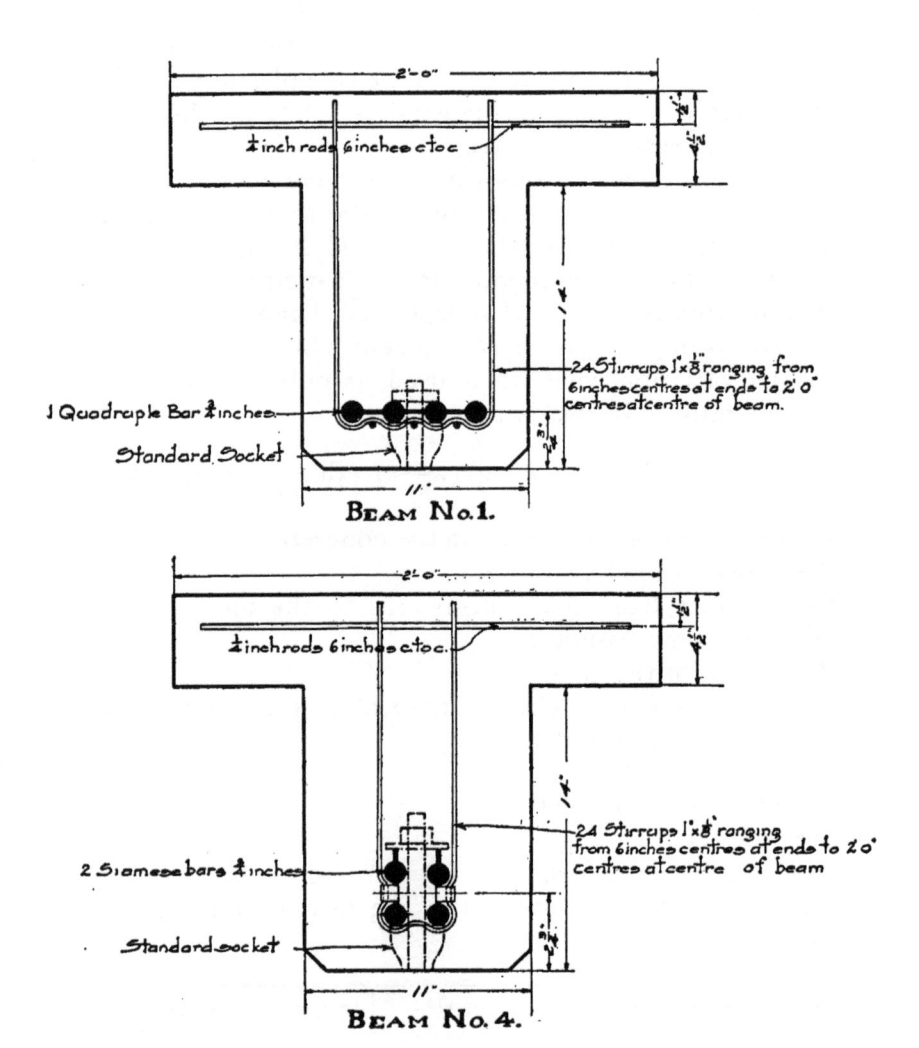

Plate III. Cross-sections

January, 1906. (*Perrot*)

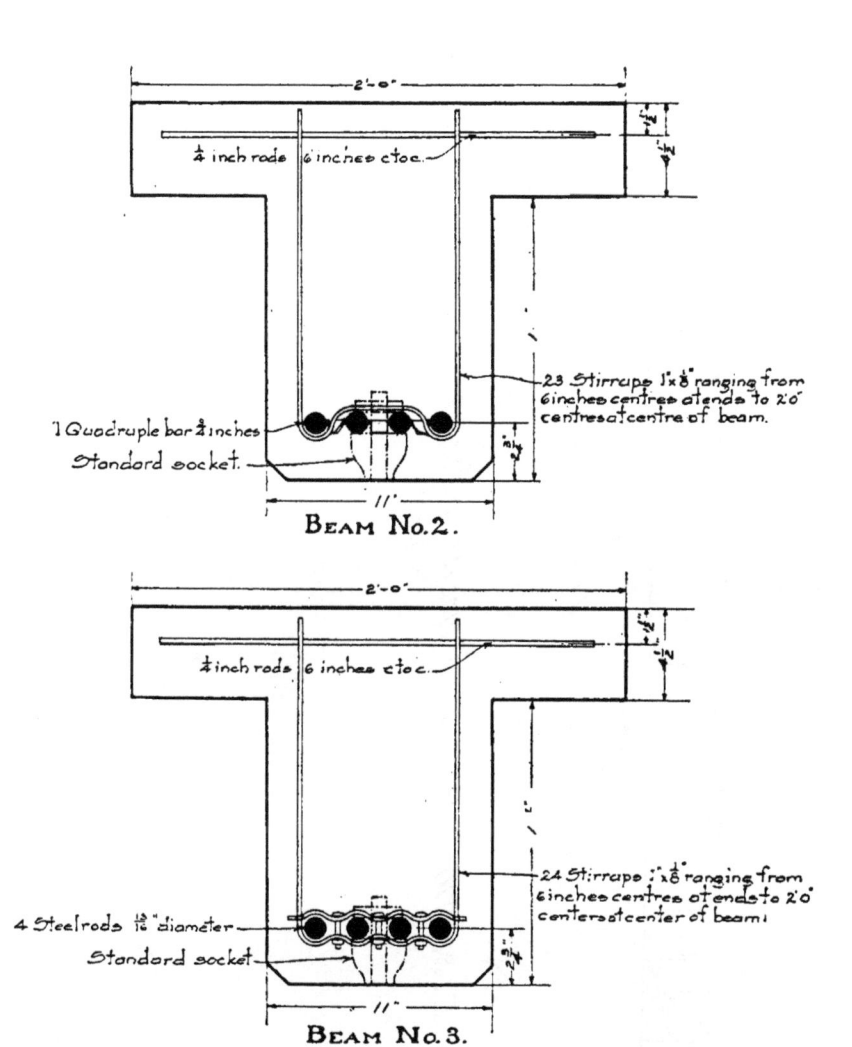

BEAM No. 2.

BEAM No. 3.

of T-beams.

CROSS SECTION.

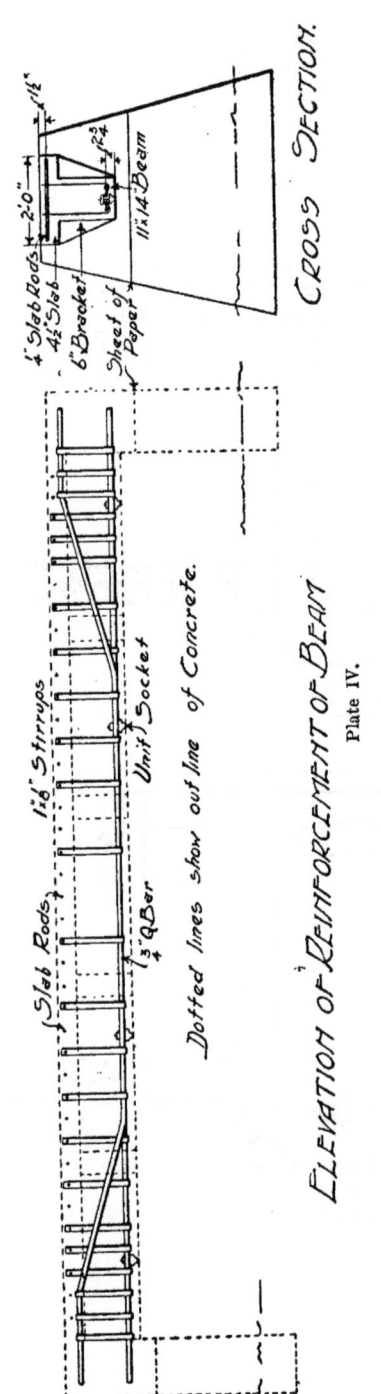

ELEVATION OF REINFORCEMENT OF BEAM
Plate IV.

stressed to their assumed values, both being stressed their proportional part, so that neither one is doing more work than the other. The formula then becomes—

$$a = \frac{h'}{1 + \dfrac{t}{pm}} \qquad (3)$$

Having obtained the value "a" in this formula, it can be substituted in Formula 1. This will give the strength of the beam.

The area of the steel to develop this strength is found by the following formula:

$$S = \frac{a^2 w - (a - g)^2 (w - e)}{2m (h' - a)} \qquad (4)$$

The safe working stress of the concrete I have assumed at 500 pounds per square inch and of the steel 16,000 pounds per square inch, these values being established by the Philadelphia Bureau of Building Inspection. This gives a factor of safety of four on the construction, as by multiplying the bending moment so obtained by four we obtain the breaking moment of the beams as shown in the table.

The approximate formula for finding the area of the steel in the bottom of a beam of tee-section, while not being entirely correct, is very convenient for checking-up purposes, and where absolute refinement is not necessary. It is based on the principle of the moment of a couple, taking the lever arm as the distance between the center of action of the steel and the center of the slab, the formula becomes—

$$S = \frac{M}{16000 \times (h' - g/2)} \qquad (5)$$

That is, the area of the steel, S, = the bending moment in inch pounds, divided by 16,000 times the distance from the center of the center of action of the steel to the center of the concrete slab.

I have called this in Table No. 2 the plate girder formula.

It is interesting to note the deflections which are shown by the following diagram. See Plate VIII.

Fig. 1.

Fig. 2.
Plate V. Breaking loads on T-beams.

No. 2

1	20	1-¾" Q Bar Punched Web	1—2—4	77	5820	3000	20	63400	39400	16000	1	¾	.8 of 1%	8½		5.68"	Total 15375 Live 9555	Def. ¼" Load 26225	52867	Concrete crushed.

No. 3

1	20	4 Plain Steel Rods $\frac{13}{16}$"	1—2—4	78	5820	3000	20	73300	53800 55500	16000	2	$\frac{13}{16}$	.89 of 1%	7½ 3.8		5.92"	Total 15909 Live 10089	Def. $\frac{5}{16}$" Load 26322	64227	Rods slipped, two rods broke at about ¼ the length of span from the end due to impact when beam dropped.

No. 4

1	20	2-¾" Simese Bars Punched Web	1—2—4	79	5820	3000	20	62550	37300	16000	1	¾	.8 of 1%	Upper 5½ Lower 11		5.68"	Total 15375 Live 9555	Def. $\frac{3}{16}$" Load 20195	52899	Same as No. 2.

NOTE—* Load includes weight of beam.
† Hand mixed stone ½ and ¾ inch size.
‡ Def. indicates deflection at center of beam.

TABLE NO. 2

Beam	A. Actual Center Bending Moment	B. Resisting Moment Christophe Formula	C. Resisting Moment Plate Girder Formula, based on 64,000 lbs.	D. Resisting Moment Plate Girder Formula, based on test strength	Variation Per Cent	Variation Per Cent.	Neutral Axis	Calculated Bending Moment at end of web	Deflection at instant of failure	Area Steel
No. 1	1820640	1928764	1848960	1900962	A 4.2 % less than D.	A 5.6 % less than B.	6"	1608400	$8\frac{13}{32}$"	2.14 sq. in.
No. 2	1586010	1844972	1607040	1591974	A 4 % less than D.	A 14 % less than B.	5.68		$16\frac{13}{16}$"	1.86
No. 3	1926810	1909056	1788480	2048368	A 6 % less than D.	A 9 % greater than B.	5.92		$11\frac{5}{8}$"	2.07
No. 4	1586970	1844972	1607040	1570630	A 1 % greater than D.	A 14 % less than B.	5.68		$18\frac{3}{16}$"	1.86

Ratio of Strength between No. 3 and No. 1.

The steel in No. 3 was 11 % stronger than that in No. 1 by actual tensile test, and had 33 % greater elastic limit, but the beam was only 5.8 % stronger by actual load, this small increase being due to the rods slipping. In No. 1 the solid web should have been at least 7′ 3″ long; instead it was only 6′ 7″. The resisting moment at the ends of the web just equaled the bending moment at these points. This is where the rods broke, so that the strength of the bar at center of beams was not fully developed.

Fig. 1.

Fig. 2.
Plate VI. Breaking loads on T-beams.

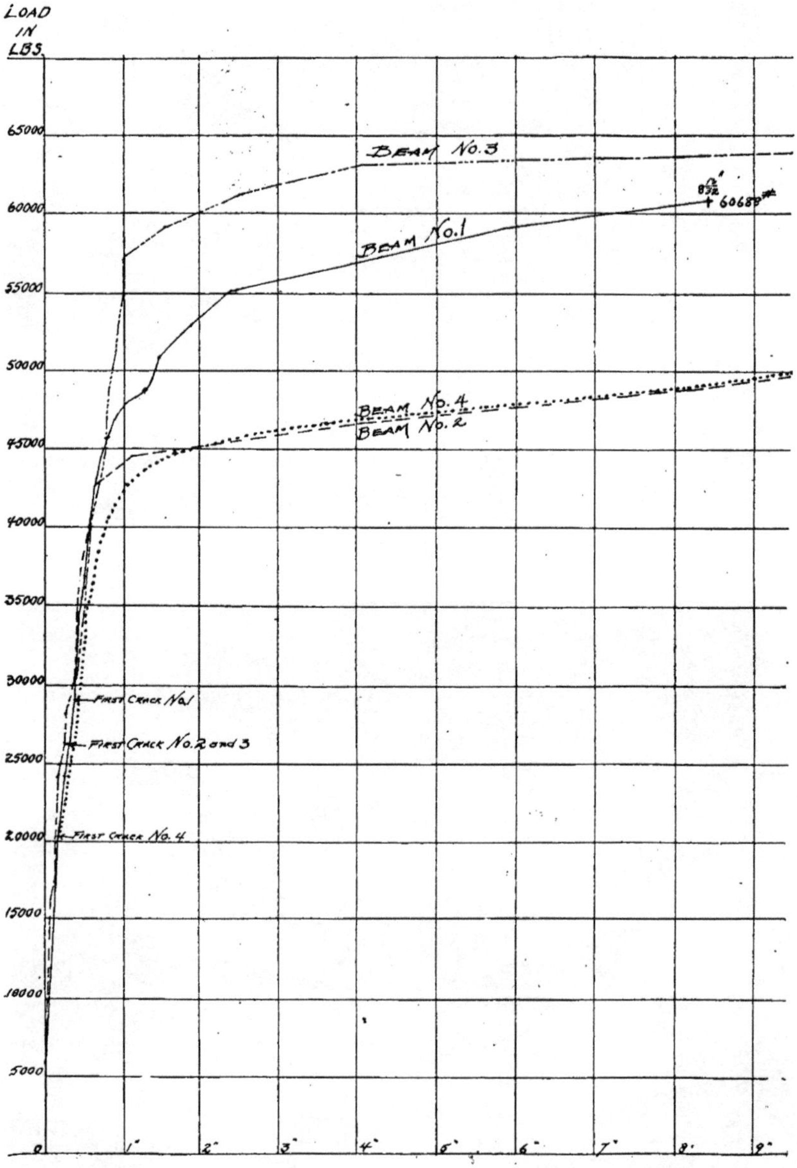
LOAD IN LBS.
65000
60000
55000
50000
45000
40000
35000
30000
25000
20000
15000
10000
5000
0
1"
2"
3"
4"
5"
6"
7"
8"
9"
BEAM No. 3
BEAM No. 1
BEAM No. 4
BEAM No. 2
60688#
FIRST CRACK No. 1
FIRST CRACK No. 2 and 3
FIRST CRACK No. 4

January, 1906. Plate VIII. (Perrot)

BEAM No. 1 — ONE ¾" QUADRUPLE UNIT BAR
WITH SOLID WEB FOR 6'-7" IN CENTER — AREA 2.14 □"

BEAM No. 2 — ONE ¾" QUADRUPLE UNIT BAR
WITH PUNCHED WEB ENTIRE LENGTH — AREA 1.86 □"

BEAM No. 3 — FOUR ⁹⁄₁₆" PLAIN ROUND RODS
CLAMPED TOGETHER — AREA 2.07 □"

BEAM No. 4 — TWO ¾" SIAMESE BARS
CLAMPED TOGETHER, WEB PUNCHED ENTIRE
LENGTH — AREA 1.86 □"

I show a number of illustrations of different methods of constructing high buildings, giving the comparison between the steel skeleton construction and the concrete skeleton construction; also pier construction in concrete and monolithic wall construction.

Various methods have been used to secure overhead shafting, fixtures, etc., by using some form of bolt or socket built in the concrete beam, but not coming in contact with the steel reinforcement, except in the case of the Unit system, where a specially-designed socket and stud is used as shown in the previous illustrations. This socket forms a part of the system and is used whether it is intended to secure overhead fixtures or not. The following views show some of the adaptations of how this socket is used to secure shafts and pipes.

The rigidity of reinforced concrete buildings is very much greater than that of any other known form of construction. The following views show the engines of the Victor Talking Machine Company, which are supported on reinforced concrete. A coin can be placed upon the cylinder of the engine and will remain there indefinitely, showing that the floor construction is so rigid as to prevent any local vibration; although it is noticed that a building of this character will rock or move as a whole, being perceptible in the top stories of such buildings; the action being somewhat in the nature of the vibration of a small end of a whip when held vertically.

[The speaker showed a number of illustrations descriptive of some of the work done throughout the country by various engineers, including the Government engineers, who have found this construction to be very suitable for the various types of buildings they design, this method of construction being largely used by them.]

Mr. WALTER LORING WEBB:—So much has been said, written and published during the last few years in regard to reinforced concrete that it is unnecessary in a short address on this subject to review the history of the development of reinforced concrete or even to attempt to establish any of the fundamental equations which are in most common use. I will therefore devote my time to a few remarks on certain phases of this work where comment seems necessary.

ELASTIC LIMIT OF METAL.

It is generally admitted by experts that the elastic limit of the metal is the virtual ultimate of the strength of a reinforced concrete structure, but the exact meaning and application of this statement is not thoroughly understood by many. In an ordinary steel bridge, which consists of but one kind of material, a stress in the structure which momentarily exceeds the elastic limit may produce a small but permanent deformation but will

Section of concrete-steel curtain dam.

not cause failure and is not immediately dangerous. Therefore the working stress may properly be permitted to approach the elastic limit much nearer than should be allowed in reinforced concrete. To put it in figures—in a bridge or a building the working stress of 16,000 lbs. is frequently permitted in metal, which has an elastic limit of 30,000 or 32,000 lbs. and an ultimate of 60,000 lbs. Experience has shown this to be safe. If a concrete structure is reinforced with this same grade of metal the same working stress of 16,000 lbs. has by no means the same *real* factor of safety that it has in the case of the steel stress, for as soon as the loading is actually increased (perhaps accidentally) to twice the working load, the structure will be at the point of actual failure. The steel will not return to its former position with reference to the concrete, and therefore the union between the concrete and the steel will have become destroyed. This is especially true if the bar is a plain bar in which the union between the concrete and the steel depends entirely on adhesion. Since the strength of a reinforced concrete structure is absolutely dependent on the intimate union of the concrete and the steel, the structure must be at the point of failure as soon as that union is destroyed. The accuracy of the above statement has sometimes been attacked by a comparison of the breaking loads of reinforced concrete beams which have been variously reinforced with plain bars, deformed bars and bars of a high and low elastic limit steel. Such tests are of almost no value for the determination of this point unless the bars of high and low elastic limit are otherwise identical. On account of the security to the union of the concrete and steel which is afforded by a deformed bar, this quality is perhaps of greater importance than the elastic limit, and it is quite probable that a plain bar with high elastic limit will have but little, if any, advantage over a deformed bar with low elastic limit, but the combination of a deformed bar with a high elastic limit gives the best form of reinforcement that can be made.

The building ordinances of cities, which have generally been written on the basis of ordinary steel construction, have been applied without any change to reinforced concrete work. This is sometimes due to the fact that inspectors have been compelled, by the literal statement of the existing building law, to follow obsolete specifications. But even this excuse does not apply to Philadelphia, which acts under a building law which was amended in April, 1903. This amendment recognizes reinforced concrete as a special method of construction and prescribes a few regulations regarding it, but these regulations are utterly inadequate and defective. Some of these regulations are not in accordance with the principles now generally recognized in reinforced concrete construction. In other respects they are inadequate and the saving clause is introduced of requiring that the plans shall have the approval of the building inspectors. Such a regulation may be all right as a makeshift, but the same reasons which justify the present elaborate details of the building law on very minor details should require a properly designed code of regulations which will enable designers to know what designs will be acceptable by the department. The initial stresses in the concrete, due to its contraction during setting, are usually ignored by designers. Tests have shown that such stress may, under some conditions, be as great as that due to the working load. The Johnson-bar sys-

Section of concrete-steel breast dam. For soft foundation and low head.

tem is based on the use of steel with an actual elastic limit of about 60,000 lbs. For calculations this is reduced to 50,000 and a factor of *four* is then used for a working stress of 12,500. Compare, for an instant, the relative safety of such a system with that of permitting a working load which will produce a stress of 16,000 lbs. in the steel, plus the unknown stress due to contraction of the concrete, when the steel is of such quality that a stress of only 30,000 lbs. will cause failure. The building laws of Philadelphia permit this very construction, and it is frequently used. It will be surprising if failure does not occur in structures built in this way.

REINFORCED CONCRETE DAMS.

One of the most remarkable applications of reinforced concrete is its use for the construction of dams. Omitting from consideration the very few dams, which on account of their exceptional location are constructed on the arch type, dams can be divided for the present purpose into two general classes. In one class the upstream face of the dam is so nearly vertical that the pressure of the water produces a tendency to overturn about the toe of the dam. The higher the flood the greater will be the overturning moment. This is resisted by the weight of the dam, which has a resisting moment in the contrary direction. The ratio of the resisting moment to the overturning moment constitutes the factor of safety of the dam. Such dams are usually built of solid masonry so as to furnish the necessary dead weight, and yet such is the cost of masonry that its volume in the design is reduced to the lowest limit which prudence can tolerate, and then some unusual flood or a weakening of the foundations will make the factor of safety (perhaps 2 or 3) insufficient and the dam gives way. The other type, of which a timber dam is the most common example, has an upstream face which makes an angle of less than 45° with the horizontal. In such a case the center of pressure of the dam is within the base of the dam. There is, therefore, no tendency to overturn. The higher the flood the greater will be the downward pressure and the greater will be the horizontal component tending to sweep the dam downstream, but the dam cannot possibly overturn. A solid masonry dam with such an outline would have a prohibitive cost. A timber dam has but a short life and a considerable maintenance charge even while it is being used. An earth dam is unreliable, especially if it has no core-wall, and if it has a thoroughly reliable core-wall it is expensive. A reinforced concrete dam is built with a comparatively flat upstream surface, so that there is never any tendency for it to overturn. Its cost is considerably less than that of a solid masonry dam. It is but little, if any, more expensive than a well-built wooden dam, and the adoption of this method has frequently permitted the utilization of water power sites which otherwise could not have been utilized without an expenditure which would not have been justified by the income.

The comparison of a hollow dam and a solid dam are illustrated in sectional view showing the actual designs made for a given location. The relative amount of masonry, the relative excavation for the foundations and the relative pressures with their resulting actions on the dam are illustrated in this figure. The limits of this article will not permit the elaboration that the subject deserves, but a few points which are easily realized from the figures here shown will be briefly stated.

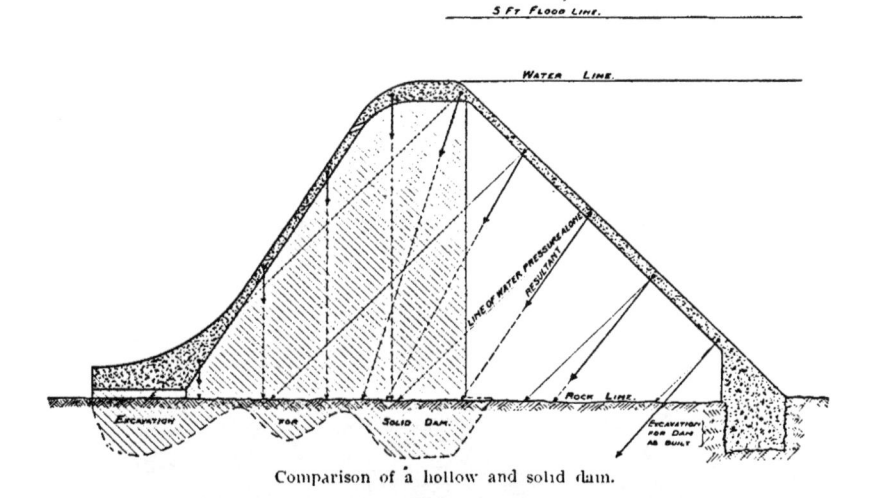

Comparison of a hollow and solid dam.

First, there is practically no danger that the dam will be swept down stream and absolutely no possibility of its being overturned.

Second, a reinforced concrete dam is "bottle tight" from the start and it remains so. This fact is easily verified by inspection, because the dam is hollow, and if by any possibility any leaks should occur they are easily repaired at a very small expense.

Third, the upward hydrostatic pressure which reduces the stability of an ordinary solid masonry dam is totally absent, for, if the soil is all porous, weep-holes are purposely made in the bottom of the dam so as to eliminate any such tendency.

Fourth, the base of these dams is so broad that a dam may be located on a soft bottom under circumstances which would be

Section of concrete-steel apron dam. For rock foundation and high head.

Section of concrete-steel apron dam. For clay or gravel foundation and high head.

prohibitive except at enormous expense for a solid masonry dam.

Fifth, a wooden dam is never any better than on the day it is finished. Its deterioration commences from that day. A concrete dam, on the contrary, gains steadily in strength. It becomes part of the geology.

Sixth—*Effect of Ice.*—Many a dam of solid masonry has been carried out by the pressure of a great ice gorge. Even a wooden dam is often carried out, since a decayed plank gives a chance for the ice to bite into the surface of the upper deck. With a few planks gone destruction follows at a cumulative

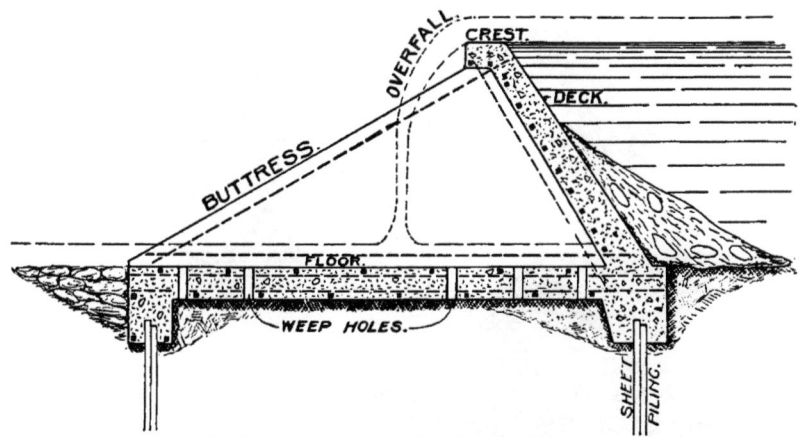

Section of dam for low head and soft bottom.

rate. In the case of reinforced concrete dams the deck rises by an easy incline from the very bed of the river, the surface is smooth, uniform and very hard. It lifts the whole mass of ice easily and smoothly to the crest of the dam and the easy incline of the apron drops it safely to the foot. If any direct expansive force of the ice exists it has no effect. The ice is merely lifted.

Seventh—*Time Required for Building.*—The time required for the construction of a dam is often a serious matter. Many a dam has been postponed for nearly a year because it would not be safe to construct it late in the season and risk severe injury by late fall floods. The record of one of these dams, 194 feet long and 10 feet high, build without machinery in twenty-two ordinary working days, is a record which probably will never be again obtained except when using a similar method.

Eighth—*Stability.*—The crushing strength of massive masonry is a somewhat uncertain quantity and therefore low unit values are used for working stresses whenever crushing strength becomes a factor in the design. In the case of a reinforced concrete structure the stresses on every part of the

Section of concrete-steel dam with half-apron.

dam are very definite and the engineer can not only tell exactly how much stress will be found in each detail of the design of the dam, but he can also design the work with any desired factor of safety. In this respect the design of such a dam becomes a very definite problem—like the design of a bridge.

Ninth—*Closing the Dam.*—The problem of closing the dam is often a very serious one which taxes the ingenuity of the engineer to the utmost. The method adopted in the Schuylerville Dam shows one of the peculiar characteristics of these dams. A water-way which would accommodate the ordinary flow of the river, was left so that the water flowed through the dam. This dam was begun September 27, 1904, and was finished December 31, 1904. For over two months the water was allowed to flow through the channels which were left. On March 11, 1905, the openings were closed up. The time required for closing was *forty-five minutes*. In any other form of

Section of concrete-steel dam—open front. For moderate head and ledge foundation.

dam used such a feat would not be even approximately possible.

Tenth.—The construction of these hollow dams has made it possible not only to avoid the usual expense of a gate house, which can be placed inside of the dam, but it will even permit the power-house to be located inside of the dam and thus effect a very considerable saving on what would otherwise be the combined cost of a dam, gate-house and power house. An il-

Schuylerville Dam. Under construction.

Schuylerville dam. Completed, except closing.

lustration of this is given in the design of the Oakdale power house at Tippecanoe, Ind.

Eleventh—*Cost.*—As intimated before, the cost is invariably cheaper than that of any other good dam. A timber dam *may* be constructed so flimsily that its cost will be lower than one of these dams, but as such a dam will ordinarily be washed out by the first unusual flood, there is no economy in it. In fact, it may be demonstrated that when you consider that the

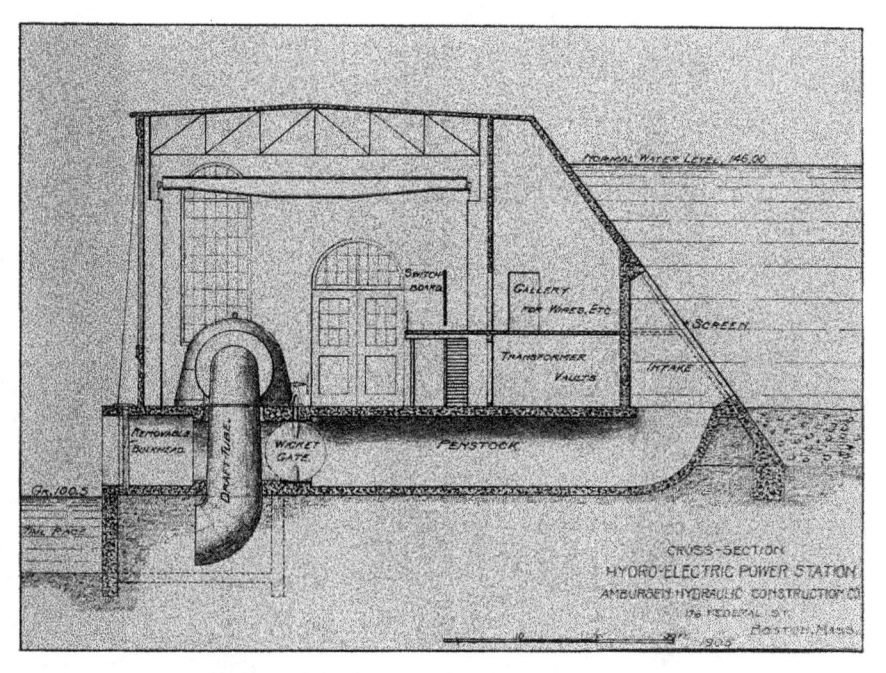

Section of Oakdale power-house, Tippecanoe, Ind.

cost of a dam (like that of any other structure) is its total cost over a long series of years, and must include all maintenance charges, the real cost of a wooden dam is even greater than that of a reinforced concrete dam. It should be strongly emphasized that this method of building a strictly first-class and permanent masonry dam at a comparatively low cost has already rendered financially practicable the utilization of many water-power sites which otherwise could not be utilized except at an expense which would not be justified by the returns.

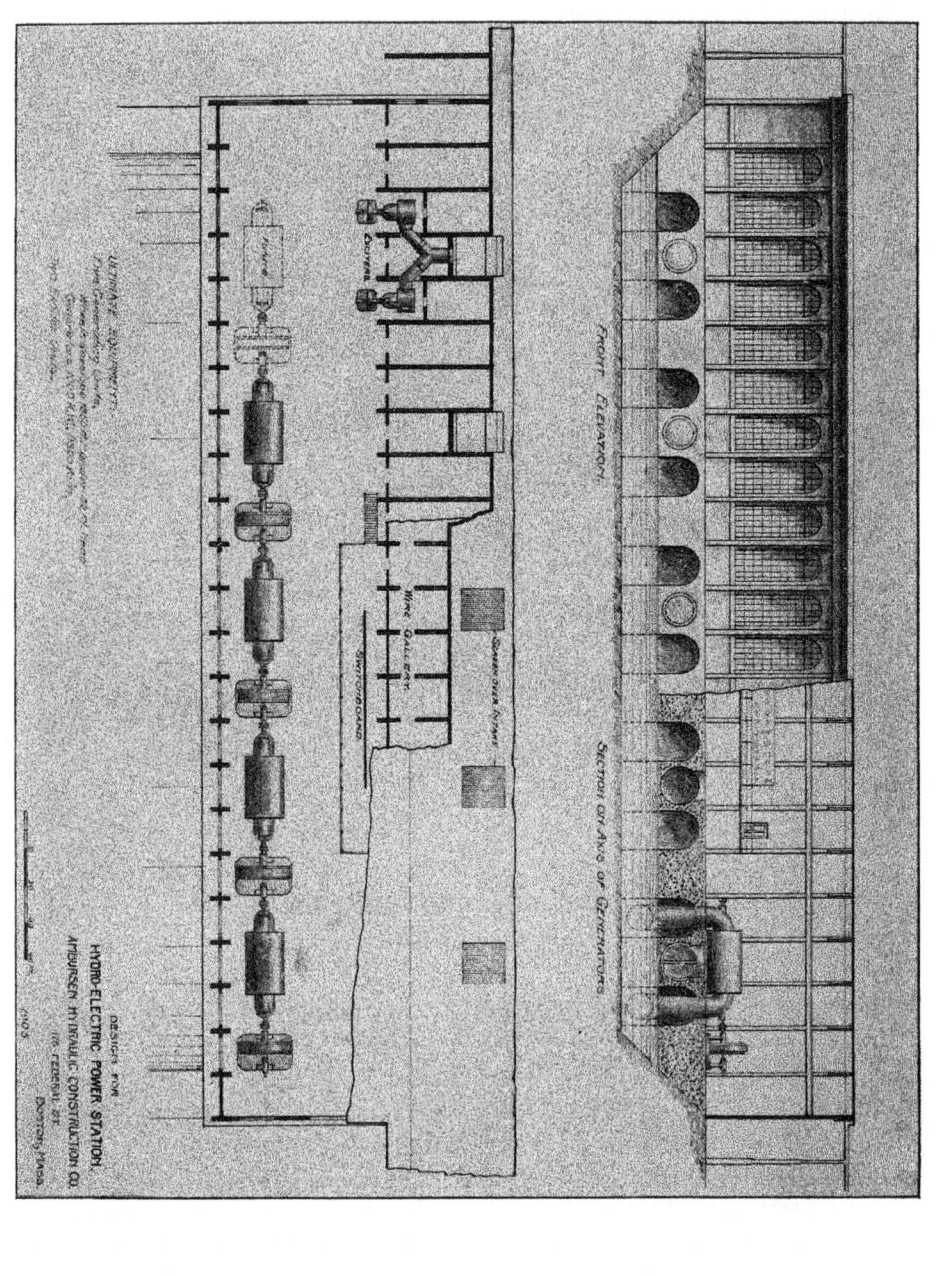

Elevation of Oakdale power-house, Tippecanoe, Ind.

Mr. C. A. HEXAMER:—The interest I have in reinforced concrete buildings is purely from an insurance viewpoint. Fortunately, there has been no serious fire involving a reinforced concrete building full of inflammable material, and therefore no actual fire test of this class of construction. Hence the insurance engineer is at present withholding his verdict. There is, however, a committee of the National Fire Protection Association at work on the question of the fire-retardent quality of this class of construction. At the annual meeting of the National Fire Protection Association, in May last, a preliminary report was presented by the committee embodying the following fundamental principles of safeguarding reinforced concrete construction:

First.—The work should invariably be designed and its entire erection supervised personally by engineers of skill and experience in this particular line.

Second.—Concrete for fire proof construction should be composed of high-grade tested Portland cement, clean sand, and broken stone, gravel, slag or cinders(?), so proportioned that the cement will completely fill the voids in the sand and the mortar thus formed will a little more than fill the voids in the aggregate.

Third.—The materials should be well mixed by machine with enough water to make the mass distinctly a "wet mixture," and should be tamped down so that no voids are left.

Fourth.—All steel members of whatever style should be imbedded at least two inches in the concrete, and in the case of important load-carrying members, three inches.

Fifth.—No cement work of any kind should be laid in cold weather without being safely guarded against freezing. This may be disputed, but is correct beyond reasonable doubt.

Mr. JAMES S. MERRITT said that on account of the lateness of the hour he would not attempt any discussion of the use of expanded metal for the Reinforcement of Concrete. but would refer to some instances in which reinforced concrete, carefully designed and properly mixed and put in place, had survived exceedingly severe treatment. Six or eight lantern slides showed concrete floors which had been cut through in the most reckless fashion to permit the passage of belts, pipes, spouts, etc.

In one case a steel beam, which had been intended to support the expanded metal and concrete floor, and come in the way of a large grain spout and had been cut entirely through, after the completion of the building. Without any additional reinforcement the floor was supporting the two halves of the beam, the load being transferred to the adjoining beams, which were about 5' away on either side.

HOLLOW CONCRETE BUILDING BLOCKS.

MR. E. P. COWELL :—That these blocks are becoming an important factor in building construction is admitted by all who have carefully investigated the subject. Concrete building blocks were practically unknown five years ago; since then 150 machines have been patented or are being made, and 1500 firms are making the blocks in this country.

The principal cause of the sudden development in this industry is the growing scarcity and consequent increase in price of lumber, and the high rates of wages forced by labor unions; also the low cost and improved quality of American cement; also the fact that concrete construction can be carried on in regions where lumber and clay are unavailable, with only cement, sand and water.

The blocks are made of a variety of sizes and shapes, and of materials and in proportions thereof innumerable, sometimes of one piece for a section of wall and again in two pieces (outside and inside).

Single blocks are made as large as 8 feet long and 8 inches wide and 10 inches thick, reinforced by steel rods. This tends towards concrete beam construction, though used as lintels. Blocks are rarely made longer that 6 feet without reinforcement. Single-piece hollow blocks are made 20 to 32 inches long by 8, 9 and 12 inches other dimensions, to make the full thickness of the wall. Two-piece blocks are made, as the name implies, for face and back wall. An argument in favor of the single-piece block is that when laid in place a section of the wall is completed, requiring no bonding to the front, containing 30 per cent. air space generally; more material and strength, therefore better and more economical than two-piece system. Advocates of two-piece system claim to secure a drier inside wall, with less material, having 50 per cent. air space and present a more even inside wall upon which to plaster.

Following are a few of the proportions of materials used by different makers of blocks:

  1 cement 4 sand.
  1 " 4 " and gravel.
  1 " 5 "
  1 " 2 " 4 cinder.
  1 " 1 " 2 crushed stone.
  1 " 3 "
  1 " 2 " sometimes used as facing of blocks.

with the view of rendering the blocks more impervious to moisture, but that seems unnecessary, as the concrete becomes practically waterproof when set, thoroughly crystallized. After the materials are mixed dry, water should be added from a sprinkling can till the mass is of a uniform color and sufficiently wet to retain shape when squeezed in the hand. The quantity of water required will vary with the condition of the sand and percentage of humidity in the atmosphere. Shovel the mixture into the mold in small quantities, meanwhile constantly tamping. Remove block on the platten to a place under cover, where it shall remain for at least one week (two weeks shows greater strength), sprinkle the block next morning and twice daily for one week, when the block may be safely used in building.

In a large city some distance from Philadelphia, one firm employing twenty-five men making hollow blocks supplied them to eighty-six dwellings and three large factories in three months this summer. These were generally sent out less than a week old, and rarely watered more than six times before placing in the yard exposed to the sun's action. No failures have so far accrued and demand exceeds production, showing that satisfaction has been given architects and builders. Other makers in the same city have furnished, in total, twice the above quantity, indicating popularity of Hollow Concrete Building Blocks. One peculiar feature of the business in that city is the uniform favor shown for blocks with an imitation rock face that is so palpable an imitation of stone work that it is repellant. A neat appearance is given by a plain or tooled face. The material we have to work with is so well adapted to many uses and designs that it deserves a better fate than being considered an imitation of something else. Let it stand for what it is—the best of building materials.

The solid wall, or so-called Monolith style of Concrete Building Construction being an established fact, why should not the block business thrive? The blocks being made in smaller units are less liable to contain defects than a large mass.

The most beautiful architectural effects may be secured by using molds for solid or monolith work, and surely the same ingenuity may be developed with hollow blocks, though it must be admitted that up to the present time we are generally confronted with the nightmare of imitation cut stone.

Adaptability to any climate, residences built in our Northwestern cities, in Lower California, along both sea coasts and in the tropics, fulfill all the requirements of a perfect structure, entirely dry, cool in summer and warm in winter. Fire protection is perfect; the material being practically indestructible. And consider the cheapness and ease of construction: with one good mason and a fair quality of laboring men, an ordinary plain building may be erected of as durable a character as the most skilled and high-priced men could build, in many cases less than a brick structure would cost, and in some cases less than frame.

For information, I append hereto the rules and regulations covering the manufacture and use of Hollow Concrete Building Blocks

APPENDIX.

RULES AND REGULATIONS COVERING THE MANUFACTURE AND USE OF HOLLOW CONCRETE BUILDING BLOCKS IN THE CITY OF PHILADELPHIA.

1. Hollow concrete building blocks may be used for buildings six stories or less in weight, where said use is approved by the Bureau of Building Inspection, provided, however, that such blocks shall be composed of at least one (1) part of standard Portland cement, and not to exceed five (5) parts of clean, coarse, sharp sand or gravel, or a mixture of at least one part of Portland cement to five (5) parts of crushed rock or other suitable aggregate. Provided, further, that this section shall not permit the use of Hollow Blocks in party walls; said party walls must be built solid.

2. All material to be of such fineness as to pass a one-half inch ring and be free from dirt or foreign matter. The material composing such blocks shall be properly mixed and manipulated, and the hollow space in said

blocks shall not exceed the percentage given in the following table for different height walls, and in no case shall the walls or webs of the block be less in thickness than one-fourth of the height. The figures given in the table represent the percentage of such hollow space for different height walls:

Stories.	1st	2d	3d	4th	5th	6th
1 and 2	33	33				
3 and 4	25	33	33			
5 and 6	20	25	25	33	33	33

3. The thickness of walls for any building where hollow concrete blocks are used shall not be less than is required by law for brick walls.

4. Where the face only is of hollow concrete blocks and the backing is of brick, the facing of hollow concrete blocks must be strongly bonded to the brick either with headers projecting 4 inches into the brick work, every fourth course being a heading course, or with approved ties, no brick backing to be less than 8 inches. Where the walls are made entirely of hollow concrete blocks, but where said blocks have not the same width as the wall, every fifth course shall extend through the wall, forming a secure bond. All nails where blocks are used shall be laid up in Portand cement mortar.

5 All hollow concrete building blocks before being used in the construction of any buildings in the City of Philadelphia shall have attained the age of at least three (3) weeks.

6. Wherever girders or joists rest upon walls so that there is a concentrated load on the block of over two (2) tons, the blocks supporting the girder or joists must be made solid. Where such concentrated load shall exceed five (5) tons, the blocks for two (2) courses below, and for a distance extending at least eighteen inches each side of said girder shall be made solid. Where the load on the wall from the girder exceeds five (5) tons, the blocks for three (3) courses underneath it shall be made solid with similar material as in the blocks. Wherever walls are decreased in thickness the top course of the thicker wall to be made solid.

7. Provided always, that no wall or any part thereof composed of hollow concrete blocks shall be loaded to an excess of eight (8) tons per superficial foot of the area of such blocks, including the weight of the wall, and no bl cks shall be used that have an average crushing at less than 1,000 lbs. per sq. in, of area at the age of 28 days, no deduction to be made in figuring the area for the hollow spaces.

8. All piers and buttresses that support loads in excess of five (5) tons, shall be built of solid concrete blocks for such distance below as may be required by the Bureau of Building Inspection. Concrete lintels and sills shall be reinforced by iron or steel rods in a manner satisfactory to the Bureau of Building Inspection, and any lintels spanning over four feet six inches in the clear shall rest on solid concrete blocks.

Provided, that no hollow concrete building blocks shall be used in the construction of any building in the City of Philadelphia, unless the maker of said blocks has submitted his product to the full test required by the Bureau of Building Inspection, and placed on file with said Bureau of Building Inspection a certificate from a reliable testing laboratory, showing that samples from the lot of blocks to be used have successfully passed the requirements of the Bureau of Building Inspection, and filing a full copy of the test with the Bureau.

10. A brand or mark of identification must be impressed in or otherwise permanently attached to each block for purpose of identification.

11. No certificate of approval shall be considered in force for more than four months, unless there be filed with the Bureau of Building Inspection, in the City of Philadelphia, at least once every four months following, a certificate from some reliable physical testing laboratory showing that the average of three (3) specimens tested for compression, and three (3) specimens tested for transverse strength comply with the requirements of the Bureau of Building Inspection of the City of Philadelphia, samples to be selected either by a building inspector or by the laboratory, from blocks actually going into construction work. Samples must not be furnished by the contractors or builders.

12. The manufacturer and user of any such hollow concrete blocks as are mentioned in this regulation, or either of them, shall at any and all times have made such tests of the cements used in making such blocks or such further tests of the completed blocks, or of each of these, at their own expense, and under the supervision of the Bureau of Building Inspection as the Chief of said Bureau shall require.

13. The cement used in making said blocks shall be Portland cement, and must be capable of passing the minimum requirements as set forth in the "Standard Specifications for Cement" by the American Society for Testing Materials.

14. Any and all blocks, samples of which on being tested under the direction of the Bureau of Building Inspection, fail to stand at twenty-eight days the tests required by this regulation, shall be marked condemned by the manufacturer or user and shall be destroyed.

15. No concrete blocks shall be used in the construction of any building within the City of Philadelphia until they shall have been inspected, and average samples of the lot tested, approved and accepted by the Chief of the Bureau of Building Inspection.

SPECIFICATIONS COVERING METHOD OF TESTING HOLLOW BLOCKS.

1. These regulations shall apply to all new materials, such as are used in building construction, in the same manner and for the same purposes, as stones, brick, concrete, are now authorized by the building laws, when said new material to be substituted departs from the general shape and dimensions of ordinary building brick and more particularly to that form of building material known as Hollow Concrete Block manufactured from cement and a certain addition of sand, crushed stone or similar material.

2. Before any such material is used in buildings, an application for its use and for a test of the same must be filed with the Chief of the Bureau of Building Inspection. A description of the material and a brief outline of its manufacture and proportions of the material used must be embodied in the application.

3. The material must be subject to the following tests: Transverse, Com-

pression, Absorption, Freezing and Fire. Additional tests may be called for when, in the judgment of the Chief of the Bureau of Building Inspection, the same may be necessary. All such tests must be made in some laboratory of recognized standing, under the supervision of the engineer of the Bureau of Building Inspection. The tests will be made at the expense of the applicant.

4. The results of the tests, whether satisfactory or not, must be placed in file in the Bureau of Building Inspection. They shall be open to inspection upon applicaion to the Chief of the Bureau, but need not necessarily be published.

5. For the purposes of the tests, at least twenty (20) samples or test pieces must be provided. Such samples must represent the ordinary commercial product. They may be selected from stock by the Chief of the Bureau of Building Inspection, or his representative, or may be made in his presence, at his discretion. The samples must be of the regular size and shape used in construction. In cases where the material is made and used in special shapes and forms too large for testing in the ordinary machines, smaller sized specimens shall be used as may be directed by the Chief of the Bureau of Building Inspection, to determine the physical characteristic specified in Section 3.

6. The samples may be tested as soon as desired by the applicant, but in no case later than sixty days after manufacture.

7. The weight per cubic foot of the material must be determined.

8. Tests shall be made in series of at least five, except that in the fire tests a series of two (four samples) are sufficient. Transverse tests shall be made on full-sized samples. Half samples may be used for the crushing, freezing and fire tests. The remaining samples are kept in reserve, in case unusual flaws, or exceptional or abnormal conditions make it necessary to discard certain of the tests. All samples must be marked for identification and comparison

9. The transverse test shall be made as follows: The samples shall be placed flatwise on two rounded knife edge bearings set parallel, seven inches apart. A load is then applied on top, midway between the supports, and transmitted through a similar rounded knife edge, until the sample is ruptured. The modulus of rupture shall then be determined by multiplying the total breaking load in pounds by twenty-one (three times the distance between supports in inches) and then dividing the result thus obtained by twice the product of the width in inches by the square of the depth in inches. $R = \dfrac{3 W l}{2 b d}$ No allowance should be made in figuring the modulus of rupture for the hollow spaces.

10. The compression test shall be made as follows: Samples must be cut from blocks so as to contain a full web section; samples must be carefully measured, then bedded flatwise in plaster of Paris to secure a nuiform bearing in the testing machine and crushed. The total breaking load is then divided by the area in compression in square inches. No deduction to be made for hollow spaces; the area will be considered as the product of the width by the length.

11. The absorption tests must be made as follows: The sample is first thoroughly dried to a constant weight. The weight must be carefully recorded. It is then placed in a pan or tray of water, face downward, immersing it to a depth of not more than one-half inch. It is again carefully weighed at the following periods: thirty minutes, four hours, and forty-eight hours, respectively, from the time of immersion, being replaced in the water in each case as soon as the weight is taken. Its compressive strength, while still wet, is then determined at the end of the forty-eight hour period in the manner specified in Section 10.

12. The freezing tests are made as follows: The sample is immersed, as described in Section 11, for at least four hours and then weighed. It is then placed in a freezing mixture or a refrigerator, or otherwise subjected to a temperature of less than 15 deg. F. for at least twelve hours. It is then removed and placed in water, where it must remain for at least one hour, the temperature of which is at least 150 deg. F. This operation is repeated ten times, after which the sample is again weighed while still wet from the last thawing. Its crushing strength should then be determined as called for in Section 10.

13. The fire test must be made as follows: Two samples are placed in a cold furnace in which the temperature is gradually raised to 1700 deg. F.; the test piece must be subjected to this temperature for at least thirty minutes. One of the samples is then plunged in cold water (about 50 deg. F. to 60 deg. F.) and the result noted. The second sample is permitted to cool gradually in air, and the results noted.

14. The following requirements must be met to secure an acceptance of the materials: The modulus of rupture for concrete blocks at twenty-eight days old must average one hundred and fifty and must not fall below one hundred in any case. The ultimate compressive strength at twenty-eight days must average one thousand pounds per square inch, and must not fall below seven hundred in any case. The percentage of absorption (being the weight of water absorbed divided by the weight of the dry sample), must not average higher than 15 per cent. and must not exceed 20 per cent. in any case. The reduction of compressive strength must not be more than thirty-three and one-third per cent., except that when the lower figure is still above one thousand pounds per square inch, the loss in strength may be neglected. The freezing and thawing process must not cause a loss in weight greater than ten per cent., nor a loss in strength of more than $33\frac{1}{3}$ per cent., except that when the lower figure is still above one thousand pounds per square inch, the loss in strength may be neglected. The fire test must not cause the material to disintegrate.

15. The approval of any material is given only under the following conditions:

    (a) A brand mark for identification must be impressed on or otherwise attached to the material.

    (b) A plant for the production of the material must be in full operation when the official tests are made.

    (c) The name of the firm or corporation and the responsible officers must be placed on file with the Chief of the Bureau of

Building Inspection, and changes in the same promptly reported.

(d) The Chief of the Bureau of Building Inspection may require full tests to be repeated on samples selected from the open market, when, in his opinion, there is any doubt as to whether the product is up to the standard of these regulations, and the manufacturer must submit to the Bureau of Building Inspection once in at least every four months a certificate of tests showing that the average resistance of three specimens to cross breaking and crushing are not below the requirements of these regulations. Such tests must be made by some laboratory of recognized standing on samples selected by a building inspector or the laboratory, from material actually going into the construction, and not on ones furnished by the manufacturer.

(e) In case the results of tests made under these conditions should show that the standard of these regulations is not maintained, the approval of this bureau to the manufacturer of said blocks will at once be suspended or revoked.

Tests made under the requirements of the City of Philadelphia on concrete blocks in the market here have developed about the following results·

Modulus of rupture, 150 to 175 pounds.

Compressive strength, f200 to 1600 pounds per square inch.

Absorption, 5 per cent.

The compressive strength is reduced little, if any, by the water absorbed.

Freezing tests show little loss.

The average compressive strength after freezing is in the vicinity of 1000 pounds per square inch.

The blocks passed the fire tests well.

A 10,000 HORSE-POWER STEAM TURBINE.

The two steam turbine sets of 10,000 horse-power each, which are being installed at the Rhenanian Westphalian Electricity Works. are the largest turbine sets. and in fact the largest stationary engines of all Europe. Each of these gigantic engines comprises a turbine running at 1.000 r. p. m., which is direct-connected to a rotary current generator of 5,000 kilowatts. 5,000 volts, and 50 periods per second, as well as to a direct-current generator of 1,500 kilowatts and 600 volts, and to a central condensing plant. The whole set is 20 meters in length, and weighs 190 tons, of' which 9.4 meters and 107 tons correspond to the turbine. The maximum height of the turbine above the floor is 2.6 meters. The turbine is of the single-cylinder type, and has only two bearings, one of which serves at the same time as a bearing to the alternator. The governor is made to compensate to 1 per cent. of any oscillations in the angular speed. with variations in the load as high as 20 per cent., while the maximum variation in the number of revolutions between running at no load and at full load is not to exceed 5 per cent. Another unit of the same size is shortly to be installed at the power station in a Westphalian mining company.—*Scientific American.*

TEN YEARS OF ELECTRIC REDUCTION AT NIAGARA FALLS.

The present month brings the tenth anniversary of the establishment of electro-chemical industries at Niagara Falls. On the developments of the decade the *Electro-Chemical and Metallurgical Industry* comments as follows:

On August 26, 1895, the Niagara Falls works of the Pittsburgh Reduction Company started operation. On October 19 of the same year the current was turned on at the plant of the Carborundum Company. Thus, it is now just ten years that electro-chemical activity started at Niagara, and in this short space of time more than a dozen varied electro-chemical industries have grown up and are flourishing within a radius of two miles from the falls. In electric furnaces are made at Niagara artificial graphite, siloxicon, silicon, carborundum, alundum, calcium carbide, phosphorus and various ferro-alloys. By electrolysis of fused electrolytes are made aluminum, sodium (for the production of various important derivatives), and caustic soda and chlorine. The latter two products are also made by electrolysis of aqueous solutions; while other products of such processes are caustic potash and hydrochloric acid and chlorates. The only example of a process using electric discharge through gases on a commercial scale at Niagara is the production of ozone for the manufacture of vanillin.

As the latest developments we may mention that ground has recently been broken for the caustic soda and chlorine works, which will use the Townsend diaphragm cell, while in the old barn in which Mr. Rossi made his pioneer experiments on ferro-titanium Mr. Rothenburg's process of agglomerating in the electric furnace magnetic iron concentrates now undergoes an experimental trial. This is certainly a splendid development, and while as a necessary concomitant to the success achieved there have been commercial failures, yet their number is remarkably small. The most notable one is probably that of the Atmospheric Products Company, on the success of which very great hopes had been founded. In the interest of the problem it it greatly to be regretted that no exact information on the causes of the failure has been published. Certainly the problem itself cannot be considered to be dead forever.

THE HEAVIEST RAILS.

The rails on the Belt Line Road around Philadelphia are the heaviest rails used on any railroad in the world. They weigh 142 pounds to the yard, and are 17 pounds heavier than any rails ever used before. They are ballasted in concrete, and 9-inch girders were used to bind them. All the curves and spurs were made of the same heavy rails, and the tracks are considered superior to any railroad section ever undertaken. The rails were made especially for the Pennsylvania Railroad by the Pennsylvania Steel Company. An officer of the railroad company states that this section of roadbed will last for twenty-five years without repairs.—*International Railway Journal.*

CHEMICAL SECTION.

(Continued from Vol clx, p. 368.)

*Mica and the Mica Industry.

BY GEORGE WETMORE COLLES.†

[In the concluding section of his exhaustive paper, the author deals with the statistics of the mica industry.—THE EDITOR.]

VII. STATISTICS.

The statistics here given have been worked up from three principal sources; namely, the charts and some pamphlets on the subject, issued by the United States Geological Survey; the annual volumes of the "Mineral Industry," published by the Editorial Staff of the "Engineering and Mining Journal;" and the statistics on imports published by the Bureau of Statistics of the United States Government.

Concerning these authorities a word here may not be out of place. With regard to the figures of the United States Geological Survey, it may be said of them, in the words of Mr. Rothwell, late editor of the "Mineral Industry," that they are "frequently but little short of absurd;" or rather it should be said, that they are in many cases *wholly* absurd, and the phrase

---

*Read by title.   †Copyright, 1905, by George Wetmore Colles.

"little *short* of absurd" may be apropriately reserved for some of Mr. Rothwell's own figures.*

However, as between the two I prefer to rely on those of the "Mineral Industry," and in justice to that invaluable publication it must be said that outside of foreign sources (for which the editors can hardly be held responsible) the discrepancies are few and the figures in general accord with known facts. The government figures are on the contrary open to such great objections that they are not to be accepted without much reserve. I have, therefore, on the accompanying charts unhesitatingly followed the former authority.

The statistics of imported merchandise published by the Department of Commerce and Labor should, it would seem, be

---

*As instances of the former. I may cite the value of asbestos, which is given by the Geological Survey as about $17.00 per ton, whereas it has been computed by a competent authority that the actual cost of mining alone amounts to about $55.75 per ton; or again, where the value of the rather common mineral *rutile* is given as $5.00 per *pound*, when the retail price quoted in the catalogue of a large New York dealer is thirty cents, and the average valuation of imports is about eleven cents per pound. As an instance of the latter, may be cited Mr. Rothwell's values for sheet mica produced in "other States," which include small amounts mined in western States, such as Idaho, Nevada and Wyoming. These he puts at an average of $2.25 per pound for 1896-7, or two and a-half times the value of North Carolina mica, while in 1898-9 it had suddenly dropped to $1.00, which is still in excess of the latter. These are for mica at the point of production, and in view of the known inferiority and lower market prices of the western mica, added to the large excess in transportation rate to the nearest market, it would seem certain that the correct value would be considerably less, instead of more, than that given for North Carolina mica. Again, the total value of imports into the United States for 1892 are given as $100,846, while upon the next page those from Canada alone are valued at $150,053, in spite of a large importation of Indian mica; and although the United States value is for the fiscal year, yet the value given for the succeeding year. 1893, is only $120,864, so cannot be said to lend an explanation.

The statistics reported for India are in no better condition, for the total exports are found to exceed the production, notwithstanding that a certain portion of the latter must be used in native industries, and notwithsanding this, the values of the exports are quoted at something like ten times the values of the product up to 1898, when the production value suddenly jumps to a figure approximately equal to the export figure. In the statistics here given both the export quantities and values have been taken as the correct ones and at all events those which concern us much more than the actual quantities and values produced.

exact as far as they go, at least they pretend to be, and there is no reason why they should not be so. Unfortunately the imports for mica have been classified only since the year 1894. These statistics are given for fiscal years ending June 30, and for purposes of proper comparison they have been reduced to calendar years in the charts by taking for any calendar year one-half the value given for the fiscal year ending on June 30 of the calendar year, and one-half of the amount for the year following.*

In like manner the statistics on Indian exports, which are taken as equal and which for our purposes are practically equivalent to the total Indian production, are given for fiscal years ending March 31, and the chart-points are figures for calendar years, found by taking one-quarter of the amount given for the fiscal year and three-quarters of that for the following year.

In addition to these sources, some fragmentary but, so far as they go, valuable figures relating to the Canadian industry are given by J. Obalski, Mining Engineer for the Quebec gov-

---

*It will of course be understood that by this method the exact amount imported during any calendar year is not found, and frequently there is a very wide variation between the amount so found and the actual amount; but this is of no real consequence in the chart, and in fact there is a reason why the average so found is better for our purposes than the true amount. The chart line or "curve" so found will be smoother and freer from jagged variations and thus show more clearly the trend through a series of years. Consider, moreover, that as a matter of fact the amount imported in any year does not agree with the amount exported from other countries, not only because of the time of transport, but also because a large fraction of the total imports is re-exported from England after lying for an indefinite period; nor does it agree with the amount of imported mica actually used, that is, entered for consumption for that year, because imported material lies stored an indefinite time after import in the Customs warehouses. These considerations will help show the futility of supposing that there can be anything like exact statistics in this industry.

†None the less it is very regretable that the native mica consumption of India cannot be given nor tabulated. What it amounts to is a mystery. By one authority it is given as several times as large as the exports; by another as about equal; while the Government statistics are so utterly discordant that nothing can be learned from them. No doubt exists, however, that India's 225,000,000 inhabitants consume an astonishing quantity of mica.

ernment, in his extended report on "Mica," 1901, which have been used as far as possible.*

The values in the accompanying charts are in all cases, except for India (those at the point of production, so that the same article will be of *less* value in proportion to the distance of the point of production from the market, and conversely, articles which appear of *equal* value on the chart would really be of superior quality in the case of the most distant point of production. With these remarks I think the diagrams will be self-explanatory.

It will be seen that the statistics at hand are not to be taken too seriously or punctiliously, but in a merely general fashion. There is unquestionably great difficulty in collecting mica statistics, more so than in the case of other minerals, owing to the reticence of the mine owners, and their reluctance to tell anything whatever with relation to their business. It is believed that in many cases the information reported is purposely falsified. It is regrettable also that the figures extend back only for ten to twenty years, and cannot be fully brought down to date owing to the delay in reporting statistics. Nevertheless I believe that such statistics as are here given will prove vastly better than none at all, as indicative of the present actual state of the mica industry and its probable future.

In all the charts, both the *quantities* and *values* are given as far as possible, both being mapped upon the same chart, and to distinguish them the quantities are drawn in light, the values in heavy lines. While both are determinative to some extent, and important as means for forming a judgment, the *values* will generally be found the most accurate guide to the state of the industry in each case.

The Canadian product as tabulated in the charts includes *all* of the product of Canada, of which trifling amounts consist of

---

*Here also it is to be noted that several departments of the Dominion government keep independent statistics of the mica production, and taken together they form a beautiful charivari of discord. To find the true values from this medley is a puzzle which would indeed tax the abilities of the proverbial "Philadelphia lawyer." The figures are dependent on the exertions of two provincial governments as well as of the Dominion government. The statistical branch of the Ontario government is evidently not all that it should be, even if the others are.

white mica, insufficient, however, to alter sensibly the general proposition that Canadian mica and amber mica are equivalent terms.

The statistics of the United States imports, as well as foreign reports, show that a great number of other countries from all the continents figure as mica producers in a small and sporadic way, but all these countries put together form in proportion a total too insignificant to be perceptible on the scale of the charts. The only three countries which have ever been steady producers for any length of time are *Germany*, whose lepidolite deposits on the Bohemian border are mined for lithia in small amounts annually;* *Australia* (South and West), which produced about $29,000 worth between 1891 and 1900, since which time production has ceased; and *Russia*, which produced about $14,000 worth between 1891 and 1898, since which time none is reported. It is perhaps worth remarking that even Great Britain produced about six tons in 1897, valued at $8,635.

Although several countries of Europe have furnished and continue to furnish exported mica in trifling quantities to the United States, this can hardly be called native mica and is presumed to be imported mica manufactured in those countries before reshipment.

World's Production of Mica. Fig. 26.

In this chart the whole mica output of the world is shown as a total, and also distributed according to the three mica producing countries. As a natural result of thus lumping together the several nearly independent branches of the industry, there is discordance between the value and quantity curves. Besides the production of sheet mica, the great quantities of scrap for the ground mica industry, and the small quantity of lepidolite mined for lithia ore are included ton for ton. As a general indication of the status of the industry, the chart will be useful,

---

*The mica statistics in the German reports are consolidated with quartz and uranium ore so that it is impossible to separate them, but the whole amount is inconsiderable, embodying some sheet mica and some massive lepidolite.

referring, however, to the values only. The value lines show a slow decrease in the total mica produced from 1890 to 1894, and from 1894 a rapid increase up to 1902, amounting to a boom, the annual production rising from less than $300,000 to $800,000; that is to say, more than doubling. This boom

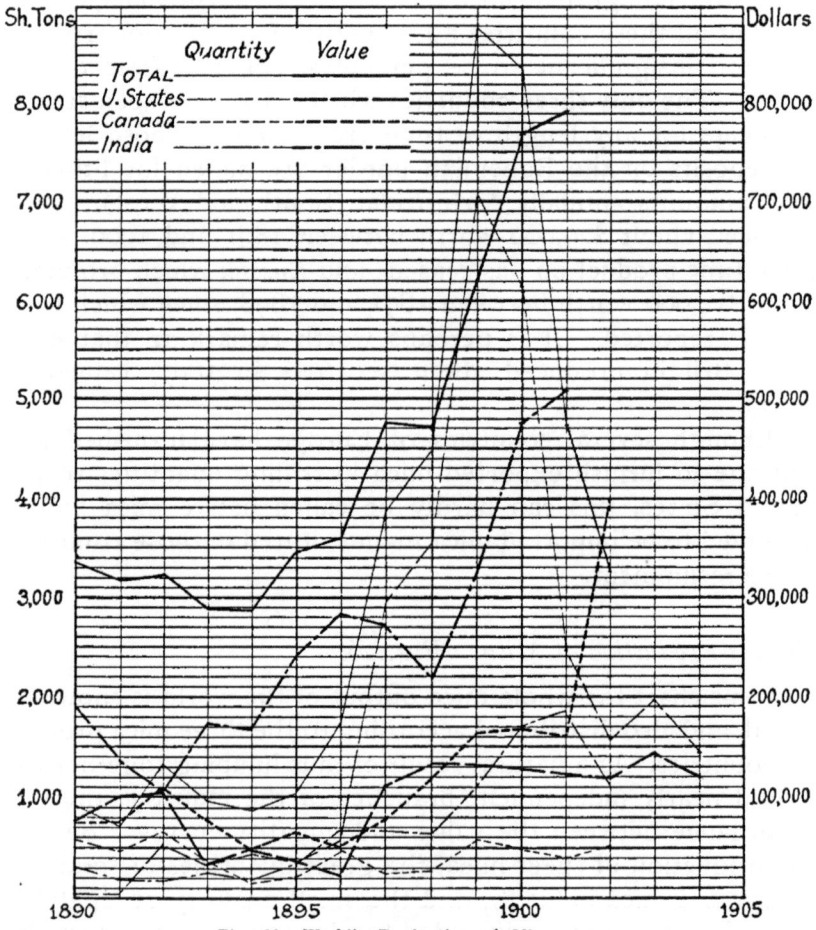

Fig. 26. World's Production of Mica.

was contemporaneous not only with the trade revival in the United States (the largest mica-consuming country), but more particularly with the great era of growth in the electrical industry, and also with the rise of the ground-mica industry to a place in the staple market. It was, however, the former and not the latter cause which was the really important factor, as will be seen by consulting the several subordinate value-lines. The United States shows a generally declining state of the industry until 1896, and from that point to a sudden rise, which took the mica industry back to its status prior to 1888 (not shown in this chart) which lasted, however, but two years, and since then has barely held its own, with a declining tendency. The increase from 1896 to 1898 was due to sheet mica principally; but less to the aforesaid ground-mica industry, which took its great rise about this time, and it will be well seen how little effect it had on the continuous increase of the line of total values.

The Canadian value line shows likewise a declining state of the industry up to 1896, when it also began to increase at a rapid rate, which increase has been maintained ever since, but unlike that of the United States, it was not due to the short-lived "boom" of a new product,—in fact the ground-mica industry in Canada came to its end just at this time; but it was due solely to the rise of the electrical industries and the increasing appreciation of the value of Canadian mica for that purpose.

The Indian value curve on this chart shows a nearly steady increase since 1892, since which time it has quintupled in amount.

The great fact, however, indicated by this chart is the comparative proportions of the mica industries of the different countries. It is to be regretted that the figures only go back to 1890. If it were possible to carry them back, we should find a time less than ten years previous when the positions of the several countries would be actually the reverse of what we find them at present. In 1892 the United States industry had declined, and the other two had increased, until they were on a par with each other, each producing one-third of the world's total supply of mica, and though the great rise of the Indian industry dates from that year, the United States and Canadian industries declined still further, keeping nearly side by side with each other, but with Canada slightly in the lead, until 1896, since which time Canada has far outstripped the United States, and bids fair to rise to a parity with India, notwithstanding the rapid increase of the latter.

The quantity-lines of this table show nothing except the impropriety of adding together scrap and sheet mica. During the years 1869 to 1902, the total curve is dominated by that of the United States, whose 7,000 tons maximum is nearly all scrap, worth a fraction of a cent a pound. Whatever is valuable in these quantity-curves can best be elucidated from an examination of the subsequent charts for sheet and scrap mica respectively.

World's Production of Sheet Mica. Fig. 27.

The statistical data at hand have enabled these figures to be stated for the United States as far back as 1880, while those for Canada run back to 1886, and those for India back to 1889. Canada produced little or no mica before 1885, but India has produced and exported mica to Europe certainly for more than a century,* and has exported to the United States since 1883. The line of total values is not very different in character from that of Figure 26, but the addition of the year 1889 is very significant, as carrying it back to a new low point in the past. It will be seen that since 1889 the sheet-mica industry has multiplied itself more than six times.

The line of total quantities in this figure is chiefly interesting as showing more violent fluctuations than that of total values. In particular it is to be noted that, while there was an advance in the total value in 1902 over the previous year, there was a considerable falling off in the total quantity.

The United States value line is probably the most interesting feature of this chart, indicating as it does a condition of prosperity in the early eighties far in excess of anything the industry has enjoyed since. It was due, however, as will be seen from the quantity line, not to a greater quantity of mica mined,

---

*Fritz Cirkel, in a memoir recently published by the Mines Branch of the Canadian Government, states that the first exports from Bengal were in 1863. This is doubtless taken from British Customs Reports and indicates the first year in which exports of mica were of sufficient commercial importance to receive separate classification. Mica was sent to China from the Indian mines centuries ago, and the name "muscovy glass" indicates a similar importation into Western Europe through Russia rather than a hypothetical native Russian mica industry.

but to the high price obtained for it in the early days. During the years 1893 to 1896 the industry of mica mining was all but extinguished in the United States, and had dwindled to insignificant proportions compared with those of Canada and India.

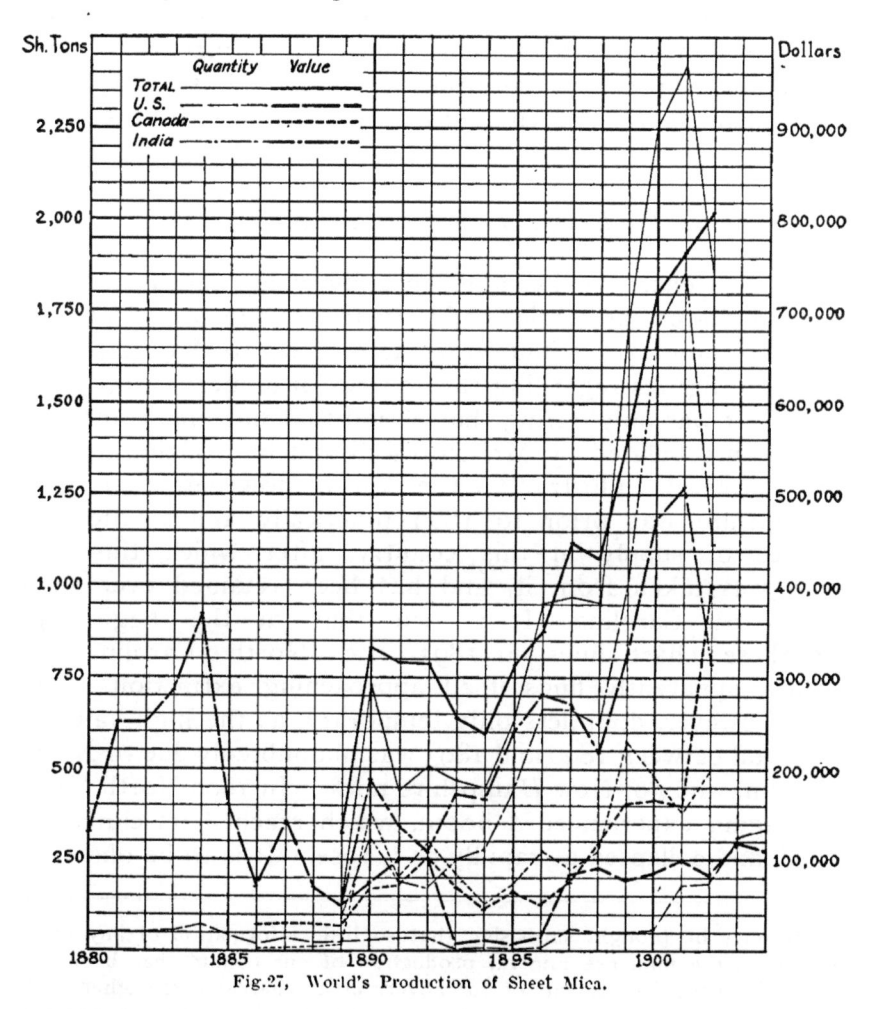

Fig. 27, World's Production of Sheet Mica.

Since 1896, however, the industry shows sure signs of a permanent revival, for while the total value of the product has fully held its own, in the face of falling prices, the volume has continuously increased.

The Canadian and Indian value lines show substantially the

same features as were remarked in considering the previous chart, and both show a marked contrast to that of the United States. The Canadian line is one especially indicative of a young industry, and one having a great future before it. The Canadian and Indian quantity lines show a tendency to separate to a constantly greater extent, indicating an increasing value per pound of the Canadian product, and a diminishing value of the Indian, as will be seen later on in connection with the price-charts.

United States' Production of Sheet Mica. Fig. 28.

On this chart we examine on a more magnified scale the United States sheet mica industry from 1890 to the present time, to show its distribution by states. Up to 1893, and since 1900, no figures regarding the sources of domestic mica were kept. It was only during that period that this important data has been tabulated by Mr. Rothwell and his collaborators in his "Mineral Industry."* We know, however, that New Hampshire was prior to 1869 practically almost the sole source of supply in this country, after which time the lead was taken from it, and has been retained ever since by North Carolina. It is true that, at the beginnings of these chart lines in 1893, the North Carolina industry apparently finds itself approaching extinction. This was, however, doubtless only temporary, as the high value of the total between 1890 and 1893 was undoubtedly due to North Carolina and not New Hampshire mica. In 1896 North Carolina again experienced a boom, which reached its climax in 1898, and collapsed as rapidly as it rose. It is to be observed

---

*It may be thought surprising that so little information of any value regarding the immense mineral production of our country has been obtained from the departments of our government, whereas every other civilized government keeps such statistics. Such is the fact, however, and the figures that are put forward by the Geological Survey deserve but little reliance. This is partly, no doubt, because the statistical force is inadequate, its organization is inadequate, its means of support are inadequate, and the means at its command for collecting information is inadequate; at all events its secrecy is probably not trusted by those making reports. I reserve the right to suspect also that its brains are inadequate.

that the rise is much greater in the value lines than in the quantity lines, indicating a boom in price rather than a boom in actual mining. It may be added (what is not shown on the chart) that the New Hampshire industry has since then sunk into a condition aproaching extinction.

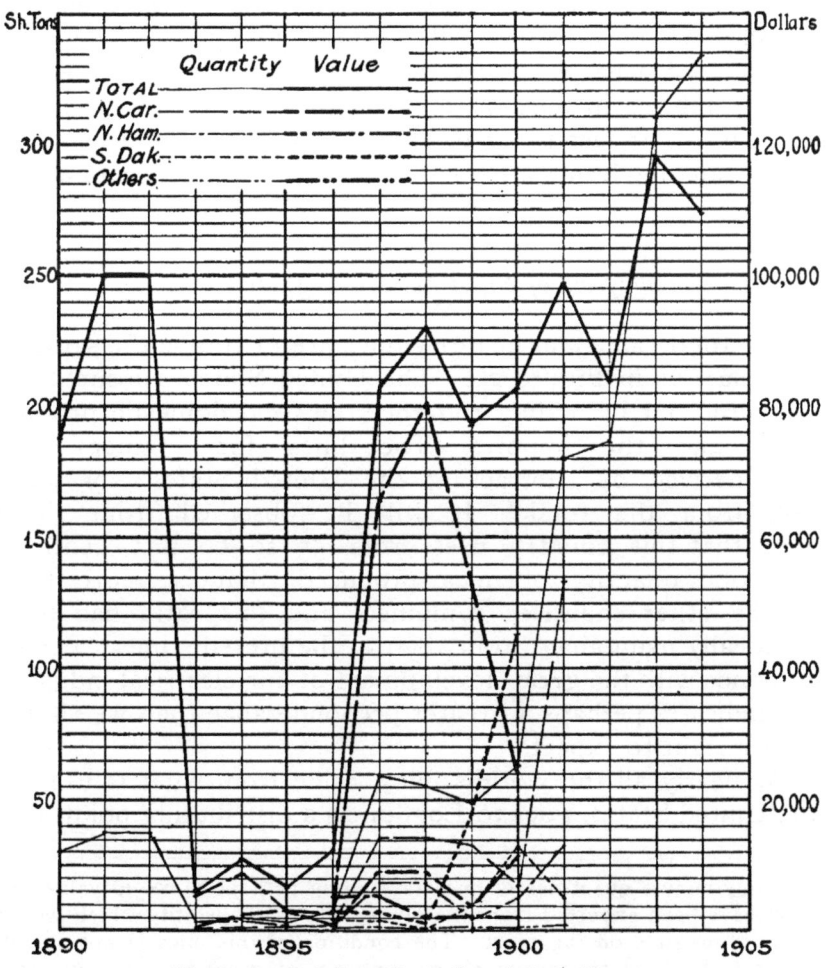

Fig. 28. United States Production of Sheet Mica.

Mica from other states first shows on the chart in 1896, in which year small quantities were produced in South Dakota, Nevada and Idaho, principally from the first. The Idaho industry has continued active in a small way up to the present time, and may perhaps grow as a source of supply for the Pacific coast. The South Dakota industry has experienced a great boom, and in 1900 took a place apaprently ahead of that of North Carolina. This is a most surprising state of affairs, considering that every physical advantage and the quality of the mica is in favor of the latter state. It is not believed that the state of South Dakota will permanently maintain this lead, but the lead it has evidently reflects great credit on the enterprise exhibited in pushing forward the product into western markets. The fact tends to show strongly the influence that energy and modern methods may have on the success of an industry.

World's Consumption of Sheet Mica. Fig. 29.

The total value and quantity lines are of course the same in this chart as in Figure 26. The United States' consumption is figured out by adding together the total domestic production* and the imports. The lines for "Others' are then derived by subtracting the United States' consumption from the total.† It will be observed from these lines that this country uses from 55 to 60 per cent. of the total quantity of mica produced, that is, more than all the other countries combined, and this ratio has been practically maintained from 1890 to the present time. It corresponds with the great development of our electrical industry, but indicates perhaps a greater preponderance than we should have a right to expect.

As nearly all the world's mica, aside from what is used in the United States, goes to London as a distributing point, it is

---

*Only minute quantities of mica (averaging perhaps $1,000 in value per year) have been exported from the United States, a quantity which would be quite invisible on the chart. The condition of this mica is not stated, but it is believed to consist almost entirely of ground mica, thus still further reducing the actual quantity.

†The figures for United States consumption thus obtained for 1892 is slightly greater than the total production reported for that year making the "Others" line a negative quantity. Of course the fact is that the amount of mica carried in stock is not taken into account in this chart, which accounts for such individual discrepancies.

hard to say precisely how the product is to be divided between the different European countries. The following summaries of shipments from India to the different countries for the six

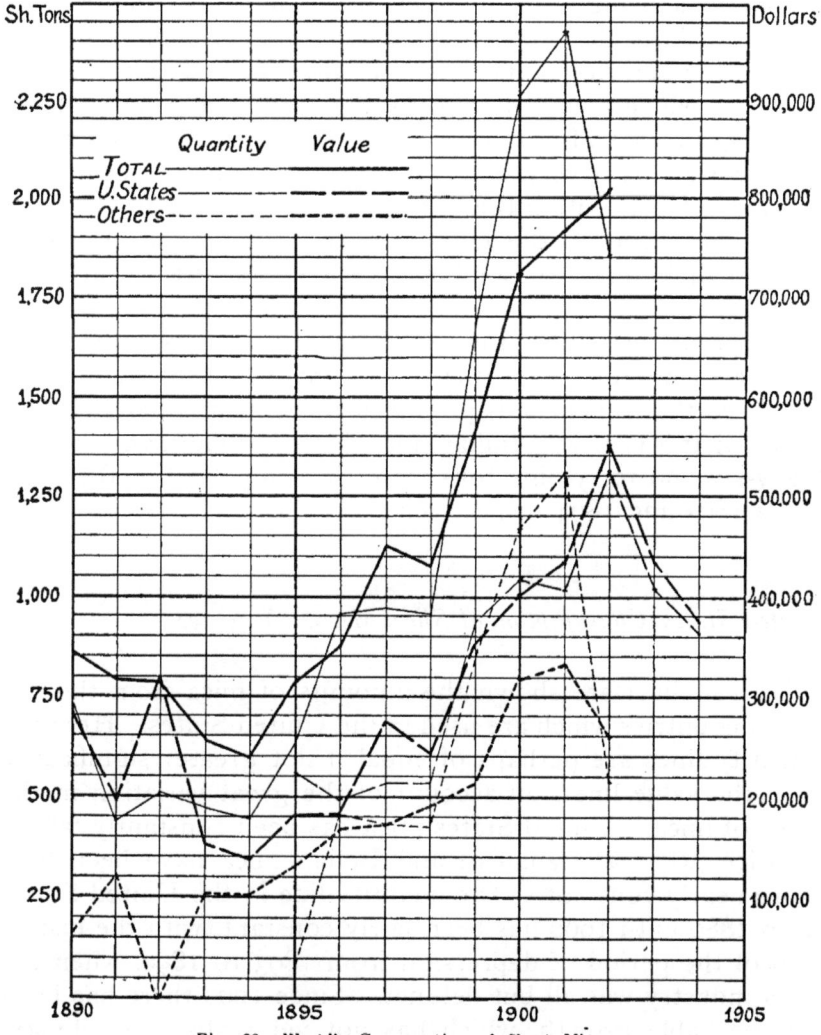

Fig. 29. World's Consumption of Sheet Mica.

fiscal years, 1891 to 1897, indicates in a general way the distribution: United States, 808 tons, 38½ per cent.; Great Britain, 1215 tons, 58 per cent.; Germany, 52 tons, 2½ per cent.; all others, 19 tons, 1 per cent.; total, 2094 tons.

Of the quantity shipped to Great Britain, nearly one-half was re-shipped to the United States. Of the balance, doubtless a considerable quantity went to Germany and France, small proportions being used by such manufacturing countries as Belgium, Australia, Italy and Switzerland. Needless to say, only manufacturing countries use mica in any considerable quantity, which means practically the United States, England, Canada, and a half-dozen European countries. The amount used by Canada is very small, and the exact amount cannot be learned. Up to within the last few years substantially all Canadian mica came to the United States, but now a large proportion of it goes to London, which is the great European mart for mica.

In recent years also the government statistics show large and increasing mica imports from Germany. As it is not to be supposed that this mica is produced there, it may safely be assumed that it is merely Indian mica floated in German bottoms. Sixty-six and forty-nine tons were thus imported in the fiscal years 1893 and 1894 respectively.

United States' Consumption of Sheet Mica. Fig. 30.

In this chart are shown the amounts of mica from the different countries which are used in the United States. Here the quantity lines are perhaps of equal if not greater significance than the value lines, on account of the great variation in the price of mica. United States mica has been constantly falling in price, whereas Canadian and Indian mica have been held more nearly constant. The quantity line in the United States from 1880 until 1900 has been fairly constant, with the exception of the period of depression from 1893 to 1896, when the mica industry was all but defunct. Since 1900 there has been a remarkable growth, which has not yet realized its climax. We must place one or perhaps several question marks after the United States figures, owing to the great difficulty in obtaining information as aforesaid.

Although 1883 is given as the first year when mica was imported from India, there were importations of mica reported long previous to that time,* and this was in all probability India mica shipped to England and re-exported from there to the United States. It has been so considered in the chart. The imports of India mica beginning with 1883 grew rapidly

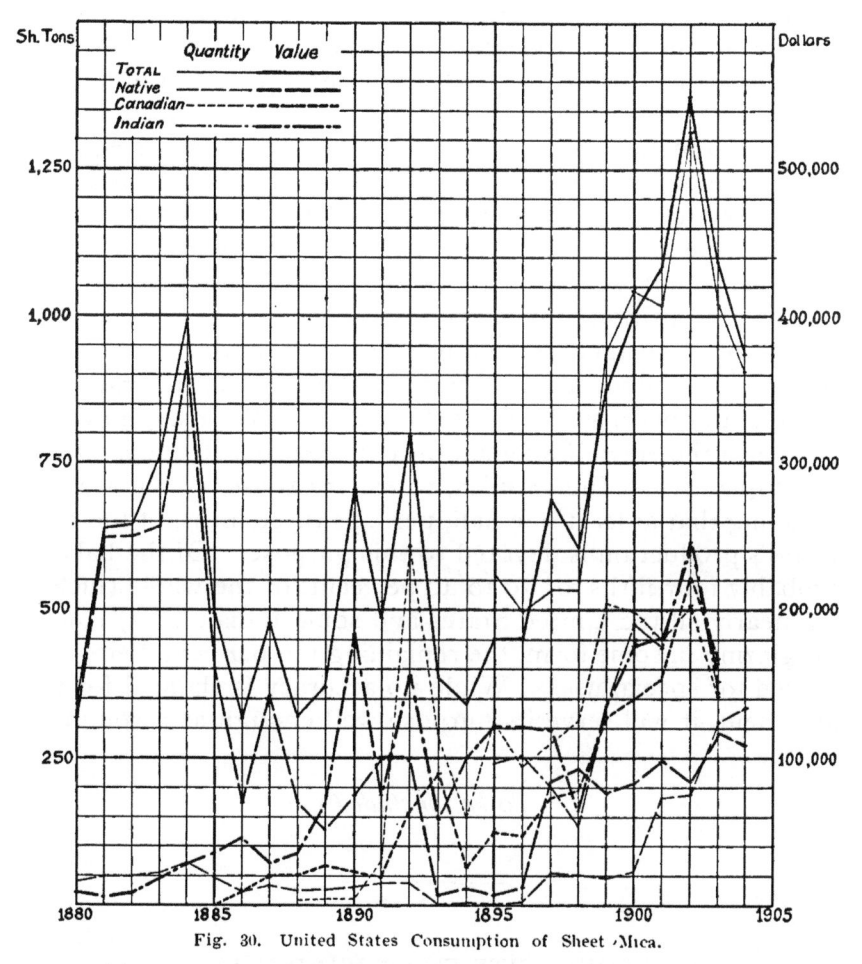

Fig. 30. United States Consumption of Sheet Mica.

up to about 100,000 tons annually from 1890 to 1898, during which time they were fairly constant around this figure, but since then the imports have practically doubled in volume and

---

*The year mica imports are first reported in the United States is 1869, and the quantity imported from then up to 1880 amount to about $40,000, or about $3500 a year.

value, and still outweigh the total of the United States production. Canada mica in the United States has experienced the same course of events, in fact the entire Canadian mica industry has, until recently, depended on the United States for a market. Beginning in 1886 imports of Canadian mica grew rapidly to nearly $100,000, an approximate equality with those of India, in 1893, but fell off during the period of depression from 1894 to 1896. From 1897 they experienced a rapid increase, and have since maintained an approximate equality with those of India. Both were for several years far in excess of the production of the United States, which in 1893 amounted to but little over five per cent. of the total consumption of mica; but the latest statistics show the native product has risen nearly to a position of equality with them.

It is of importance to inquire how the total production of sheet mica is divided according to uses. It is naturally impossible to obtain statistics of this nature. We may, however, put down the entire Canadian output, 500 tons in 1902, and an equal quantity of Indian mica, to electrical uses. We are safe in saying that at least one-half, perhaps three-fifths, of the total world's production is devoted to this purpose. Of the balance probably four-fifths, or 30 to 40 per cent. of the whole (including nearly all the United States and some Indian mica) is used for glazing in some form, the remaining ten per cent. being employed for special uses. While no accuracy is claimed for this allotment, it will serve its purpose well enough as a rough estimate.

(*To be concluded.*)

PROPER SITE FOR A WIND MILL.

The safest and most secure situation for a wind mill to avoid danger of being blown over has been generally supposed to be a place sheltered by trees or barns. Such, however, is not the case. The safest place for such a tower is on a hill, where the wind can strike it equally from all directions. In such a location shifting winds are less pronounced than behind buildings or hills, and it is also found that there is less lifting force to the wind in the open than behind structures.

THE FRANKLIN INSTITUTE.

(Stated Meeting, held Wednesday, October 18, 1905.)

Etching by Machinery.

By Louis Edward Levy.

Photo Chemist, Member of the Institute.

[Presentation of an Etch Powdering Machine, a newly invented device for applying resinous powder to metal plates in preparation for etching. The machine is designed to supplement the Levy Acid Blast in the process of etching by machinery. A number of plates powdered for various stages of the process were exhibited. The description of the apparatus is preceded by a consideration of the subject matter in general.—The Editor.]

It is only sixty-five years since the discovery of the Daguerreotype process of photography suddenly widened out the narrow confines of the Graphic Arts into a new and larger field. And it is only about fifty years since this field was expanded by combination with the lithographic processes; scarcely more than thirty years since it was extended into the region of the typographic arts by combination with the etching process and but little over twenty years since this, in turn, has been further enlarged into a practically boundless domain by the autoglyphic, or so-called half-tone process of photo-engraving.

When the illustrative value of the half-tone process, as shown by Ives in the early eighties, and its practical possibilities as indicated by the Meissenbach method of the same period, had been realized by my invention of the engraved line screen in 1887 and its practical perfection by my brother, Max Levy, in 1891, the way was opened for the general use of halftone as well as line etchings in the illustration of daily newspapers. It was not long before commercial competition, along with the exigencies of modern journalism, led to various efforts to expedite the etching process, generally through the use of stronger mordants, and though something was gained in this direction, frequently at the expense of the workman's health, there still remained a further need of greater brevity in the process and the inevitable demand for greater brevity in the cost. These requirements were sought to be met by my invention of the Acid Blast method of etching, which I had the honor of bringing out before this Institute in February, 1899. Much time and effort had, however, yet to be devoted to the perfection of the machine for the application of this process. The main obstacle that presented itself from the beginning was the necessity of making the apparatus proof against the nitric acid carried by the air blast, and the structural difficulty of working a thoroughly acid-proof material into the requisite form with an adequate degree of accuracy. The machines first put into practical use were made entirely of aluminum, that metal affording sufficient resistance to the action of nitric acid to make it practically proof against the dilute solutions used with the blast. In time, however, it was found that while parts of the machine which were subjected most to the acid blast, but which were intermittently washed by the water spray, such as the cover and plate-carrier, remained intact, the aspirators and other interior features of the machine were strongly affected by the resultants of the process of etching zinc. It developed that the aluminum gave way through the formation of its hydroxide, the trihydrate $Al_2(HO)_6$, doubtless due to the presence of free hydrogen in the course of the decomposition of the zinc by the acid. The use of vulcanized india rubber was also found objectionable, as that material is apt to become brittle from the effects of the nitric acid, and the atomizers to become gradually enlarged by the abrasion of the acid blast, thus changing the character of the spray. Eventually the entire apparatus, except the cover, carrier-plate and outer mechanism, has been made of a vitrified material which is permanently unaffected by the acid, or its products. In this form the machine has gone into extensive use and the blast method of etching, through its economic and sanitary advantages, has become firmly established in practice.

In the course of early experience with the acid blast it became apparent that the "powdering" of the plate, a preliminary operation of the etching process, could also be advantageously effected by means of a mechanism adapted for the purpose. This powdering process consists, at the outstart of the work, in a hardening of the soft ink composing the features of the design, so that the metal surface under the ink be thoroughly protected from the action of the acid while the rest of the surface is being etched by it. This is effected by covering the plate with a resinous powder, some of which adheres to the viscid body of the ink when the surplus is brushed away. The plate so "topped" is then heated to the point where the resin melts into combination with the ink, making, when cold, an effective resistant to the acid. This operation is comparatively simple, requiring only a reasonable degree of care in

Fig. 1. Etch Powdering Machine. Front and Working Side.

the distribution of the powder, the clearing of the surface and the heating of the plate.

But after the plate has been etched to a depth where the sides of the projecting lines or other features of the design require to be protected as well as their surfaces, the proceeding becomes more complicated. This may be effected by the so-called "rolling-up" process, which consists of passing over the wetted surface of the plate a roller carrying a fatty ink, which, while rejected by the moist surface, adheres to the resinous covering of the design. After being dried, the plate, with its fresh supply of ink, is again covered with the resinous powder, so prepared as to be limpid when fused; the surplus is blown away with a bellows and the adhering powder melted and allowed to run down over the edges of the projecting lines upon their sides. The plate is then again etched and the process repeated, usually five or six times, until the required depth has been reached. This method, known in this country as the European process, is still extensively in vogue on the Continent, where it was originally worked out. In 1881, while I was employing this process commercially, I hit, or to be, perhaps, more accurate, stumbled on the idea of laying the powder against the sides of the lines with a flat brush, melting it into place, and repeating the operation in the four directions at right angles across the plate. This method of protecting the sides of the lines proved more expeditions in itself, and by packing a considerable bank of the powder against them afforded an amount of resistance to the acid that permitted the successive etchings to be continued to a greater depth, and so reduce their number by half. With the growth of the photo-engraving industry at that period, when the swelled and wash-out gelatine methods were beginning to give way to the etching process, the four-way powdering method gradually spread into general use in this country, and came to be known in Europe as the American process.

The four-way powdering operation, being essentially mechanical in its nature, clearly lent itself to the application of machinery. As a supplement to the acid blast particularly, such a mechanism became the more desirable as the action of the blast, being mechanically uniform over the surface of the plate, required that the powdering of the design to prepare the plate for the action of the acid, be likewise accurately uniform over the entire surface.

On a small plate a trained and skillful hand can lay the powder evenly enough for all practical purposes, but when the plate is as large as a newspaper page, and especially when the brushing has to be done in the hurry and rush of a newspaper etching room, the work is inevitably more or less notably defective. Such plates, whether etched in the tub or in the blast, have here and there to be "cleared," that is, the inequalities of the powdering have to be reduced by dint of local etching with a brush.

Besides this technical consideration in favor of powdering by machinery there was also the want, not to say the need, of some means of relieving the operator from the strain of handling these large plates and from having to work in a powder-laden atmosphere while brushing them. The saving of time in the work, particularly with the larger plates, and still more where these are of double thick gauge or thicker, and correspondingly heavy, afforded but another reason for the use of machinery in powdering them.

Accordingly, in 1898, I began the construction of a mechanism for the purpose. The problem presented by it appeared simple enough at the start, but proved to be complicated by the necessity of making provision for the variations in the "packing" quality of powdered dragon's blood (which is the preferable resinoid for the purpose), under changes of the temperature and the moisture of the atmosphere. The first machine, completed in 1899, was succeeded by one after another of several modifications of the original design, but since October, 1903, a perfected machine has been in successful operation in the works of the Boston *Herald*. A number of others have since been put to work, notably at the St. Louis *Republic*, the New York *Journal* and the Philadelphia *Press*, each, however, modified in minor details to meet some special requirement.

The machine is represented in the accompanying autoglyphs and diagrams. It takes a plate of any size up to 24 inches square, which, if not thicker than the customary sixteen gauge, can be "topped," that is to say, prepared for the first etch, in one and one-quarter minutes. Thicker plates require correspondingly more time in the heating and cooling.

The "banking" of the powder, whether "wide" or "close," being primarily dependent on the nature and condition of the powder in use, is controlled by means of a movable handle, which determines the position of the brushes in relation to the plate and the degree of their pressure upon it. Apart from this preliminary adjustment, the process is entirely automatic.

The plate to be powdered is laid on the receiving table, face up, against the projecting prongs of a carrier bar; the machine is then started; a small handle is next pressed down to bring the cylindrical feed-brush into operation; the plate is now carried forward by the carrier-bar under the feed-brush, thence onward under a gang of elliptically-moving flat brushes which pack the powder on the plate and clear away the surplus, still forward through a gas furnace which melts the powder in place, and finally on to a cooling table, where the hot plate is swept on both sides by an air blast, which quickly cools it in readiness for

Fig. 2. Etch Powdering Machine. Back and Driving Side.

a repetition of the proceeding in another direction through the machine.

Putting the plate on its proper position on the receiving table, adjusting the gang brushes if necessary, starting the machine to go and bringing the feed-brush into operation is all the etcher has to do. When the plate passes the feed-brush the latter is automatically lifted out of operation; when it reaches the furnace the gas is automatically turned on and lighted; when it arrives at the cooling table the air blast is automatically started; when it emerges from the furnace the gas is automatically cut off, and when the plate is cooled to a proper temperature the air blast ceases and the machine automatically stops. By this time, seventy-five seconds from the starting, the carrier-bar is back in its place at the front of the machine and the plate is ready for the next operation.

The furnace of the machine can be used by itself, independently of the rest of the mechanism, by turning on the gas through a lever-handle cock on a by-pass. The cover of the furnace is hinged, so it can be raised to take in a plate to be warmed or dried.

DETAILS OF THE MACHINE.

The plate is carried from the receiving table, A, through the powdering compartments, B, B, by means of the carrier-bar, C, attached to two sprocket chains, D, D. The carrier pushes the plate into the gas furnace, E, upon a train of sprocket chains, F, which, along with another similar train, G, carry the plate out upon the delivery table at the further end of the machine.

The powdering apparatus embraces principally the powder feed hopper, 1, the sweep-brush chamber, 2; the surplus powder receptacles, 3, 4; the gang-brush chamber, 5, and the fan chamber, 6. One side of the hopper-wall is movable on a pivot, 7, and held in place by a clamp, 8. This movable side abuts below against the surface of a feed screw, 9, which closes the outlet of the hopper.

The feed screw delivers the powder against the cylindrical feed-brush, 10, which revolves in a direction against the movement of the plate. It is operated by the worm wheel, 11, fastened on its shaft end and engaging with the worm, 12.

This worm wheel is normally out of contact with the worm, leaving the feed-brush out of operation; it is brought into operation by pushing down the small lever, 13, which operates cams that pull the worm wheel into engagement with the worm.

The small levers, besides operating the cams as above, also actuate a shaft to which are attached two dog-levers, 14, one on each side of the mechanism. These levers are engaged by lugs, 15, on the carrier chain, 16, placed just in front of the carrier bars, 17, the lugs in passing pull the dog-levers forward and thus turn the shaft and the attached cams sufficiently to lift the worm wheel of the feed-brush out of engagement with the worm, placing the feed-brush out of operation and allowing the carrier bar to pass freely under it.

After receiving powder from the feed-brush, the plate passes under the sweep-brush, 2a, which packs the powder against the projecting features of the plate and sweeps back the surplus until it falls off into the drawers, 3, 4, below. From these receptacles, it is to be put back into the feed-hopper for further use.

Fig. 3. Etch Powdering Machine. Powdering Compartment and Furnace.

The plate is carried beyond the feed-hopper into the gang-brush chamber, where the powder placed against the sides of the lines by the feed- and sweep-brushes is further packed in place and the surplus brushed off by the gang brushes, 18. There are six of these brushes, carried on two link belts around sprocket wheels in an elliptical orbit over the plate. The sprockets on the first shaft, 19, have recessed depressions, which permit the brushes, in their first contact with the plate, to drag over it at an angle, while on their last contact, under the second sprockets, 20, they are held over the plate vertically.

The shaft of the first sprockets is movable vertically, 20a, to adjust the pressure of the brushes in their first contact with the plate, and it is also movable laterally, 20b, so as to adjust the tension of the link belts and thereby the angle of the brushes as they pass over the plate. This tension, however, is not to be changed except as the wear of the brushes may make it necessary. The shaft of the second sprockets is movable up

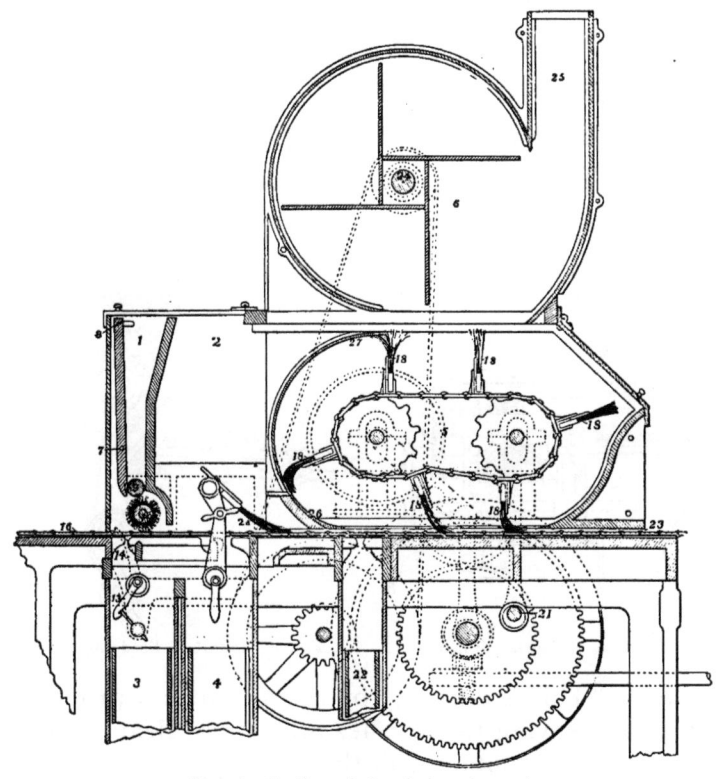

Fig. 4. Section of Powdering Apparatus.

and down, and adjustable by means of a hand-lever, 21, so that the "bank" produced by the final contact of the brush on the passing plate can be varied as desired.

The surplus powder brushed off from the plate by the gang brushes falls down into a drawer, 22, whence it is to be taken, sieved and used over again in the hoppers.

The revolution of the gang-brushes draws a current of air inwards through the exit opening, 23, over the surface of the plate as it emerges from the brush chamber. This draft of air prevents any fine particles of powder from settling on the plate and leaves the ground perfectly clear. The air thus drawn in is carried around by the brushes and is taken up, together with its floating powder particles, and driven out by the rotary fan, 24, up through the ventilating chimney, 25.

The gang-brushes in their course over the passing plate are lifted from its surface in their inclined position by contact with the brake surface, 26, and carried around against the curved extension of this surface to its end at 27. At this point the brush being suddenly freed from restraint, the hairs spring forward, spread out and throw off any surplus powder remain-

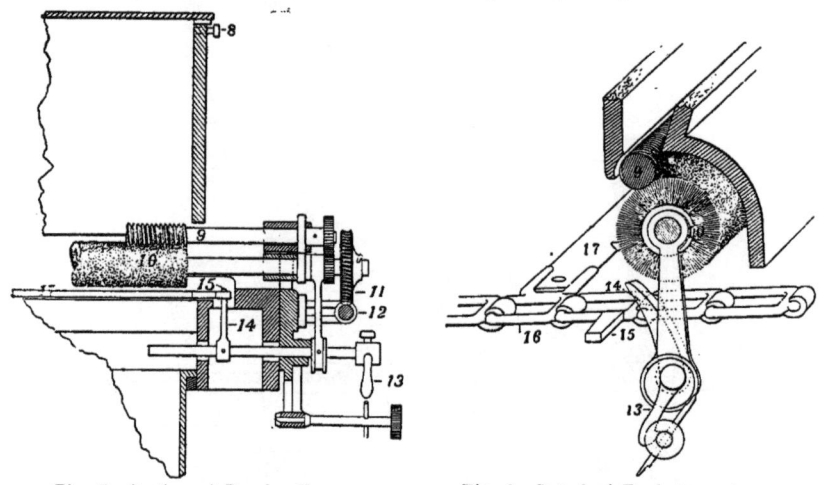

Fig. 5. Section of Powder Hopper.   Fig. 6. Detail of Feed Apparatus.

ing on them. This is taken up by the air current produced by the rotary fan, leaving the brushes practically clean as they revolve into renewed contact with the passing plate.

To prevent an accumulation of powder against the edge of the advancing plate, the table over which the plate is carried forward under the gang-brushes is provided with two depressions, traversing the surface diagonally in divergent directions, like a V. As the moving plate passes over these depressions, the powder collected against the front edges of the plate is brushed down into the hollows, and after the plate has passed the powder is swept out of these back to the opening under the brake and down into the drawer, 22, below.

The control shaft, 28, makes one revolution while the machine completes its cycle. It carries a mutilated gear wheel, 29, which engages a pinion on the gas cock, 30, and alternately turns on and shuts off the gas supply; also a cam, 31, which actuates the belt shifter that alternately starts and stops the air blower; also the pivoted latch lifter, 32, which, in due course, lifts the belt shifting bar, 33, and lets the spring, 34, pull it back to stop the machine. These appliances are adjusted on the control shaft in such relation as to perform their functions in proper sequence.

The remaining features of the machine, such as the gas pipes, 35, the pilot light, 36, the by-pass and cock, 37, for independent use of the furnace, the rotary air blower, 38, and its distributing connections, 39, the ribbed delivery table, 40, &c., are sufficiently indicated in the illustrations and need no detailed description.

THE PRODUCTION OF BORAX IN 1904.

The production, imports, uses, and technology of borax are interestingly described by Mr. Charles G. Yale in a report on the borax industry in 1904, written for the United States Geological Survey.

All the output of borax in the United States comes from California, and the larger part of that from the extensive colemanite deposits in San Bernardino County. The total product for the year 1904 amounted to 45,647 tons crude, valued at $698,810. Of this amount 38,000 tons, valued at $508,000, came from San Bernardino County, Cal., the remainder coming from Ventura and Inyo Counties. In 1903 the returns gave an aggregate production of crude amounting to 34,430 short tons, valued at $661,400. The production in 1902 was 17,404 short tons of refined borax, valued at $2,447.614, of which 862 short tons, valued at $150,000, were stated to be boric acid, and 2600 short tons of crude borax, valued at $91,000, a total of 20,004 short tons, valued at $2,538,614.

The amount of borax imported in to the United States in 1904 was 153,952 pounds, valued at $10,569. Borates, both the calcium and sodium, were imported to the amount of 89,447 pounds, worth $6,630. The quantity of boric acid imported is given as 708,815 pounds, valued at $27,658.

An interesting feature of Mr. Yale's report is a description of the Blumenberg sulphur-dioxide process, which is used by the American Borax Company at Daggart, Cal. Other processes for the manufacture of boric acid from colemanite have been described in previous Survey Reports. These include the Moore process, hydrochloric-acid process, sulphuric-acid process, and Bigott process.

Mr. Yale's paper is published as an extract from the Survey's annual volume "Mining Resources of the United States, 1904."

THE PRODUCTION OF BROMINE IN 1904.

The superior activity of American manufacturers has made for the bromine industry of the United States a place on the European market. The great deposits of haloid salts at Stassfurt and Leopoldshall in Germany are capable of supplying an almost unlimited market, but the American product has nevertheless forced itself into recognition. The result is that German manufacturers have been obliged to offer their goods in America at a price far below that usually current. Hence the price of bromide of potassium has fallen from twenty-five cents to fifteen cents a pound.

American bromine is obtained chiefly from salt brines in Michigan, West Virginia, Ohio and Pennsylvania. The manufacture of bromine in the United States was begun in 1846, at Freeport, Pa., but subsequently has been carried on chiefly in certain areas of brine production, which are mainly at or near Lake St. Louis, Mich., Pomeroy, Ohio, and Malden, W. Va.

To produce bromine the residual liquids or bitterns from the processes of salt manufacture are treated with sulphuric acid, thus forming hydrobromic acid. From this the bromine is separated by the use of an oxidizing agent which removes the hydrogen. For this purpose either chlorate of potash or binoxide of manganese is used.

Bromine is used by manufacturing chemists, who make from it the bromides of potassium, sodium, and ammonium used for medicinal purposes and as photographic reagents. A small amount of bromine is also used in the preparation of the coal tar colors, known as "Eosine" and "Hoffmann's blue." It is employed also as a chemical reagent for precipitating manganese from acetic acid solution, for the conversion of arsenious into arsenic acid, and for detecting nickel in the presence of cobalt in a potassium cyanide solution. Bromine dissolved in water may also be used as a disinfectant. Interesting metallurgical results have been obtained from its use in the bromination and bromocyanide processes of gold extraction, which may, in a measure, become substitutes for chlorination and cyanidation.

The total output of American bromine in twenty-five years has been 10,499,625 pounds, valued approximately at $2,887,917. During 1904 the total output amounted to 897,100 pounds, valued at $269,130. Germany furnishes annually about 300 tons of bromine.

The above facts are taken from a report on the production of bromine in 1904, which Mr. Frederick J. H. Merrill has written for the United States Geological Survey. It is published as an extract from the annual volume "Mineral Resources of the United States, 1904," and may be obtained on application to the Director of the Geological Survey, Washington, D. C.

STEEL, with one-quarter per cent. C, one per cent. W, two-fifths per cent. Mn and one-quarter per cent. V, broke at 112,540 pounds per square inch.

Rondinella's Photo-Printing Machine.

[*Being the Report of the Franklin Institute, through its Committee on Science and the Arts, on the invention of Prof. L. F. Rondinella, of Philadelphia. Sub-Committee: L. E. Levy, Chairman; Henry R. Heyl, Waldemar Lee, Thos. P. Conard.*]

No. 2357.

The Franklin Institute, acting through its Committee on Science and the Arts, investigating the merits of the "Photo-Printing Machine," by Prof. L. F. Rondinella, of Philadelphia, Pa., reports as follows:

This is an apparatus for producing photographic prints in continuous form from tracings or other flexible transparencies of unusual length.

The form of the apparatus under present consideration, to which the inventor has given the name of "Star Photo-Printing Machine," is designed especially for the practice of the blue-print and paper-negative processes, and was patented March 19, 1901, No. 670,349. The machine is adapted to print by sunlight or artificial light as may be most convenient, and accordingly comprises two independent parts, the printing machine proper and the electric lighting apparatus upon which it is supported.

The printing mechanism is contained in a casing which is provided with ball casters so as to be easily rolled out from its support upon tracks arranged for the purpose through a window for sun-printing. The casing is made of well-finished oak and contains all the requisite materials of the printing process throughout and also all the mechanism of the apparatus except the small electric driving motor and its reducing gears. These are fixed on the exterior, on one side of the machine. The casing is curved on top, whence it runs down into a slant of about 45° and then projects to form a receiving compartment at the front. The covers of the slanting and horizontal parts are hinged together and to the front edge of the casing.

forming a two-leaved lid which, when lifted and brought forward, opens the machine and at the same time forms a projecting work table. This hinged cover may also be brought to rest in two other positions, leaving the machine only partially open and the work table out of the way. The curved section of the casing is covered with a roll-top shutter which serves as a covering slide over the exposure opening, permitting this to be varied in extent up to 105° for rays from the sun and up to 120° or more for those from the electric lamps.

The printing is effected through a transparent covering-sheet which holds the tracing and sensitive paper down upon the surface of a felt-covered drum that revolves under the ex-

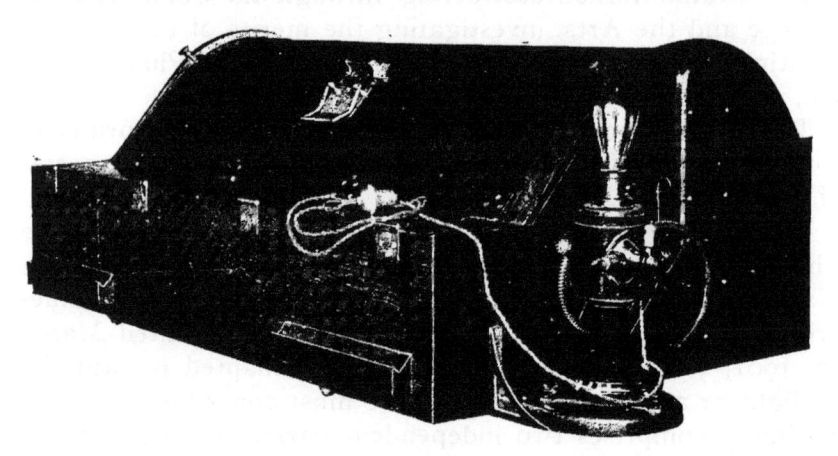

Fig. 1.—Star Photo-Printing Machine, with Casing Closed.

posure opening concentrically with the curved top of the casing. The transparent cover consists of a strip of the best tracing cloth over seventy feet in length, permanently fastened at one end to the drum and passing around this to a winding roll under proper tension. The tracing cloth is carefully prepared so as to wind true from roll to drum and back again, and its edge is spaced off into feet and marked with numbers which indicate the maximum length of print which may yet be made when part of the cover has been wound off. The cover-strip passes from its winding-roll up under an idler and then around this down to its contact with the drum, the material thus forming an inclined feed-apron down which the tracing and sensitive paper are carried into contact with the revolving drum, and

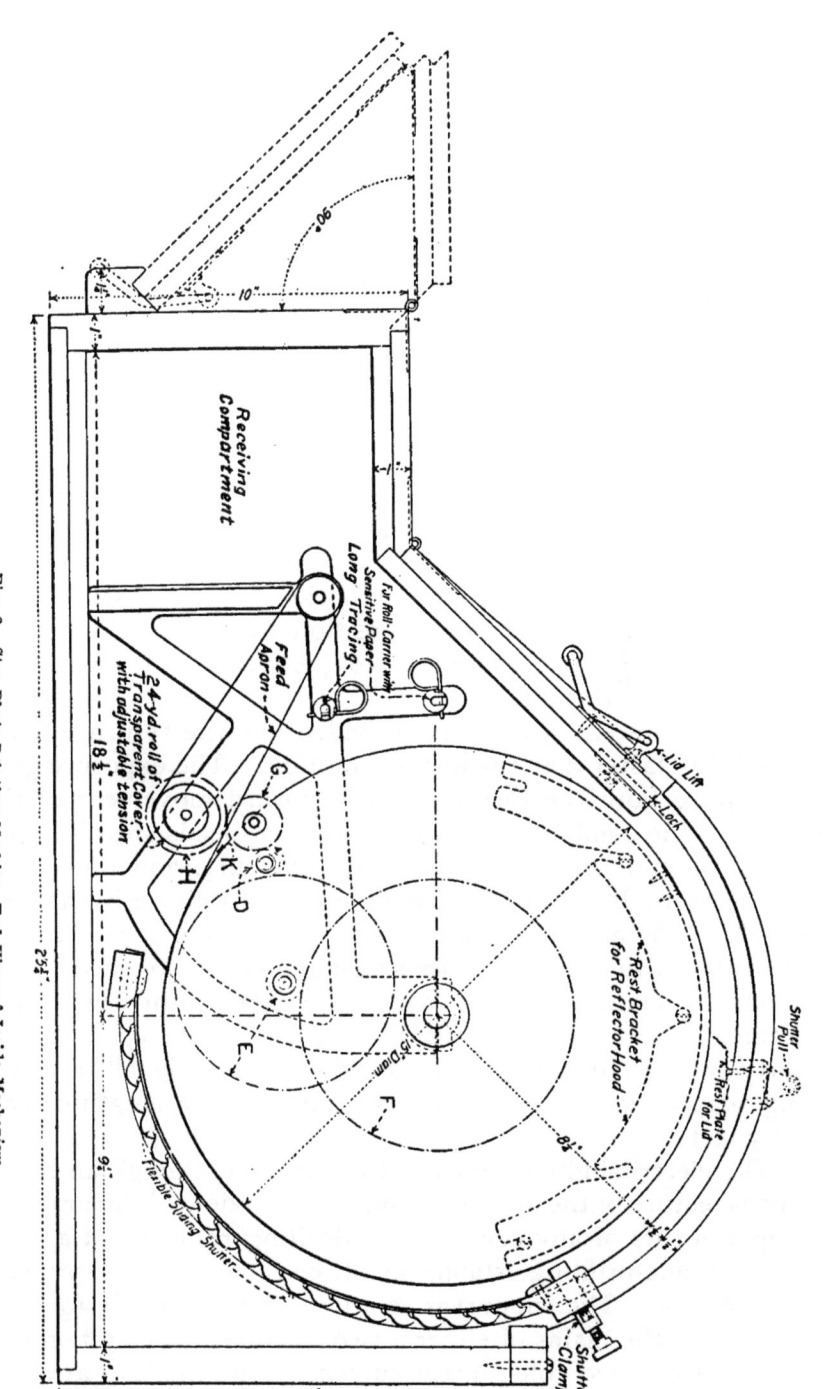

Fig. 2.—Star Photo-Printing Machine, End View of Inside Mechanism.

thence around with it under the exposure opening. The printing is continued on the return of the drum, and after the printing the tracing and prints are carried up the inclined plane and delivered over the idler into the receiving compartment in front. The tension of the transparent cover can be regulated by means of a friction brake at one end of the winding roll and can be effectively controlled so as to insure a close contact of the sensitized paper with tracings on thick or rumpled tracing cloth at any desired printing speed.

The drum is actuated from the outside by means of a reducing gear-couple from the motor to a driving spindle which passes into the casing and carries a small pinion on its inner end. This spindle is movable laterally so as to carry the pinion into mesh with either of two sets of reducing gears on the inside, one set serving to move the drum forward at a certain speed and the other to move it backward at a faster speed. The spindle is held in either position by means of a spring catch which fits into either of two grooved collars carried by the spindle on the outside.

Three different speeds are provided for, the return in such case being faster than the forward motion. The exposure goes on during both the forward and backward movement of the drum. The combination of these various speeds affords a gradation of nine different periods of exposure, and these may furthermore be varied by changing the lamp resistance attached to the motor. The use of cone pulleys or of an external rheostat for this purpose is thus advantageously avoided.

The machine, together with its electric-light supporting frame, is built in three widths, affording prints up to thirty, forty-two and forty-eight inches wide respectively, all three being adapted to make continuous prints up to a length of seventy feet.

The electric-light support of this machine, though an extraneous feature of the apparatus, must yet in view of the practical requirements of any considerable drafting room or of a commercial blue-printing establishment, be regarded as of primary importance. The support is so wired as to be ready for connection to the mains of either a two- or three-wire system carrying 110 or 220 volts, direct current. The lighting arrangement consists of enclosed-arc lamps, four, five or six of them for the three respective widths of the machine, or of a set of three or four Coper-Hewitt mercury lamps of the requisite length.

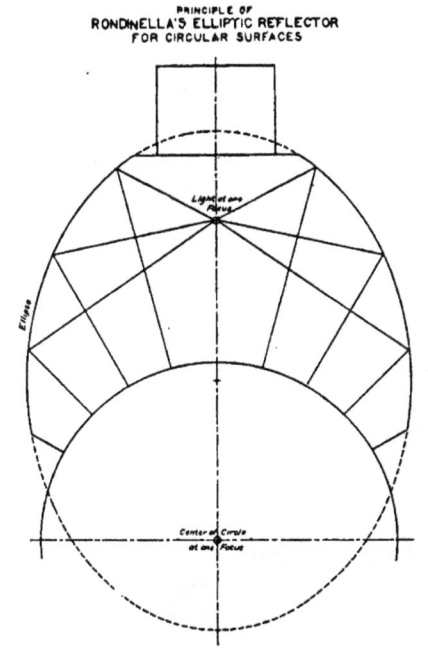

The most actinic rays of light from an arc lamp radiate in a beam outward and downward from the crater formed by the arc in the upper carbon, and the idea is to so reflect all the rays as to bring about a zone of uniform actinic force over the entire surface of the exposure opening of the machine. To this end the lamps are placed directly over the axis of the drum, under a reflector-hood especially designed for the purpose upon the principle of reflection from the surface of an ellipse, as first demonstrated by the present inventor before the Franklin Institute, at its meeting on December 21, 1904. The lamp and the enclosing reflector-hood are both hung from a suspension beam adapted to be raised and lowered on the support over the machine, the lamps being held to the beam in a fiixed position and the hood by chains that permit its being raised and lowered about the lamps. Through sight-holes placed for the purpose the hood can be adjusted over the lamps so that the arcs coincide with the focal line of the elliptical inner surface of the hood. In this position all the rays that strike the inner surface are reflected towards the opposite focus of the ellipse. The suspension beam with its lamps and hood is then lowered until the ends of the reflector hood rests upon the brackets at the two ends of the machine casing. Thus placed, the reflected light is intercepted by the cylindrical surface of the drum, reaching it and its overlying tracing and sensitized paper in rays of equal length and in directions uni-

[Fig. 4—For Uniform Distribution of Electric Light

PRINCIPLE OF RONDINELLA'S ELLIPTIC REFLECTOR FOR CIRCULAR SURFACES

formly normal to the surface, thus producing the desired area of uniform illumination. For the long tubes of the mercury lamps, the same principle is applied in a form modified for the purpose. In either case the uniformity of illumination which is produced is such that tracings coming within the area of the opening can be effectively printed without moving the drum.

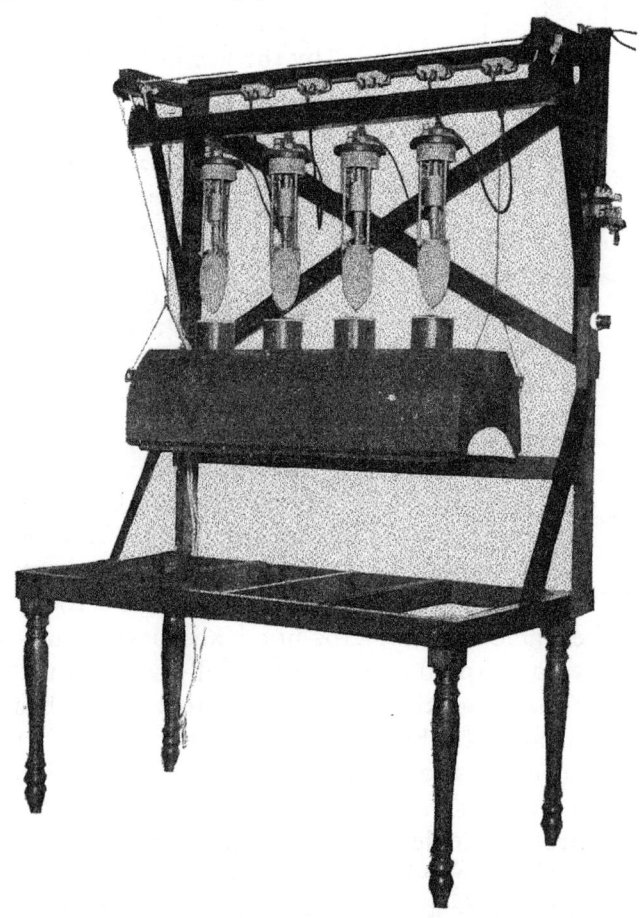

Fig. 3.—Support for Star Photo-Printing Machine, and Electric Equipment.

The Rondinella Photo-Printing Machine marks a distinct advance in the practice of this art. Long prints, whether from negatives or from tracings, were formerly produced by simply pasting shorter ones together or by the obviously difficult and unsatisfactory method of making the continuous print by successive exposures of shorter sections. To obviate the imperfections inevitable in either of these processes, various expedients have been resorted to, one method being to fasten the sensitized paper with its over-lying flexible transparency on a board of sufficient size in width and length, springing the board lengthwise to force the material into the best possible contact

Fig. 5.—Star Photo-Printing Machine, in its Support, as Used for Printing by Electric Arcs.

over the convex surface and then exposing to as even light as possible. An effort to replace these crude procedures with a mechanism for the purpose was made some fourteen years ago by Paul Heinze, of Chicago, Ill., whose invention was patented No. 469,244, on February 23, 1892. In that device the tracing and sensitized paper were each to be separately rolled upon spools from which they were then drawn off together by the rotation of a drum to which the outer ends of both were fastened. These spools were geared so as to make the same number of revolutions, but on account of the different thicknesses of the two materials, the diameters of the two rolls would be or would soon become different, and unequal lengths of the two materials would be drawn off in the same time. In the course of the rotation, the two sheets were drawn under a succession of scrapers spanning the exposure opening, the scrapers serving to smooth out the sheets and to force them into printing contact. The machine seems to have been devised especially for printing by daylight alone and appears never to have been brought into practice. The Rondinella machine of 1901 is the next on the record, and the first to meet the practical requirements of the occasion. It has been followed by several others, most of them utilizing a rotating drum on which to effect the exposures, but manifestly retains its leading place as an efficient solution of the problem.

In view of the scientific accuracy and mechanical thoroughness and simplicity with which all the various requirements of the process of continuous photo-printing have been fulfilled in this machine, the Franklin Institute recommends the award of the John Scott Legacy Premium and Medal to the inventor, Prof. Lino F. Rondinella, of Philadelphia, Pa.

Attest,   WM. H. WAHL, Sec'y.

Adopted June 7, 1905.

Sections.

MECHANICAL AND ENGINEERING SECTION. *Stated Meeting*, held Thursday, November 16, 8 P.M. Mr. James Christie in the chair. Present, thirty members and visitors.

Prof. H. W. Spangler read an illustrated paper describing some experiments on heating the Dormitories of the University of Pennsylvania.

Mr. Arthur Falkenau followed with some Notes on the Use of Materials in Hydraulic Machinery.

Both papers were freely discussed. The meeting passed a vote of thanks to the speakers of the evening and adjourned.

FRANCIS HEAD, Sec'y.

SECTION OF MINING AND METALLURGY. *Stated Meeting*, held Thursday, November 23, 8 P.M. Prof. A. E. Outerbridge, Jr., in the chair. Present, thirty-seven members and visitors.

The chairman introduced Mr. Clifford Richardson, of New York, who presented an interesting and instructive communication on the Constitution of Portland Cement.

The paper was discussed by Mr. Robt. W. Lesley, Dr. W. J. Williams, Dr. Wm. H. Wahl and the author.

The thanks of the meeting were voted to the speaker of the evening. Adjourned.

WM. H. WAHL, *Sec'y pro tem.*

Franklin Institute.

Proceedings of the Stated Meeting held Wednesday, December 20, 1905.

HALL OF THE FRANKLIN INSTITUTE,
Philadelphia. December 20, 1905.

President JOHN BERKINBINE in the chair.

Present, fifty-two members and visitors.

Additions to membership since last month, fifteen.

The Secretary presented the resignation from the Board of Managers and the Committee on Science and the Arts, of Dr. Coleman Sellers. Mr. H. R. Heyl, Mr. L. E. Levy, the President and the Secretary referred in complimentary terms to the long and eminent service which Dr. Sellers had rendered to the Institute during his long association with it; whereupon the following resolution, offered by Vice-President Washington Jones, and seconded by H. R. Heyl, was unanimously adopted:

Resolved, That the Franklin Institute in accepting the resignation of Dr. Coleman Sellers from membership on the Board of Managers, the Committee on Science and the Arts, and the Committee on Publications, directs that a minute be entered on its records expressive of its high appreciation of, and grateful thanks for his long and valued services to the Institute, as President, manager and committeeman.

The Secretary was directed to transmit a copy of this resolution to Dr. Sellers.

The following nominations were then presented for officers, managers and committeemen to be voted for at the annual election, to be held on Wednesday, January 17th, 1906, viz.:

For President	(to serve one year)..........	JOHN BIRKINBINE.
" *Vice-President*	(" three years).........	JAMES M. DODGE.
" *Secretary*	(" one year)..........	WM. H. WAHL.
" *Treasurer*	")..........	SAMUEL SARTAIN.
" *Auditor*	(" three years).........	W. O. GRIGGS.
" *Auditor*	(" one year)..........	SAMUEL P. SADTLER.

For *Managers* (to serve three years).

CYRUS BORGNER.	JAWOOD LUKENS.
JAMES CHRISTIE.	LAWRENCE T. PAUL.
PERSIFOR FRAZER,	HORACE PETTIT.
HARRY W. JAYNE.	OTTO C. WOLF.

WM. H. LAMBERT (to serve for two years).

For *the Committee on Science and the Arts* (to serve three years)

RICHARD L. BINDER,	ARTHUR FALKENAU,	W. N. JENNINGS,
HENRY F. COLVIN,	JOHN M. HARTMAN,	H. F. KELLER,
THOS. P. CONARD,	C. C. HEYL,	LOUIS E. LEVY,
GEO. S. CULLEN,	HENRY R. HEYL,	TINIUS OLSEN,
CHARLES DAY,	GEORGE A. HOADLEY,	CHAS. E. RONALDSON,
J. M. EMANUEL,	RICHARD L. HUMPHREY,	SAM'L P. SADTLER,
ERNEST M. WHITE,		W. J. WILLIAMS.

C. A. HALL (to serve for two years).   WILHELM VOGT (to serve for one year).

The President then introduced Mr. Howard DuBois, Mining Engineer, Philadelphia, who presented a communication, profusely illustrated with lantern photographs, on "Hydraulic Mining in British Columbia and Alaska."

The thanks of the meeting were voted to the speaker. Adjourned.

WM. H. WAHL, *Secretary.*

Committee on Science and the Arts.

(Abstract of Proceedings of the stated meeting held Wednesday, December 6th.)

DR. EDWARD GOLDSMITH in the chair.

The Committee adopted the following reports:

(No. 2372.) *New Physical Diagnostic and Surgical Instruments.* Dr. Henry Emerson Wetherill, Philadelphia.

ABSTRACT: These instruments are in part protected by U. S. Letetrs patent (No. 780,315, January 17, 1905, and No. 798,938, September 5, 1905). They consist of a medical hygroscope, called by the inventor the Hygromed, which has for its object the detection of the amount of moisture in the skin; an improved precipitating device, called generically the Hæmatokrit, for the separation and precipitation in a graduated tube of the contained solids in the liquid treated—a compact, portable and handy device for its intended use; an ingenious pair of scissors, combining in one seven different instruments.

The report awards the Edward Longstreth Medal of Merit to the inventor. (*Sub-Committee*, Dr. W. O. Griggs, Chairman; Dr. W. J. Williams.)

(No. 2377.) *Automatic Momentum Car-Brake.* Wm. L. Barker, Llanerch, Pa. An advisory report.

The following reports passed first reading:

(No. 2365.) "*Speed-Jack.*" C. J. Reed, Philadelphia.

(No. 2375.) *Quartz-Glass Mercury Lamp.* W. C. Heraeus, Hanau, Germany.

(No. 2336.) *System of Wireless Telegraphy.* Dr. Lee De Forrest, New York.

The following protests were disposed of:

(No. 2216.) *Narnst Lamp.* Protest of F. M. F. Cazin, Hoboken, N. J.

(No. 2360.) *Ives's Diffraction Grating Replicas.* Protest of R. James Wallace, Yerkes Observatory, University of Chicago.

The protests were not sustained.

WM. H. WAHL, *Secretary.*

JOURNAL

OF THE

FRANKLIN INSTITUTE

OF THE STATE OF PENNSYLVANIA

FOR THE PROMOTION OF THE MECHANIC ARTS

Vol. CLXI, No. 2   81st YEAR   FEBRUARY, 1906.

The Franklin Institute is not responsible for the statements and opinions advanced by contributors to the *Journal*.

CHEMICAL SECTION.

(*Concluded from vol clxi, page 58*).

*Mica and the Mica Industry.

By George Wetmore Colles.†

Average Prices.

An important element in statistical values is the average *prices* of the product, more especially when taken in connection with charts for total values and quantities. *Average* prices of miscellaneous products of assorted sizes, can of course be of no service to those requiring to purchase or needing for service a particular size or sizes, but they are of real importance to the mining industry, because they designate the real value which may be counted for the run-of-mine product. We will next consider, therefore, two charts, Figs. 31 and 32, showing the

---

*Read by title.   †Copyright, 1905, by George Wetmore Colles.

average prices of the sheet in the respective countries of production, and also the prices received for the different varieties of our native product. In Figure 31 the first thing that strikes the eye is the character of the United States price-curve,

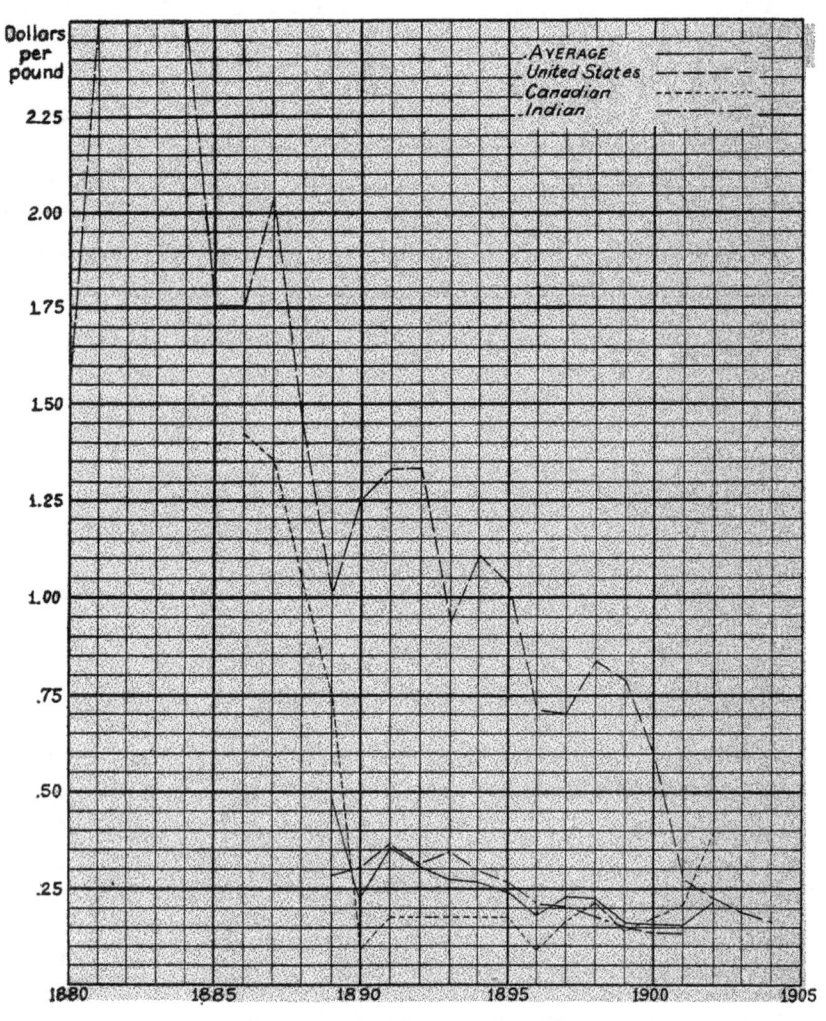

Fig. 31. Average Prices of Sheet Mica.

which slants steeply downward from a generous $2.50 a pound in the early eighties to a beggarly 16.4 cents in 1903,—not to mention the $8.00 a pound which was received in the sixties.

Oh, what a fall is this, my countrymen! Can there be any wonder if mica-mining has ceased to be profitable?

True, not alone United States mica has suffered this fall. So has Canadian mica; but it is clear that the $1.42 rate for the latter was very far above its proper level. This rate was merely assumed when Canadian mica began production in 1886, because of the prevailing high price in this country; but as soon as production seriously began, the price rapidly fell to nine cents per pound, and pulled our domestic product down with it to half-price.

While United States mica has fallen off continuously in value, Canadian mica has been, since 1896, in the course of recovery, and in 1902 ran higher than United States mica. And there is reason that it should do so; its superior merits for electrical purposes have been slow of appreciation by users, and it has only begun to take the position in this respect that it has a right to claim. During the same period, the average price of Indian mica has fallen steadily from 30 to 35 cents in 1891 to about 14 cents in 1901, and is now reckoned at a lower value than either of the others.

The general price of mica has likewise fallen during the entire period under consideration, keeping pretty close to the India line, and is now reduced to less than half its former value. This is undoubtedly to be explained by the increased competition, and that more particularly in India. The increase of competition has been doubtless called forth by the increasing demand for mica for electrical purposes. The opening of the new Madras mica district of India in 1897 has undoubtedly also had a strong effect on the reduction of the price, adding an annual output of some 300 tons to the supply. The great increase in the supply can in fact readily be seen by referring to Figures 26 and 27.

A most interesting point in connection with this chart is the question of the effect the increase of the tariff on mica has had on the price in this country. Prior to 1892 mica came in free. From that time until July 24, 1897, the tariff was 35 per cent., and subsequent to that date the combined specific and ad valorem duty has averaged 45 to 60 per cent. We cannot, however, observe any permanent effect on the price from these tariff duties, although perhaps a temporary effect may be discerned in the high levels of 1892 and 1898. The tariff, if it was its object (as it really was) to exclude the foreign product, has

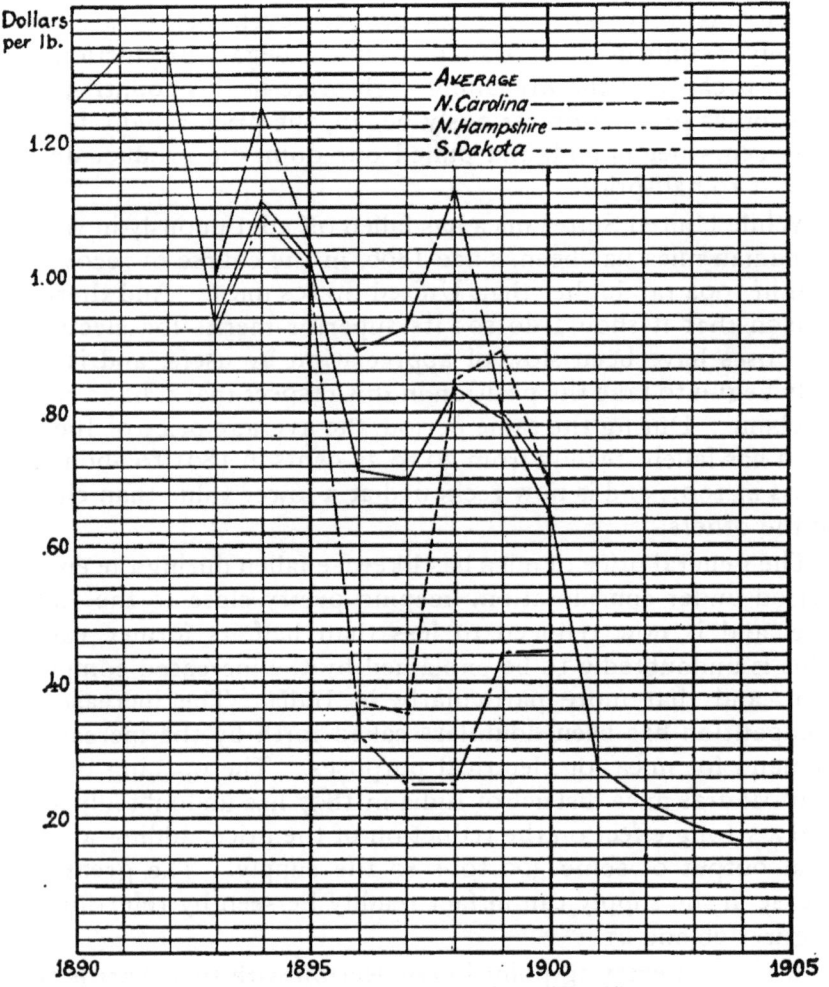

Fig. 32. Average Prices of United States Sheet Mica.

totally failed of effect, and this we shall see even more clearly in considering one of the following charts.

In Figure 32 we can compare the average market prices of North Carolina and New Hampshire and South Dakota mica, as deduced from the reports collected by the "Mineral Industry." North Carolina is, and always has been, rated at a higher value than New Hampshire mica, and probably always will be, for it is of undoubtedly superior quality for the purposes for which it is used, namely, glazing, to any elsewhere produced. Both New Hampshire and North Carolina mica have necessarily fallen in value with the course of the average curve; but New Hampshire has fallen the most, and although values are only given to 1900, they are believed to have sunk still lower since that date, and have practically put an end to the New Hampshire industry. It is surprising that South Dakota mica should show such a great increase in average price from 1897 to 1898, and it is thought that these figures are not to be relied on.

There is a very important point to be noted in connection with these price diagrams, however, and which renders the wide differences observed more apparent than real. It is this: That the value of the mica will naturally depend on the condition of manufacture, and this is different in different localities. Practically all the North Carolina mica is shipped, as it always has been, cleaned and cut into rectangular panels, ready for use by the stove manufacturers; whereas a large proportion of that from New Hampshire is sent to market as trimmed mica, whose value *per pound* is not more than 40 per cent. of the value of what is left after cutting, on account of the waste. But aside from this, the average value is largely influenced by the sizes obtained. It is less because of its inferior quality than because of its small sizes that the *average* price of New Hampshire mica is lower. Indeed, much of what is sent to market as sheet in New Hampshire to be cut into small electrical forms, is termed scrap in North Carolina, and, if not suitable for grinding there, never leaves the dumps.

These points are also to be considered in connection with the foreign products. At the present time and for some years past none of the product of Canada, and only a small proportion of that of India, is cut into panels or other shapes in the mines; so that the quantities shipped are really much larger than the quantities of corresponding value in cut mica shipped in the United States, and show correspondingly in the price charts, and on the quantity lines of the production charts.

Valuation of United States Imports. Fig. 33.

This chart is of great interest and importance as indicating the actual valuations at which the foreign product comes into competition with that of the United States, and the effect, or rather the lack of effect, of the imposed duty. Two sets of lines are shown, one indicating the invoice-price of imports in dollars per pound, the other the invoice-price with the addition of the duty; and independently of these is shown a duty line

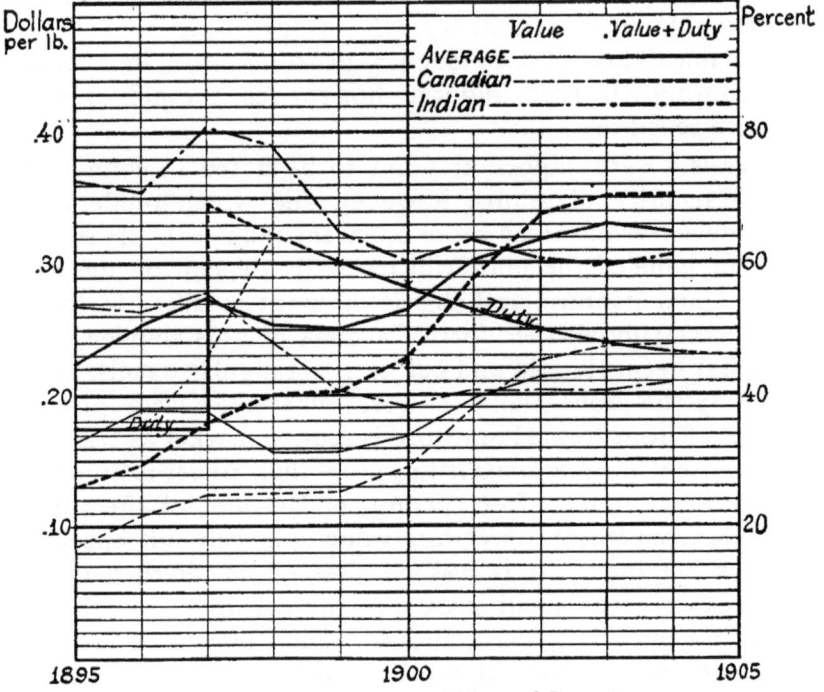

Fig. 33. Valuation of Imported Mica, and Duty Thereon.

indicating the *equivalent* ad valorem rate. This latter shows that from a fixed rate of 35 per cent. previous to 1897 the duty rose in that year to 46 per cent. and in the following year to about 64 per cent., since which time it has been regularly declining. The reason for the decline is to be found in the increased average price per pound on the invoices, the specific proportion of the duty being of course equivalent to a lower ad valorem rate as the price increases. It is not explained by any change in the proportion of manufactured to unmanufactured* mica, the former kind having remained fairly steady since 1899† at 3 to 5 per cent. of the total quantity, and 9 to 12 per cent. of the total value.

Practically all the imported mica, therefore, comes to us in a merely trimmed condition, in which condition it is more suitable of electrical purposes, owing partly to the irregularity of the shapes in which it is used in the simple form, and partly because a large if not the greater proportion is used in the manufacture of mica board or compound forms, such as rings, sleeves, tubes, etc., in which the individual laminæ of mica are irregularly cemented together.

The increasing appreciation of Canadian mica is readily seen from this chart. It has been increasing steadily in value from 1895 to the present time, and that of India has been almost as steadily decreasing in value; so that, where in 1895 the imported value of Indian mica was 36.3 cents, or nearly three times that of Canadian mica, in 1904 Canadian mica stood at 35.1 cents and Indian at 30.7 cents. Even more significant, perhaps, is the fact of the well-defined and fairly steady increase in the average value of imported mica from 22 to 32 cents per pound, while during the same period United States mica was falling from $1.00 to 16 cents.

It is clear that the change in the rate of duty has had no perceptible effect either in keeping up the price of the United States production, or in keeping out imports, the former having decreased, and the latter increased quite without regard to such changes. The conclusion to be drawn from this state of affairs will be set forth in the following section.

---

*The specific rate on unmanufactured mica is six cents, and on "cut or trimmed" mica twelve cents per pound. The actual line of division between the two seems to be uncertain in the Treasury itself, for I am informed that both knife-trimmed and thumb-trimmed mica, even when the product is split into thin sheets and cleaned, is rated as "unmanufactured." It is certainly true that thumb-rimmed mica is and always has been so classed. It fact, it would be difficult to draw any line between raw mica and thumb-trimmed, which would permit any mica to be imported at the lower rate.

†In 1899 it was 6 per cent. of the total quantity and 18.1 per cent. of the total value.

Proportions of Sizes in the Total Production. Fig. 34.

The only complete statistics on this subject are those furnished for Canadian mica, and these illustrate very well the proportion of the output of the Canadian mines, though of course those of the Indian, North Carolina and New Hampshire mines and other mines would be quite different in each case. The output of Canadian mica classified in sizes for the three years, 1899, 1900 and 1902, are as follows:

From the percentage-figures for total quantities and values the chart shown in Figure 34 has been prepared.

Prices of Different Sizes.

From the preceding table price-curves have been also charted in Figure 35, for the three years above given, which will indicate pretty clearly the wholesale rates for Canadian mica. In the United States a price-schedule is determined on by the mica-dealers and published annually. This schedule for the years 1897, 1898 and 1902 is given below. There are discounts of 50, 10 and 5 per cent. from the schedule, and the prices with all of these discounts deducted, are also shown in the table.

LIST PRICES OF CUT INDIA AND NORTH CAROLINA MICA.

Size	List Price 1897	List Price 1898	List Price 1902	Net, with Discount 1897	Net, with Discount 1898	Net, with Discount 1902
1 x 4	.90	1.15		.384	.492	
2 x 2	.60	.85		.256	.363	
2 x 4	1.20	1.45	.30	.513	.620	.128
2 x 6	3.00	3.25		1.28	1.39	
3 x 3	2.50	2.75	.80	1.07	1.18	.342
3 x 4	5.00	5.00	1.50	2.14	2.14	.641
3 x 6	8.00	8.00		3.42	3.42	
4 x 4	7.00	7.00	2.00	2.99	2.99	.855
4 x 8	9.75	9.75		4.16	4.16	
6 x 6	10.00	10.00	3.00	4.27	4.27	1.28
6 x 8	10.75	10.75		4.60	4.60	
8 x 8	12.00	12.00		5.73	5.73	
8 x 10	13.00	13.00		5.56	5.56	

It will be understood that the "nominal size" of a piece of trimmed mica is the size of the largest rectangular sheet without flaws which can be cut from it. Indian mica is not classified in this way, but according to numbers and grades. The numbers are specified in available square inches as follows:

PROPORTIONS OF SIZES OF CANADIAN MICA, 1899–1902.

Nominal Size	Equivalent Sq. Inches	1899 Quantity Pounds	% of Total	1899 Value Dollars	% of Total	Avge Price per lb.	1900 Quantity Pounds	% of Total	1900 Value Dollars	% of Total	Avge Price per lb.	1902 (Quebec only.) Quantity Pounds	% of Total	1902 Value Dollars	% of Total	Avge Price per lb.
1x3	3	284,036	42.9	$18,926	17.5	$ .067	338,200	62.5	$31,860	30.3	$ .094	64,463	46.5	$ 7,364	20.5	$ .114
2x3	6	136,054	20.5	19,146	17.7	.141	92,359	17.1	18,534	17.6	.201	27,861	21.0	7,201	21.0	.258
2x4	8	179,113	27.0	32,721	30.3	.183	71,332	13.2	24,953	23.7	.350	27,296	20.6	10,756	31.4	.394
3x5	15	37,284	5.6	16,720	15.5	.449	25,637	4.7	15,706	14.9	.611	11,772	8.9	7,578	22.1	.646
4x6	24	17,937	2.7	10,908	10.1	.608	11,762	2.2	11,451	10.9	.974	890	0.7	820	2.4	.921
5x7 5x8 & over	35 40	7,767	1.2	9,642	8.9	1.240	1,995	0.4	2,696	2.6	1.351	540	0.4	585	1.7	1.183
		662,191		108,063		.163	541,285		105,200		.194	132,822		34,304		.258

1. 36 to 50 sq. in.
2. 24 to 36 " "
3. 16 to 24 " "

4. 10 to 16 sq. in.
5. 6 to 10 " "
6. 4 to 6 " "

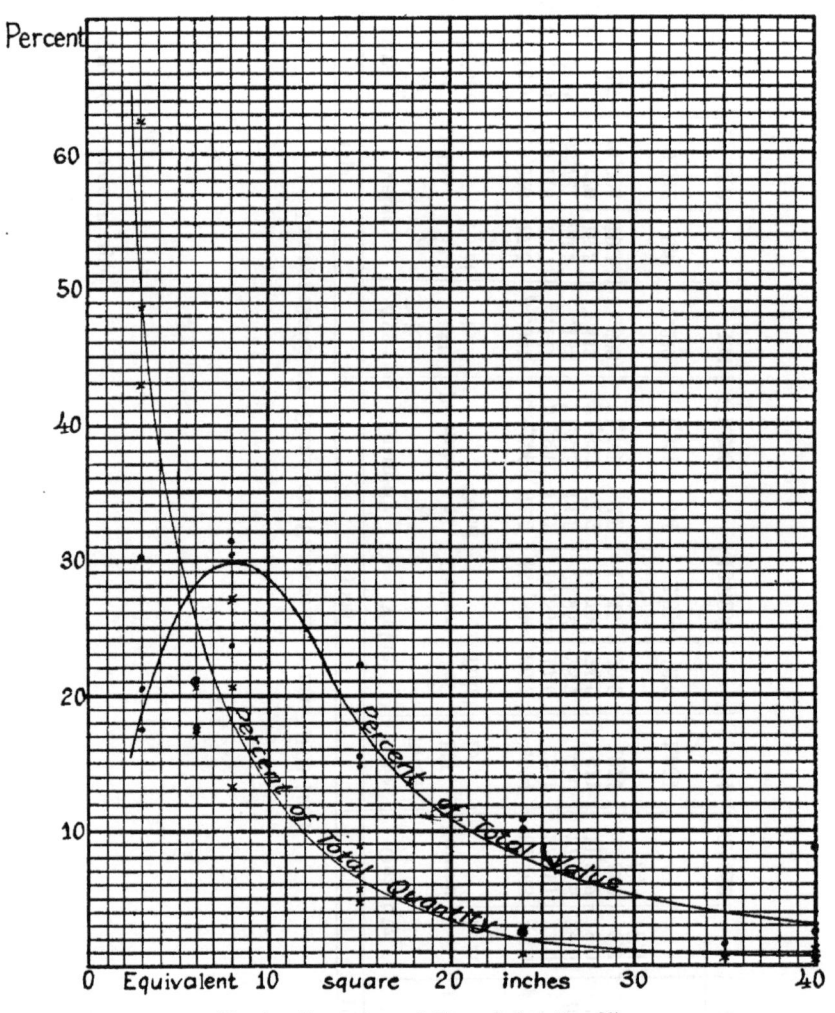

Fig. 34. Proportions of Sizes of Canadian Mica.

Sheets over fifty square inches in available area are termed "specials."

The grades are as follows: (1) Ruby mica, hard and tough; (2) white, transparent; (3) discolored and smoked; (4) black and flawed. Each of these grades represents about double the value of a sheet of the same size in the next lower grade.

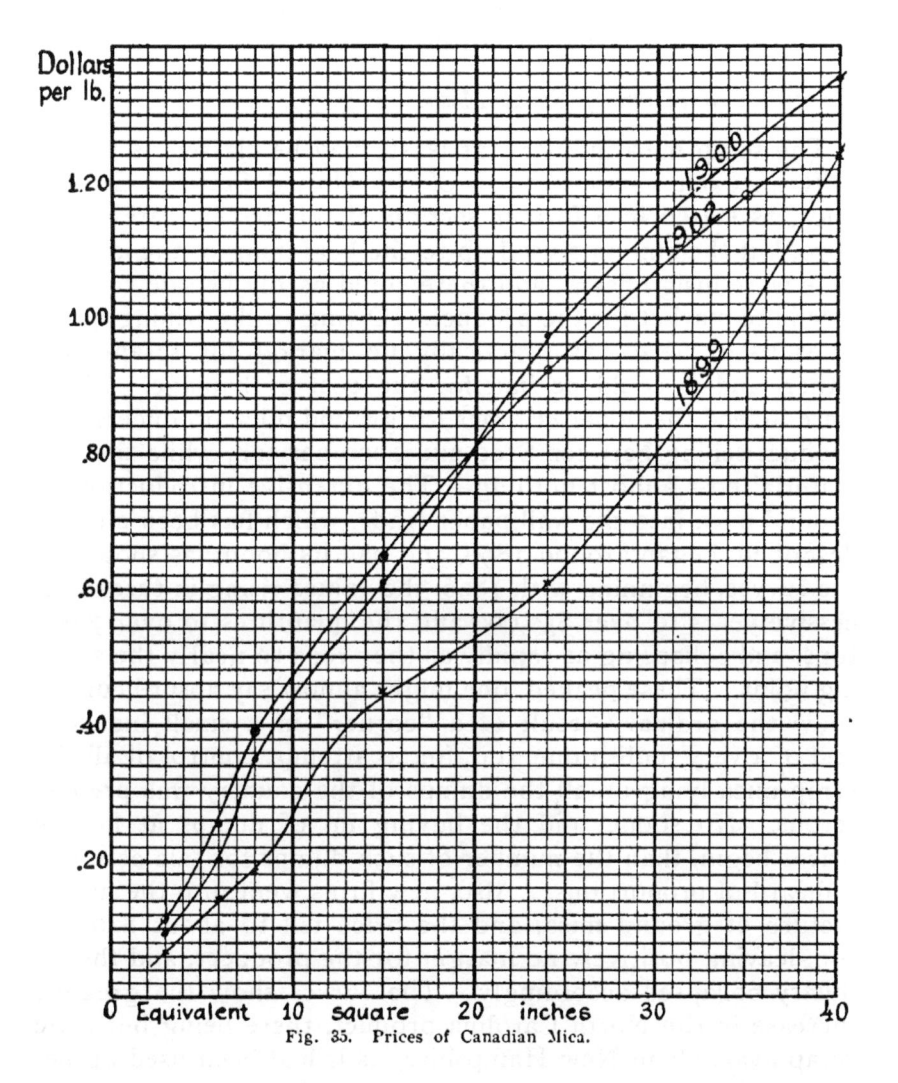

Fig. 35. Prices of Canadian Mica.

The following quotations on best ruby, No. 1 grade, were made by auction in London, according to A. Mervyn Smith:*

---

*Proceedings Institution of Mining and Metallurgy, Volume VII, page 171.

No. 1. 6s 8d. = $1.62   No. 4. 1s = $ .24
 " 2. 4s   =   .97   " 5. 4d =   .08
 " 3. 2s   =   .49   " 6. 2d =   .04
Specials bring as much as £1 per pound.

World's Production of Scrap Mica. Fig. 36.

The scrap mica industry, at least, may be classed as practically an exclusive possession of the United States. Ground mica, as stated in a former section of this paper, dates from 1870, but the amounts used up to the beginning of the last decade were insignificant, and there could hardly be said to exist a definite market for the product. Owing to the trifling quantities used there was practically no market for scrap-mica, either, and it could be had for the cost of haulage. These were indeed the bonanza times for ground-mica producers, the product selling between $100 and $200 per ton, while the raw material cost almost nothing. The extensive introduction of ground mica into the wall-paper trade, which has been already described, began around 1890, and had its boom, as shown by the curves, beginning with 1896, the values rising in three successive years to over $50,000 and the quantities to over 7,000 tons, and collapsing as rapidly as they rose to nearly the starting point. These peaked, one might almost say absurd curves, show the ordinary result of a "corner" in a small industry, and of a very foolish one, too, for to attempt control of all the scrap mica lying on all the dumps of the country was preposterous. No sooner did the buying up of dumps at famine prices begin, than the number of such dumps for sale assumed unheard-of proportions. Also, new firms went into the manufacture of ground mica, and the price fell to about $40 per ton, leaving positively no margin for the producer, and the industry collapsed accordingly. The curves show the greatest increase in the North Carolina product, there being but little scrap available in New Hampshire, as it had been used up before. The increase, however, was not really due to actual production, but merely to the throwing on the market of the aforesaid dumps, which were the product of the previous years' work and had suddenly become valuable. The writer has made a vain effort to secure statistics of prices of scrap and ground mica, which would enable charts of prices to be drawn. The industry being a small one, the "friction of trade," as it is called, is very great, and the prices slide from one level to another by big jumps from time to time, keeping constant,

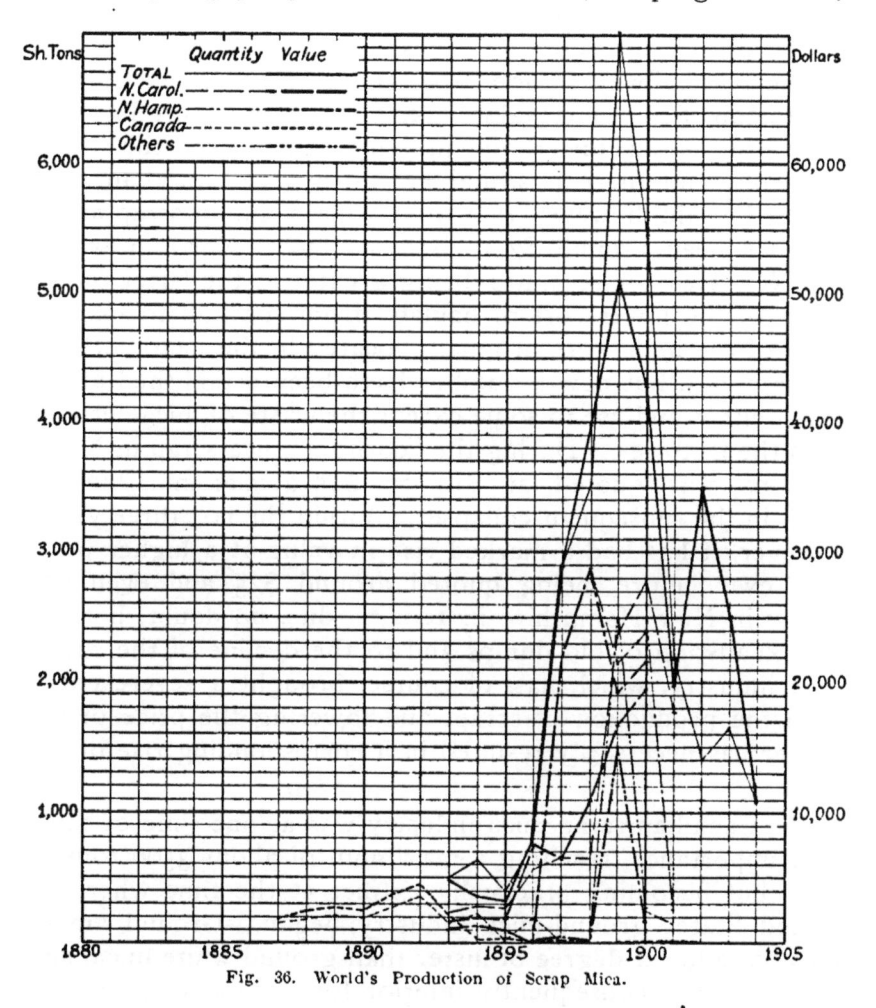

Fig. 36. World's Production of Scrap Mica.

nominally at least,, in the intervals. Actually the prices are more or less subject to secret discounts, which renders any quotation chart which might be formed practically valueless. The value of scrap, moreover, varies widely with the place of production, not only because of the difference in proportion of scrap mined in the different districts, and the different proportionate values of the sheet product, but also, and more especially, because of the great differences in the cost of hauling to market, which may equal or exceed the value of the scrap. The value of scrap on a dump in North Carolina, for example, might be $1 or $2 per ton; at a near-by mill, or town, $5; at the nearest railway station, $10; and at New York, Chicago or Boston, $20. The value even at "place of production," therefore, is an uncertain quantity, unless on the dump is meant, and there it would be more uncertain than ever. It may be said, however, that the normal value of scrap at the mill has been around $7 to $10 at most North Carolina and New Hampshire mines, being greater at the latter on account of the lower cost of haulage from the mill to market. Most of the mills were originally located in the neighborhood of the mines, but now a number have been established at the larger cities, and use to some extent the waste of mica cuttings from the factories. During the "corner," mica-scrap in North Carolina was quoted at $12 to $15 at the mill, but fell again in the subsequent collapse. Within the last three years, however, it has had a marked rise in value at a higher point than ever before, being quoted around $25 and even $28 to $50 in New Hampshire in 1902,*—due, no doubt, both to the exhaustion of old dumps and to the decline of the sheet-mica industry, which has, of course, brought a severe check upon the production of scrap, and lowering the price of the former has necessarily resulted in a corresponding rise of the latter.

The Canadian scrap-mica industry was at one time of some little importance, reaching a maximum of about 3500 tons in 1892, but becoming practically extinct at the time when the United States ground-mica industry began. Ground amber mica has a lower degree of luster than ground white mica, and the product is consequently inferior for decorative purposes, which, added to the duty of 20 per cent., has practically excluded it from the United States, its only market. It has had some use in lubricants, paints, and the like, but the amount

---

*Albert F. Hoskins, in "Mineral Industry," 1903.

used of these materials in Canada is insufficient to support a ground-mica industry.

Of course it is to be anticipated that, should the United States mica industry decline again as in the early nineties, the price of domestic scrap will rise to such an extent as to render the importation of Canadian ground mica necessary. Under the present tariff, however, it would be impossible commercially to import Canadian scrap on account of the six cents per pound specific duty, equivalent to $120 per ton. The importations of the ground product are taxed at 20 per cent. ad valorem. thus rendering its importation less difficult. However, it can hardly be supposed that this product can ever wholly, or even largely, replace that of the United States. No other country produces any ground mica, and consequently, aside from the United States, Canada is our only outlook for either the raw material or the finished product under the present tariff, as it can hardly be supposed that there would ever be an industry in Europe exclusively for supplying the United States, and it is a strange and interesting fact that Europeans apparently make no use of this material except for the small quantities that are exported from the United States.

It has been assumed in this discussion of the scrap mica production that all of it was used for the purpose of grinding. Of course there are certain amounts of mica classed as scrap which are actually used for such things as small electrical insulating plates and washers, but most of the material used for such purposes is classed as sheet mica.

VIII. CONCLUSIONS.

From a consideration of the facts hereinbefore cited we are in a position to draw certain conclusions of more or less importance.

We have seen that there is an important natural division line between the micas, dividing them into two classes. having not only different physical and chemical characters, but with most important commercial differences in respect to hardness, flexibility and coloration. The harder and more brittle—igneous—micas are the only ones suited for glazing; the softer and more elastic best suited for most electrical purposes.

This line, commercially speaking, is not absolute, however, for much of the former class, especially that containing lithia, has a softness and flexibility nearly if not quite equal to the latter.

Moreover, the line of division is not drawn through the center, so to speak, but on one side. Only one locality furnishes mica of the second class;* but mica of the first class is found in commercial deposits in most countries of the world.

The two classes also differ very much in their respective facilities of mining, that of the first class being found in hard, flinty and feldspathic rocks, or their derivatives, and then in isolated bodies which must be recovered individually; while mica of the second class occurs not only in softer rocks—calcite, apatite and pyroxene— but also in dense masses, rendering the mining cost but a fraction of that for the first class of mica.

In neither class, however, is there any deficiency in the natural supply, which acts as a factor of any importance in the price. In this respect mica stands apart from the generality of minerals, the supply of which is limited to a greater or less extent through a scarcity of the deposits; it stands rather on a par with materials like building-stone, slate and cement, but differs from these latter in that the price per pound is so much higher that the cost of transportation is generally of small account. Not only are there in many countries unlimited deposits of mica which have never been seriously worked, but even in the producing localities the surface of the deposits has scarcely been scratched. In Canada, North Carolina, New Hampshire and even in India there are great numbers of deposits of good quality which are not worked, and that which is worked, as in India, is worked very wastefully and a large proportion of the total output is spoiled.

The reason for the limitation of the production to three countries is due to special conditions. India naturally takes the lead in production for three reasons: First, the excessively low cost of the unskilled labor necessary; secondly, the decomposed nature of the rock in which most of the mines are

---

*It is understood that amber-colored mica is mined in India; this may or may not be phlogopite, but it does not make its appearance in our markets.

located; and thirdly, of course, the great abundance and good quality of the mica itself. These conditions practically place other countries out of competition with it.

Canada, however, is an exception to the fact just noted, in that it alone possesses magnesian mica, and this mica is mined with so much greater facility than the other that even with a higher cost of labor, it can be placed on the market quite as cheaply as that of India. It has, besides, an added advantage of transportation, but the most important fact is that Canadian mica is *better* than Indian mica for certain purposes, and is therefore preferred, and will continue to be used without much reference to price differences.

The last remark is also applicable to North Carolina mica and explains, I think, sufficiently its reason for existence at a time when in the matter of price it cannot approach the other two varieties. The addition of 25 or even 50 cents a pound is a matter of small consequence in the glazing art, where it is desired to have the best, and where the total cost of the mica used is but a small element in the cost of the manufactured article, and detracts little from the profit.* For this reason it is believed that North Carolina will also continue to be a mica producer to some extent, whatever the course of future prices may be. The same, however, cannot be said in the same degree, of localities like South Dakota and New Hampshire.

There is another feature, however, to be considered in this connection as helping to enable the domestic industry to hold its own: namely, the production of scrap, which will undoubtedly continue to increase from year to year, although it can hardly rise to any very voluminous proportions. Now the scrap and sheet-mica product together form the support of the industry, and it follows that one must rise as the other falls. and conversely. If then the price of sheet mica falls from 50 to 25 cents per pound, it is natural that the price of scrap should rise from $7 to $25 or even $50, as we see that it has done. We have seen, moreover, that the United States.

---

*An ordinary sheet of stove-paneling. 4x4 inches, weighs about $\frac{1}{100}$ to $\frac{1}{125}$ of a pound, and would cost from one to two cents or less. For such purposes as lamp-chimneys it is absolutely necessary that perfectly colorless mica be used.

ground mica industry is practically dependent on the United States for its supply of scrap, and is likely to continue to be, and this alone will form a source of support for mica mining in the United States. A price of $25 to $50 a ton for scrap looks high indeed from our former condition, but there is really no reason why it should not remain there if the industry requires it, for the grinding can be done at a profit for $40 per ton, and there is no question but that the ground mica trade can endure the price of $80 to $100 per ton.

The mica industry is influenced fundamentally by the uses to which the product is put, for as we have seen, these uses are special in their nature. In consequence the different classes of mica come only slightly into competition with each other, and the corresponding branches of the industry are semi-independent of each other. Electrical mica is not suited for glazing, nor is glazing mica best suited for electrical uses. If, indeed, either class of mica should become very scarce, we might expect a considerable influence upon the mining of another class, but so long as prices remain within reasonable limits, there is no special inducement to substitute one class for the other. This is convincingly shown by the absolutely imperceptible effect of changing tariffs upon the imports and prices. Of course, India mica to a large extent competes with both Canadian and domestic, and owing to a general ignorance of the subject, prices have doubtless been due largely to prejudice; but the enormously increased use of mica in the past ten years have served to bring each more nearly to its respective level.

During the whole of the last decade the future of our mica industry hung in the balance; but in the last few years it has assumed a new aspect, due undoubtedly to the stimulus of enterprise in the western fields, and now, even with the prevailing prices lower than ever before, it is rapidly forging ahead. It must be assumed that this new condition is due to a change in the method of mining rather than to new possibilities of exploitation or a new use for the product.

The New Hampshire industry is practically extinct, and must become absolutely so in a few years unless a growth of the ground-mica industry should take place beyond what it is reasonable to expect.

In spite of the recent great relative increase in importance of the western mica, the greater possibilities of the future undoubtedly lie in the state of North Carolina. A mica field equal in quality of its product to that furnished by this state for glazing purposes is yet to be discovered.

There seems to be no likelihood of any new districts becoming producers on an important scale. The industry is so well established in India under European control that it unquestionably has a lead which it would be difficult for a new district to overcome without some special reason; and there is no doubt that by the introducing of more system and the prevention of waste in the Indian mines, the mica could be produced at lower cost even than at present; and there is no doubt of the ability of India to supply the world with all the mica that it needs.

Large sizes, both in glazing and electrical mica, more especially the latter, have undergone a permanent decline. The increasing use of artificially-compounded board, and its manifest superiority from an electrical standpoint to the large sheets, indicate that these will never rise again to their former value. Precisely how great an effect on the price per pound has been caused by the proportionate increase of the use of small sizes and the decrease in large sizes, it is difficult to say, but it undoubtedly has had an influence in determining the average price per pound of marketed mica, by increasing the amount of utilizable material per ton of rock mined.

Finally, it may be said that the mica industry at large is certain to increase in the future as it has in the past. Its industrial position is for all practical purposes impregnable. In its two principal uses no substitute for it is known, nor is one in the least likely to be found, which can have any market effect on its production. More especially the electrical industries, it is well understood, are on the increase, and that of mica must increase concurrently. True, the percentage of installations which use direct current (for which commutators are necessary) is on the decrease, but it is hardly to be supposed that the actual number of direct-current machines in use can decrease; these machines must use commutators, and the commutators must be insulated with mica. Mica has been and can be artificially prepared; but the conditions under which natural mica is produced seem to place its artificial manufacture on a commercial scale beyond the range of reasonable possibility; and should any new art arise requiring the use of mica on a large scale, it would of course give a large additional impetus to mica production.

[THE END.]

TITANIFEROUS IRON ORE.

The blast-furnace is not suitable for the reduction of titaniferous iron ore. But in the electric furnace it is possible to obtain a final product containing a large or small percentage of titanium as desired. The advantage of the treatment of titaniferous iron ore, vast quantities of which exist in the United States, lies in the value of the by-products, particularly the ferro-titanium and titanium carbide.—*Eng. and Min. Jour.*

PHILADELPHIA FIRE ESCAPES.

A type of fire escape has been developed under the Building Laws of Philadelphia primarily for use in factories, which is so remarkably efficient and so far ahead of safety of anything else that exists, that we may wonder why it has not been copied in other cities. It is somewhat expensive, but the safety it gives is well worth the extra cost. The fundamental idea is that the stairway tower is absolutely cut off from the various rooms and floors which it serves. One must go out from the room into the open air and then enter the stairway. Once within this, he can proceed without danger to the bottom. The same idea can be applied to the fire escapes from a theatre.

ALUMINUM-ZINC ALLOY.

An alloy of two parts of aluminum and one part of zinc is equal to good cast iron in strength, and superior to it in elastic limit. Its color is white. It takes a fine, smooth finish and does not readily oxidize. It melts at a dull red heat, or slightly below, and is very fluid, running freely to the extremities of the mold and filling perfectly small or thin parts; in that respect it is said to be superior to brass, but it is brittle, and hence unsuited to pieces which require the tougness possessed by brass. The tensile strength of the alloy was found to be approximately 22,000 pounds per square inch, and its specific gravity 3.3.—*Eng. and Min. Jour.*

Section of Physics and Chemistry.

(*Stated Meeting held Thursday, November 9th, 1905.*)

Labor-Saving Appliances in the Laboratory*

By Edward Keller, Baltimore, Md.

[The author in this paper describes and illustrates a number of improved methods and apparatus introduced in the newly equipped assay laboratory of the Anaconda Copper Mining Co. in Baltimore.—The Editor.]

Having been authorized by the Anaconda Copper Mining Company to build a new laboratory in Baltimore, Md., a works-laboratory in which the chief duty is to determine the values of copper, silver and gold in crude copper, the writer determined to make its equipment superior to any he had seen in this country and abroad. The result of this effort is given in the subjoined description:

CONVENIENCE FOR HANDLING BEAKERS AND ACIDS.

Assaying forms the bulk of the work at the Baltimore laboratory. Scores of silver-determinations and hundreds of scorifications and cupellations for gold are often made. Silver, and sometimes gold, is determined in the metallic copper-material by what is commonly known as the combination method, which consists in dissolving both copper and the contained silver in nitric acid, leaving gold as a metallic residue. The silver is reprecipitated in the form of chloride and, with the gold, is separated from the copper solution by filtration. The incineration of the filter, scorification with metallic lead, and cupellation with a subsequent parting of the two precious metals, completes the assay. We have simplified this operation first by the introduction of metallic trays to hold the No. 5 Griffin beakers, thus avoiding handling them either singly or in pairs. The trays, shown in Fig. 1, hold nine beakers, and as soon as the copper has been dissolved the tray may be placed directly on the fire in order to expel the nitrous fumes by boiling the solu-

---

*The *Journal* is indebted to the American Institute of Mining Engineers for the use of the illustrations to this paper. [Ed.]

After the copper solutions have become cold and the proper quantity of sodium chloride requisite to precipitate all the silver has been added, it is absolutely necessary to stir the mixture in tions. The acids used to dissolve the copper is kept in a five-gallon bottle, having a glass cock, or spigot, with a 0.25 in. orifice, through which the outflow is very rapid. From this bottle the acid is tapped into a specially-designed measuring-cylinder shown in Fig. 2, provided with a stop-cock similar to that of the bottle, which allows the acid to be tapped into the beakers, without dripping, in a neater and cleaner manner than can be done by pouring from a bottle, beaker or cylinder.

Fig. 1. Metallic trays holding nine No. 5 Griffin beakers.

order that the reaction shall take place throughout, and that, after adequate settling, none of the silver chloride will run through the paper during the subsequent filtration. The stirring of a large number of solutions by hand occupies much time and is very tedious, but omitted, the determination becomes faulty. Figs. 3, 4 and 5 illustrate a stirring-machine which obviates all the difficulties incident to the manual operation of stirring and, by its use ten solutions can be stirred in the same length of time occupied in stirring one by hand, and there is no splashing of solution or breaking of beakers. Fig. 3 shows the machine ready for use; Fig. 4 shows the position after the stirring-rods have been lifted from the solutions and are ready to be rinsed; and Fig. 5, the position assumed after the rods have been rubbed off, the disks holding them and forming a cover to the beakers, now being ready to be washed. This latter operation, however, is generally superflous, for very seldom does any of the solution reach the covers.

The characteristics of this stirring-machine are the three based feet, the convenient driving-gear on the right, which admits of the ready application of mechanical power; the slack take-up for the belt on the left; and the rubber disks which hold the rods and form covers to the beakers. The action of the machine is made universal by having several holes in each disk so that the position of the rods will conform to any size of beaker. The separate construction of both beaker-stand and stirring-stand permits a ready change of a set of beakers and allows the rods to remain a permanent part of the whole.

Fig. 2. Cylinder for measuring acid.

Descriptions of stirring-machines are given in some catalogues of manufacturers of chemical apparatus, but they seem to lack the essentials necessary for successful manipulation.

FILTERING APPARATUS.

The construction and operation of our filtering or decanting apparatus are illustrated in Figs. 6, 7 and 8; and by its use twenty filtrations can be performed with perfect ease, and without the least danger of loss by splashing or breakage. The beaker-rack is tilted by means of a hand-wheel on the right, retrograde motion being prevented by a ratchet. The point of

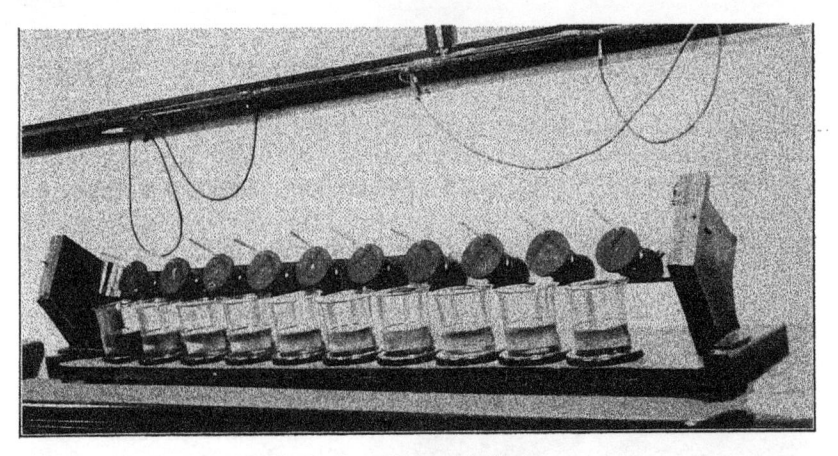

Fig. 3. In position ready for use.

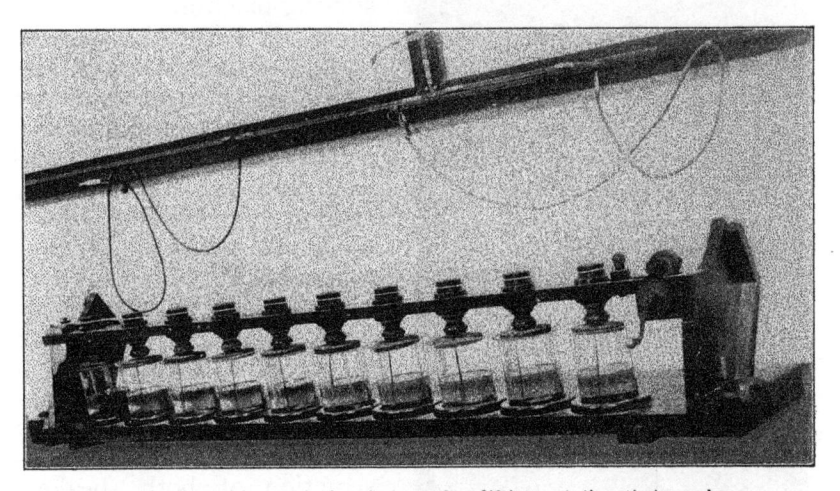

Fig. 4. In position ready for rinsing, after lifting out the stirring-rods.

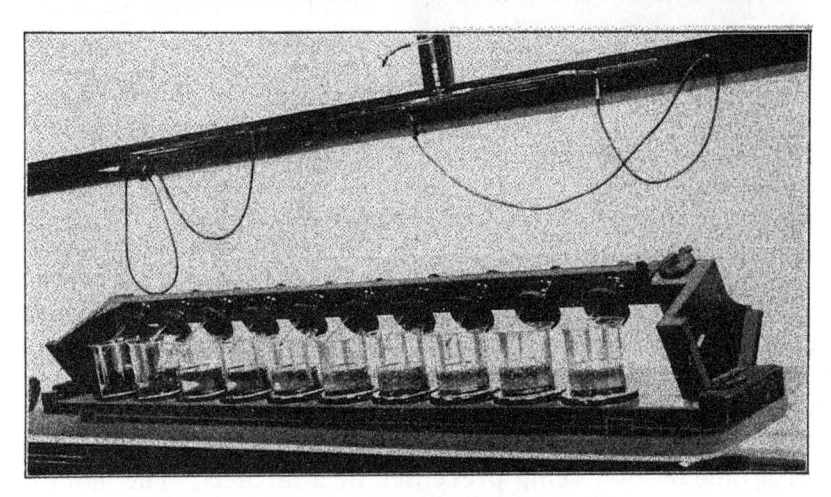

Fig. 5. In position ready for washing off the disks.

THE STIRRING MACHINE.

Fig. 6. Original machine, ready to begin filtration or decantation.

Fig. 7. Original machine, position at end of filtration.

Fig. 8. Original machine with new locking device.

THE FILTERING OR DECANTING APPARATUS.

rotation of the whole series of beakers lies some distance from their lips, about at the end of the glass rods that guide the stream of liquid to a definite point in the filter, an arrangement which is essential for steady pouring. The lifting of the load is aided by a counterpoise on the left. By the use of this apparatus the filtering can be done leisurely with one hand, so steadily that the precipitate remains undisturbed until all the clear liquid is poured off, and the time of the whole operation is greatly shortened. Figs. 6 and 7 show the machine as originally constructed, with a locking-device for each beaker, and rods, consisting of an arm pressed downward by a spring, and free to rotate vertically within an angle sufficient to clamp, as well as to allow it to sweep horizontally over the rod and beaker, the horizontal rotation also being entirely free. Fig. 8 presents a locking-device on the same machine by means of which ten beakers and rods are locked simultaneously. It is a matter of taste which arrangement is the more satisfactory. The beaker-rack and the filter-stand are adjustable to several sizes of beakers and funnels, and the machine can be made for any number of beakers. Light wooden trays are provided to carry the beakers, in sets of ten, to or from the apparatus.

Table I., containing the results of a test recorded by one of my assistants with twenty beakers, shows the difference in time between the new method of mechanical filtration and the old method by hand. The quantity of solution contained in each beaker varied between 275 and 300 c.c.

TABLE I.—COMPARISON OF RESULTS OF MACHINE FILTRATION VS. HAND FILTRATION.

Machine Filtration.			Hand Filtration.		
		Time Consumed.			Time Consumed.
10.21 a.m.	Started.	Minutes.	11.13 a.m.	Started.	Minutes.
10.22 a.m.	Placed rods	1	11.14 a.m.	Placed rods	1
10.30 a.m.	Finished decanting	8	11.38 a.m.	Finished decanting	24
10.41 a.m.	Washed beakers and rods	11	11.58 a.m.	Washed beakers and rods	20
10.45 a.m.	Washed filters	4	12.04 p.m.	Washed filters	6
10.55 a.m.	Rubbed out beakers	10	12.13 p.m.	Rubbed out beakers	9
10.59 a.m.	Folded filters and placed in scorifiers	4	12.17 p.m.	Folded filters and placed in scorifiers	4
	Total	38		Total	64

Difference in favor of machine, 26 minutes

I have decanted ten beakers, each containing from 275 to 300 c.c. of solution in 4 min. 30 sec., the solutions being run through a double S. & S., No. 597, 12.5 cm. filter.

In connection with the work as just described we have entirely dispensed with the well-known wash-bottle, on account of the unnecessary physical strain involved in its use, as well as its unsanitary character in many instances. The distilled water contained in carboys placed about 4.5 ft. above the floor-level is siphoned into a system of glass-tubing, extending throughout the laboratory wherever the water may be needed. This glass-tubing is provided with numerous "T" connections, to which are attached rubber-tubes, having pinch-cocks and glass nozzles at their ends, which permit the direction of a stream of water to any desired point. The water in the overhead carboys is replenished by forcing a new supply from another carboy in which it has been condensed, placed on a carriage on the floor, up through a glass siphon by means of compressed air.

When the filtration is complete and the precipitates have been washed out of the beakers the latter must be rubbed out with paper in order to be sure that neither silver chloride nor gold is retained on the walls of the vessels, perhaps by a little oil often contained in the samples, or by the drying and hardening of small particles of chloride above the surface of the solution. In order to facilitate this cleansing, we have designed a machine in which the beakers are rotated. This ma-

Fig. 9. The "Policeman" in beaker in rotating machine.

chine is on rollers, and runs on rails fastened to the table in front of the filtering-apparatus, and may be locked at any desired point. The rubbing is done by means of a "policeman," Fig. 9, which consists of a rod having a cork fastened at one end, over which is clamped a piece of filter-paper, held in place by a conically-cut ring. The paper thus fastened is run along the bottom and side of the rotating beaker. For each beaker a fresh piece of paper is quickly clamped to the cork.

NEW ASSAY-FURNACE TOOLS.

The implements used in performing the furnace-work in assaying are shown in Fig. 10. A and B are the traditional tongs, universally used for handling singly the scorifiers and the

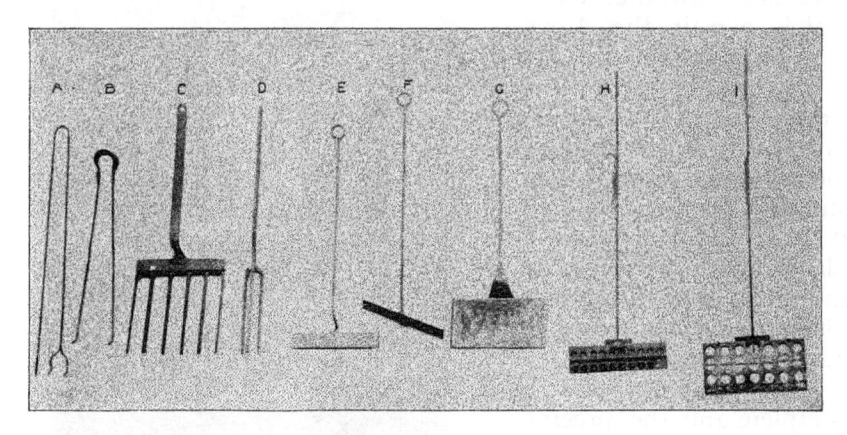

Fig. 10. Furnace tools.

cupels. The former has been entirely replaced with a fork, C, with which a set of twenty scorifiers can be handled at one time. In silver-assaying each set of scorifiers is placed in the muffle twice and taken out twice; first put in the muffle for the incineration of the filters, then taken out for the addition of the test-lead, then returned for scorification, and finally taken out for pouring the slag and molten lead into the molds. By the use of the fork, which works perfectly if the muffle be properly supported so that it will not sag to any marked extent, sixty handlings are reduced to three. Four scorifiers constituting a longitudinal row in the muffler are poured at one time by means of a pair of tongs, D, and with a little practice the pouring is made just as easy as with a single scorifier. It is necessary only that the mold correspond to the arrangement of the scorifiers and that the pockets are shallow. The cupels are placed into, or taken from, the muffle in sets of one or more rows, by means of the tools, E, F and G, an idea which, I believe, was first put into practice by my brother, Richard Keller, of Durango, Colo. E and G are sharp-edged shovels, the latter having up-turned sides. F is a rabel, with which the cupels are raked onto the shovel, removed therefrom to the place where they are to be deposited by placing it behind them and withdrawing the shovel. The tool, H and I, is entirely new, and by its use one or more rows of cupels in the muffle may be charged with the lead-buttons from the scorifiers. Fig. 11 shows the idea of the device more clearly. It comprises a top sliding-plate with openings corresponding exactly to the position of the cupels. The openings in the lower plate correspond exactly with those of the upper one; the plate, however, rests on two adjacent sides

Fig. 11. Device for charging scorifier buttons into cupels.

extending downward at right angles to the plate and to each other, thus forming two closed sides of the instrument; one at the front, and the other at the right-hand side. The height of these sides is such that, when resting on the bottom of the muffle, the bottom plate will be some distance above the cupels and, by a slight pull forward and a push to the left with the handle of the instrument, the set of cupels will be perfectly aligned in both directions, and the apertures in the lower plate will exactly cover the tops of the cupels. The lead-buttons are placed in the apertures of the upper plate and rest on the lower plate before introducing the instrument into the furnace, and when it is placed over the cupels, which have been properly aligned in the muffle, the upper plate is pushed forward to a stop-point, bringing the apertures of the two plates to register, thus causing the lead-buttons to drop down into the cupels. The handle of the upper plate runs through guides fixed to the handle of the lower plate; both handles are connected by a spring, which acts as a brake when the upper plate is pushed forward to drop the buttons, and also serves to bring it back into its original position, in which the buttons cannot drop through the apertures in the lower plate.

Charles Tookey* first recommended the use of hydrochloric

Fig. 12. Convenient parting-bath.

acid (1 HClaq, 2 H_2O) instead of a brush for cleaning the buttons, and for this purpose a small silver dish and tray having perforated pockets give excellent satisfaction. By the use of this device fifty or more beads at a time can be treated, washed and dried without transfer.

Fig. 12 shows a very handy parting-bath; which, though old in principle, has not been in general use. The vessel is a constant-level water-bath and the tray an ordinary test-tube holder. The silver beads to be parted are dropped into the

---

*Journal of the Chemical Society (London), vol. xxiii., p. 366 (1870).

test-tubes, and the latter filled with dilute nitric acid of a strength of one of acid (sp. gr. 1.42) to 9 of water. The water in the bath is first brought to the boiling point before the tray with its contents is set into it. Treated in this way the gold almost invariably remains in the form of a small coherent bead, even from an alloy as low as one part of gold to 500 of silver.

I have recognized that the system of handling everything in sets was incomplete as long as I was unable to take a whole set (twenty) of scorifiers from the muffle and pour their contents simultaneously into the molds. Recently a tool for that purpose, shown in Figs. 13 and 14, has been perfected. It is composed of quintuple tongs, corresponding to the five longitudinal rows of scorifiers in the muffle. The lower part of each pair

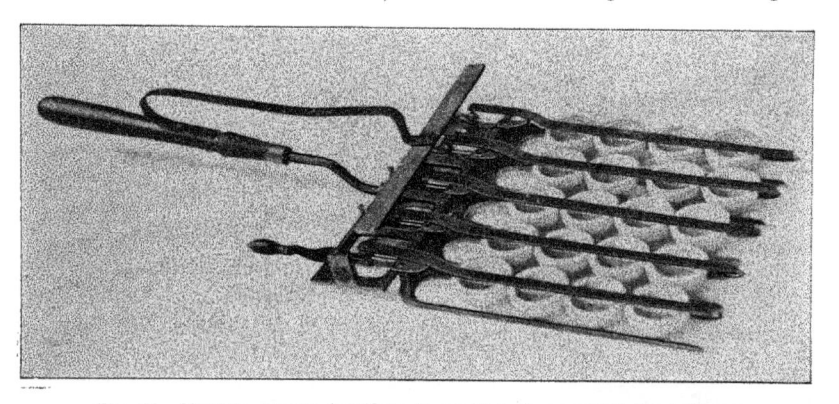

Fig. 13. Multiple tongs; Scorifiers in position for removal from furnace.

of the tongs consists of a fork, on which the scorifiers rest. and one of whose prongs is retilinearly extended through two bearings in a frame, and held in position by collars. This extension is free to revolve in the bearings, and it is the axis of rotation of the tongs, To each of them is attached, at right angles, a lever, extending upward at an angle of 45°, and all the levers are connected by slotted joints to a cross-rod. Therefore, if by means of a crank, fastened to the end of one of the extended prongs, one of the forks is turned and the scorifiers tilted to the desired angle, the others performing the same rotation. The center of gravity of the scorifiers lies to one side of the rotation-point, and they would therefore, on being lifted, tilt in that direction; this, however, being prevented by the cross-bar resting against a post at that end of the frame toward which the inclination tends.

The scorifiers are clutched by the upper prong of the tongs, which is fastened to a spring on a post of the fork below, and which is free to move in a vertical plane; the pivotal point lying over the spring and post. By bringing pressure on the extended ends of these clutch-bars behind the pivot, their other end will rise from above the scorifiers, and thus release these, or permit the placing of them onto the tongs. The pressure exerted on the rear ends of the clutches is accomplished by means of a cross-bar fastened to a spring-bar, which is itself riveted to the handle of the instrument; all of which may be plainly discerned in the illustrations. In pouring the contents of the

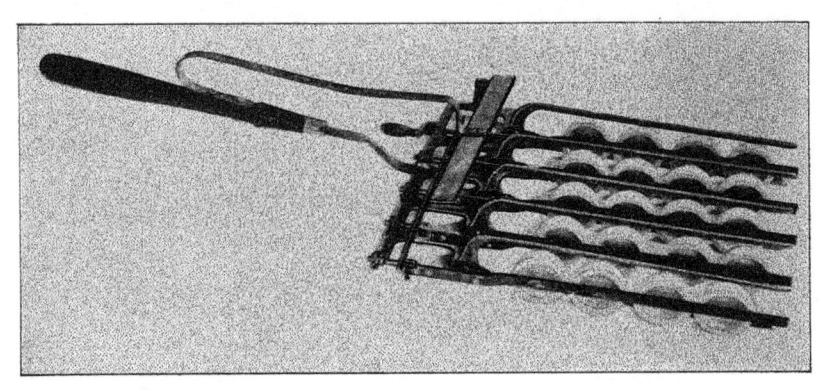

Fig. 14. Multiple tongs; Scorifiers tipped for pouring contents.

scorifiers, the frame of the tool rests on the edge of the mold, leaving the tongs free to turn.

The introduction of the new system of manipulation described in this paper has resulted in economy in several ways. Much labor has been saved; breakage of expensive glass-ware has been very largely eliminated; and the time of the furnace-work, and, consequently, the consumption of gas, have been much reduced. Furthermore, the gain has been a moral one, and work formerly regarded as tedious has become more of a pleasure; especially has the sojourn in the furnace-room during the hot summer months been rendered cooler by being greatly shortened.

The appliances described in this paper may eliminate the

laboratory boy, but if not, they will make him more reliable. They also increase enormously the quantity of chemical work that one man can do in a day. In the small domain of our laboratory the change from the old system to the new is considered to bear about the same relation as the change from ancient horse-cars to the modern rapid transit.

In conclusion, I take pleasure in acknowledging the efficient callaboration of my assistants, Mr. Albert Ferrell and Mr. K. W. McComas, the former having shown himself particularly useful as a skilled mechanic in the construction of various improvements.

CAUSE OF FAILURE OF BOILER FURNACES.

The collapsing of boiler furnaces is almost always the direct result of scale or of oil in the feed water, the latter being a particularly prolific source of trouble, according to a recent paper before the Northeast Coast Institution of Engineers. No ordinary furnace fails for lack of strength if clean and covered with clean water. A very thin smear of oil, however, has an effect totally out of proportion to what might be expected. In a furnace having a normal factor of safety of five, this factor rapidly decreases after the temperature reaches 650 degrees F., and entirely vanishes at a red heat. Steam at a pressure of 200 pounds has a temperature of about 380 degrees, or 270 below the point at which the tenacity of the steel begins to be affected, but a clean furnace, rubbed over with a very clean and thin coat of mineral oil, will soon rise above 650 degrees, even under light duty, and often reach 1200 degrees, at which point 75 per cent. of the strength has departed. With the use of high-grade mineral oils the danger is less than with low-grade oils, due to the fact that the latter emulsify and hence cannot be removed from the feed water except by chemical treatment.

PHYSICAL PROPERTIES OF TANTALUM.

Warner v. Bolton in the *Zeit. f. Elektrochem*, 1905, pp. 503-504, gives details of tests upon the hardness of tantalum in the purest form that has yet been obtained. The results obtained show that tantalum, containing only very small amounts of oxide as impurity, after hammering into sheets at a red heat, is equal in hardness to the best and most carefully finished steel. Tantalum, however, greatly exceeds this hard steel in toughness, for while these hard tool steels are brittle, tantalum can be rolled into sheets without injury. The new metal in its purest state, combines the hardness of the best steel with a greater toughness and ductility than is known to be possessed by any other metal.— *Eng. and Min. Jour.*

THE ZUÑI SALT DEPOSITS, NEW MEXICO.

Forty miles south of the pueblo of Zuñi, in the west-central portion of New Mexico, there is a deposit of salt which is not only of great geologic interest, but promises to prove of considerable economic importance. It is briefly described by Mr. N. H. Darton in the year-book of the United States Geological Survey, which is entitled "Constitutions of Economic Geology, 1904."

The location is eighty miles south of Gallup, on the main line of the Santa Fe Railroad, and about the same distance west of Magdalena, on a branch of the same railroad system. This deposit has been a source of supply for the Indians and Mexicans for several centuries, and of late the salt is hauled to ranches in a wide surrounding district. The present output averages only approximately 1000 tons a year, valued at about $2.50 a ton. A small colony of Mexicans at the locality collect the salt in a very crude manner. Ordinarily, persons desiring a supply go to the place and help themselves.

The deposits occur in a lake, which occupies a portion of the bottom of a deep depression in a plain of Cretaceous sandstone. This depression is about a mile in diameter, and has walls of sandstone, in part capped by lava, averaging about 150 feet in height. The lake is about 4000 feet long, east and west, and about 3000 feet wide, and is apparently shallow. The water contains about twenty-six per cent salt, mostly chloride of sodium. The region is arid and the evaporation causes the crystallization of the salt, especially in the shallow water.

A large amount of salt could be obtained by properly conducting solar evaporation of the lake water. By washing the salt with a small amount of the lake water the more soluble foreign salts are removed and almost pure chloride of sodium remains.

FORCED DRAFT.

Forced draft appeared some years ago in danger of being abandoned for all ordinary work because of frequent complaints regarding its effects in burning our the grates, injuring the boilers and blowing gas and smoke from the fire doors into the fire room. It has since been demonstrated that these difficulties were due to the fans being too small for their work, requiring them to be operated at far above their normal speed. This caused an ash-pit pressure of from five to ten inches of water where only ¾ to 1¼ inches is required. Later and more accurate knowledge regarding the proper application of the fan blower for this purpose has resulted in restoring the system to its former measure of popularity. When it is used the air passing from the ash pit to the combustion chamber is greatly reduced in pressure, due to the resistance of the grates and fuel, while the stack tends to create a partial vacuum in the furnace It is thus impossible to maintain more than a slight excess of pressure in the combustion chamber and this should not be forced, as was formerly the case.—*Iron Age.*

The Problem of Gravitation.*

By Charles Morris.

Of the varied problems with which modern science has had to deal, that of gravitation has proven the most difficult to solve. While nearly every other agency of nature has in some measure yielded up its secret, as regards this, apparently the most universal of them all, we remain almost completely in the dark. The scope of the problem has been somewhat narrowed, but this is all that can be said. Physicists are satisfied that attraction is but a name for something else, that "action at a distance" does not exist, that some agency unknown is at work upon matter, and even surmise that this agent may be the luminiferous ether. But there they stop, in ignorance of how this agent operates, or how it produces its effect.

It is certainly very probable, in view of the universality of the force of gravitation, that it is dependent upon some simple principle, and possibly this simplicity is the chief cause of its elusive character. There is so little to take hold of, such a lack of those varied elements of action which aid us elsewhere. If, indeed, we could accept attraction as the actual force which many believe it to be, there would be no problem. Newton's law of gravitative action is based on the proposition that every atom in the universe attracts every other atom; the force of attraction being inversely proportional to the square of distance. But Newton accepted attraction only provisionally; as a convenient working tool, not as a credible fact. He expressly says:

"That gravity should be innate, inherent, and essential to matter, so that one body can act upon another at a distance, through a vacuum, without the medium of anything else, by and through which their action and force may be conveyed from one to another, is to me so great an absurdity, that I be-

---

*Read by title December 7, 1905.

lieve no man, who has in philosophical matters a competent faculty of thinking, can ever fall into it."

This view of the impossibility of "action at a distance" is widely held by physicists to-day. It is unthinkable to the logical mind that one substance can act upon another through a void, or through space occupied by material that takes no part in the action; as, for an extreme instance, an atom in the sun exercising a pulling force upon one in the earth, by virtue solely of some property dwelling within itself. Recognizing this conclusion as a finality, no matter how small may be the interval between the atoms, all scientists who have attempted to explain gravitation have done so on the basis that the seeming attraction is due to the agency of an external medium, acting through some system of impact, pressure, or tension, stress or strain.

Various hypotheses have been offered in pursuance of this view. In some the active medium is the luminiferous ether; in others it is a special ether confined to this duty alone. Newton suggested the agency of a special medium in some way rarified by material substances, the rarity decreasing outwardly in due proportion to distance. Le Sage invented a system of "ultra-mundane corpuscles," flying in all directions through the universe, impinging forcibly upon the spheres and moving them towards each other. Lord Kelvin imagined an incompressible fluid, filling all space, flowing in from infinity and absorbed by each particle, or momentarily created in each particle and flowing outward to infinity. It is not necessary to go into the details of these hypotheses, or any one demanding a special ether, for none of them is to-day accepted. The tendency now is to look upon the luminiferous ether as in some way the agent at work in gravitation; but in what way remains unknown. A considerable number of hypotheses have been offered in the past few centuries in which gravitation is ascribed to some action of this ether, some of them quite ingenious, but few of these attained even momentary acceptance and none of them is now considered of any potency.

Such is the status of the gravitation problem as it exists to-day. Its scope, as above said, has been narrowed. Gravitative force is believed to be due to some agency of the ether, but this is as far as anyone has gone. All efforts to show in what this agency consists have failed. This is all, but must it remain all? Is there no clue to the labyrinth? May it not be that the failure so far is the result of a radically false conception of the scope of the problem? Has science put the real nature of the question at issue before it; studied its initial elements before seeking to reach its solution? May not one source of error be the following? In popular acceptation matter, acting through the force of attraction, is the sole active agent in gravitation. In scientific acceptation the ether, or some other universal substance, is the sole active agent. There is no hypothesis in which the possible agency of both matter and ether, each taking a direct part in the issue, is considered. Yet may it not be that matter and ether are jointly active in the process, and that the seeming attraction is the result of conditions of energy directly interacting between these two elements of nature?

Again, gravitation appears to many to stand apart; to have no traceable relation, direct or indirect, with the other forces; to be an isolated phenomenon, which must be dealt with by itself. It is, no doubt, maintained by scientists that all the forces of nature are interrelated, and some hold that there is a close affinity between gravitation and magnetism, but no one has succeeded in demonstrating the existence of such a relation, and so far as we actually know the gravitative force is isolated. Thus the case stands at present; but few physicists doubt that this universal force is definitely related to the other forces of nature, and efforts to find the line of connection have often been made. That they have been so far unsuccessful is no evidence that such a connection does not exist, in view of our ignorance of many of the underlying principles and conditions of nature.

In dealing with the subject of attraction it must be premised that it is not confined to gravitation, but exists also in magnetic, electric, and chemical action. Instead of being a single, it is, in reality, a four-fold phenomenon. But this by no means indicates that there are four separate attractions, unlike the origin and character. Such an idea would infinitely complicate the problem. A single attraction is difficult enough; four undelated forms of attraction would be insuperable. The force we call attraction is in all probability single and simple: an unique phenomenon, not a composite enigma. That variation in force and method of attraction may arise from variation in conditions is easy to understand; that it arises from primary differences in the force itself is infinitely less probable. We certainly cannot admit this of a peculiar force like attraction unless it be proved that the diversities in its manifestation are irreconcilable. Yet if we view all displays of attraction as due to one general couse, the fact becomes important that attraction does not constitute all we know of electricity and magnetism, as it does in the case of gravitation. Attraction, for instance, is but one among many electric phenomena. It appears to be a temporary one, subject to passing conditions, and a survey of these conditions may throw some light upon the hitherto sealed mystery of gravitation. The path to the unknown leads through the gateway of the known. It is a path that seems never to have been trodden in the present inquiry, but it is one well worth traversing.

It certainly seems a judicious course to study the problem of gravitation at a distance, beginning with the somewhat transparent electric attraction and leading up to the opaque gravitative attraction. This work has been largely done for us in the case of electricity, though no one has carried the effect there visible into the domain of gravitation. A cursory examination of electric attraction seems to render it evident that, in this case at least, attraction does not manifest itself as an inherent property of matter. It comes and goes in response to changes in conditions, and apparently without regard to the matter involved. It may even be transformed into the opposite force of repulsion by a simple reversal of conditions.

We are not aware of what takes place in static electricity, other than that attraction and repulsion appear when the electric state is produced and disappear when electricity is neutralized or conducted away. In current electricity we can with some ease discover a general principle of action. Let us take two metallic wires placed in general parallelism at a moderate distance apart. They do not apparently affect each other. Their gravitative force is too slight to yield any visible result. But if we send currents of electricity through them in the same direction a marked effect appears. They show a tendency to approach, to move together. In common parlance we say that they attract each other—the seeming attraction aroused being far more powerful than that of gravitation. Reverse one of these currents, cause them to flow in opposite directions, and a new force manifestation appears, that of repulsion. The wires show a tendency to move apart. Cut off the currents and the wires return to the passive state. Every visible indication of attraction of repulsion in them disappears.

We have here a very significant fact in any study of the cause of attractive force. Electric currents seem capable of generating attraction when they move in parallel or accordant directions and repulsion when their directions are reversed. As these forces disappear when the currents are cut off, they may be looked upon as in some way generated by these currents, the part taken in them by matter being, so far as appears, confined to its power of conducting electricity.

It is well to say here that exact parallalism in the conducting wires is by no means necessary, it being simply requisite that the currents should have a general consonance in direction. If the wires form an angle, starting from a common point, and both currents move toward or from this point, attraction will appaers, no matter how wide the angle, even if it be a broadly obtuse one. On the other hand, if one current moves toward, the other from, such an angular point, the wires will repel each other without regard to the width of the angle. The tendency will be to bring the wires into parallel positions and then to draw them together or force them apart as the relations of the currents may demand. All that is requisite in the one case is accordance, in the other discordance, in the direction of the currents.

In seeking to deduce a definite conclusion from these facts, let us take the most recent theory of electricity. This is, that the current is conveyed by electrons the minute constituents of atoms, which fly with extraordinary rapidity from atom to atom, so that practically a swift stream of them flows through the wire. If this theory is correct it gives us some warrant to conclude that attraction and repulsion are dependent upon, not matter in itself, but matter in motion. These forces seem to be in some way generated by moving matter, or movement in matter, attraction by parallel or accordant motions, repulsion by reverse or discordant motions.

The phenomena of magnetism leads us to a similar conclusion. In fact, the magnet can be exactly imitated by sending an electric current through a spirally-wound wire. Two such wires act like two magnets, attracting and repelling, displaying north and south poles. The results are the same as when straight wires are used, the coils attract when their spiral currents flow in parallel directions, repel when their directions are reversed. These electric phenomena are so significant of what is taking place in the interior of magnets that they have led to Ampere's theory, everywhere accepted, that every particle of a magnet has closed currents flowing round it in fixed directions, and that to these currents its attractive and repulsive forces are due. Without going into the details of the magnetic phenomena, the one point with which we are here concerned is that magnetic attraction, like electric, gives very strong indications of being a function of matter in motion, not of matter in itself.

What conclusions shall we draw from all this? While we are ignorant of what takes place in static electricity or in chemical action, the facts of dynamic electricity and of magnetism lead irresistably to the inference that attraction is not an inherent property of matter. It is dependent upon special conditions, and these conditions can be readily so changed that attraction will disappear or will be replaced by repulsion, with no evident change whatever in the character or position of the matter concerned. The one thing necessary for the display of these forces, so far as appears, is the development of special conditions of motion in the material employed.

Here is a fact of extraordinary significance, and one we must take into account in dealing with attraction in any of its manifatesitons, general and local alike. It is, stated broadly, that attraction is not a property of matter in itself, but arises from a special condition of matter, this condition appearing to be that of motion—not an irregular and constantly reversed motion like that of heat, but a motion in fixed directions, as in the electric current. This inference seems to hold good whether or not we accept the recent theory of electricity. The current may not be due to the motion of electrons, but few doubt that in electric conduction motion of some kind and in some material is taking place, and the appearance of attraction or repulsion seems immediately dependent upon this motion. The query naturally arises, is gravitative attraction similarly conditioned? Is motion in spheral matter necessary for its manifestation?

This is certainly a very probable conclusion, far more probable than the opposite, that attraction is due to more than one cause and dependent upon more than one condition in matter. It is difficult to imagine that there can be two or more separate causes for so mysterious a force. On the other hand, the condition predicated as necessary is existent in all spheral matter. Motion is everywhere present. The earth, for instance, is circling around the sun at the extraordinary speed of more than eighteen miles a second, and through general space, in company with the whole solar system, at a speed of twelve or more miles per second. The same is true of all spheral bodies, many of them moving with a far greater speed than that named. If now, instead of looking upon the earth as a unit, we regard it as a vast collection of independent atoms, it becomes evident that all its contiguous atoms are rushing in parallel lines with immense rapidity through space. Thus the condition which seems requisite to attraction in local instances is present in the spheres, and we have some justification in assuming that it may be the exciting cause of attraction there also.

To return now to the view so widely entertained, that the luminiferous ether is the active agent in what we call attractive force, let us recall what was above affirmed, that attraction may not be a result of the action of either matter or ether considered separately, but may arise from a close interaction between the two, each playing its part in the result. As to the part taken by matter, we have indications of it in electric and magnetic action. It remains to try if we can discover and coördinate or resultant action in ether.

As a preliminary, something must be said about the nature and conditions of ether, so far as known or conjectured. It is held to be an universal substance, occupying all open space and penetrating all the spheres. It is possibly the basis of matter itself, since each atom may be an ether aggregate. Whether it is composed of particles or consists of substance in a state of infinite division; whether it is a plenum, filling all space absolutely, or permits of void intervals; whether it is compressible or incompressible, are questions unanswered, perhaps unanswerable. We do not know if cohesion exists within it, for the pressure of a plenum may produce the effect of cohesion. We know, from the phenomena of light and other radiations, that it is capable of transmitting vibrations in extraordinary variety and complexity. And the phenomena of light have given rise to the hypothesis that the ether is an electric solid, it being maintained that only a solid could transmit the transverse vibrations of light. To explain the passage of the spheres through solid ether without obstruction the example of an ordinary jelly is adduced. As a weight will sink through a jelly and leave no trace of its passage, so it is held that the spheres can pass freely through the almost infinitely thin jelly of the ether. But the weight does not pass through the jelly without obstruction. Its speed of fall is retarded, and it is hard to avoid the conclusion that the spheres would be similarly retarded by the ether, however thin, if it have such a constitution as here conjectured. Certainly this hypothesis does not appeal to us as probable, however necessary physicists may deem it, and we have before us a problem of very difficult aspect.

We know nothing about the innate constitution of the ether, but from its remarkable readiness to vibrate in response to material impulses we may deduce that it is highly mobile, and are justified in conjecturing that it is a reservoir of intense motor energies and capable of exerting vigorous pressure if these energies are brought to act in one direction. The belief that matter may be derived from the ether is an argument to this effect, since the energies of the ether would then be the source of the very active motions possessed by matter, while energy may be retained much in excess of that given. There are other problems connected with the ether. For instance, is it quiescent or is it adrift through space? Have its particles onward movement or only vibratory or other local motions? This problem is unanswerable, but for our purpose it is simply necessary that the ether should be a storehouse of potent energies.

Returning now to the subject of the relations between ether and matter, it may be said that it has long been a vexed problem how the earth could rush at its immense speed through the ether without obstruction. It was formerly supposed that the earth dashed the ether aside in vast eddies as a ship does the ocean waves. It is now believed that the earth is readily permeable to ether, which fills its innumerable pores and drifts through it almost as freely as though no matter were present. But this view merely leaves us confronted with the same problem in a different form. It is simply transferred from the earth as a whole to its constituent atoms. The transfer of the problem from earth to atoms does not change the probable result. The question of obstruction remains the same. The difficulty seems an insuperable one if we hold that the ether is a solid. It is a very awkward one even if we hold that it is an almost infinitely thin gas, especially if this gas be a reservoir of intense energies.

We have now reached a critical point in our inquiry. The problem before us assumes a triple aspect. Does the ether oppose and obstruct the atoms in their swift movement? Does it yield and flow past with no effect upon them? Or does it react upon them in a way differing from that of obstruction? The first of these seems the most probable, but all the evidence goes to prove that it does not exist. The second, while possible, does not appeal to us as probable. The third is a new point of view, never yet taken, but one of great interest, since it may possibly be the missing element in gravitation.

Let us consider the case of a group of atoms moving at a speed of eighteen miles per second through a mass of quiescent but highly mobile ether. In view of the fact that the atoms appear to move without obstruction, we might predict that the ether is non-coherent and therefore non-resistant, and is dashed aside in a broad eddying whirl, curving backward so instantaneously that the atom group is past before its internal energies are brought into play. But the atom group is no more solid than is the earth. It is permeated everywhere with pores, through each of which the ether will drift backward. But if the formation of the ether is disturbed and an eddying movement is produced, it seems probable that it would flow backward more freely outside the group than through its pores, and more freely through its larger than its smaller pores. In other words, if there is a disturbance of the ether formation and interference with its drift through the atom pores, this interference increasing as the pores grow more minute, we should have a degree of rarefaction in these pores, a partial ether vacuum growing more declared as the pores decrease in size. The natural result from this would be a pressure of the external ether similar to that seen when an air vacuum is produced.

The point to be made is this:—If the ether, despite its internal energies, makes no frontal attack upon the atoms, but flows back in unresisting eddies, seeking in preference the larger channels, it seems fair to presume that it may make a lateral attack, crowding the atoms into the spaces in which there is a partial ether vacuum. This is the crux of the situation. If the whirling ether reacts laterally instead of frontally, forcing the atoms inward instead of backward, we are led to an explanation of two mysteries, that of the lack of obstruction to the motion of the spheres and that of the pressure of gravitation. The conceivable former is converted into the actual latter. If the possibility of this be admitted, all forms of attraction will be brought into conformity, that of gravitation coming into harmony with those of magnetism and electricity, to the extent that the motion of matter in each case is the instigating cause.

Such an inward pressure in ether of course would produce condensation, and it might be imagined that the atoms would be driven together until the spaces between them were obliterated and vacuus intervals become impossible. But the atoms have other motions than their spheral ones. Their heat agitation exerts an energy sufficient to keep them permanently apart, despite the gravitative force, and the degree of separation in any case would be the resultant of the opposing forces of external pressure and internal energy.

There is a variation in results between the cases of the movement of a body through air and of a body through ether. The resistance in the former case is due to the cohesion of the air particles. However feeble this be, it would seem likely to cause some resistance, and we know that such a resistance takes place to the motion of meteorites in the exceedingly rare upper atmosphere. The fact, then, that ether makes no resistance seems to indicate that, as above suggested, it is quite destitute of the condition known as cohesion. Its mission may be to produce in other matter the cohesion it lacks in itself.

If there be such a non-coherence, this may be the explanation of its non-sisistance to moving matter; while its internal energies, and its rarefaction in the pores of moving matter, may be the explanation of the compressing pressure known as attraction of gravitation.

So far we have dealt merely with atom groups and with the very slight pressure exerted upon each of these. Let us now consider the effect upon the sphere as a whole. Its innumerable swarm of atoms, each yielding a minute effect, must in the aggregate yield an immense effect. If the vacua produced in each group of atoms gives rise to an inflow of ether reaching beyond the immediate vicinity, or a pressure making its effect felt for a slight distance outward, then the whole ether impulse would presumably extend to a great distance outward from the earth, the ether moving inward or its energies being directed inward toward the earth throughout a broad field of space. If gravitation is due to this cause, the ether action must extend to and far beyond the sun. In the case of the latter it is supposed to extend to the fixed stars.

In the LeSage hypothesis it is held that the earth and the sun exert an indirect or "shadowing" influence upon each other, each cutting off a part of the stream of corpuscles moving upon the other. The result is held to be a diminished pressure in their connecting line, with the result that the external pressure is the stronger, the effect being to force them toward one another. In the case of an ether pressure, such as we have suggested, the same result would appear. But the present hypothesis suggests a different agency, which may be of still greater effect. The ether between the sun and the earth would be affected by opposing influences and would tend to move in both directions, vigorously toward the sun, more feebly toward the earth. Thus its action toward each orb is resisted by the other and is thus greatly diminished. There must be a limiting zone in which there would be a balance between these tendencies and the ether fail to exert energy in either direction. In this zone the ether, seeking to act in both directions, would tend to thin out or become rarified, a partial vacuum being produced into which both spheres would be forcibly driven by the external pressure. Thus the cutting off of the pressing force of ether from each sphere by the other, thus immensely reducing its range of action in this line, and the rarefaction of ether between them, would form a double agency forcing them together and holding them in mutual relations of position. If this be the case, the influence of a vacuum must be the active agency not only between atom and atom, but between earth and sun, and perhaps between distant orbs of space.

If the present hypothesis is based on correct premises, and gravitation is a result of the action of moving matter upon ether and the reaction of ether upon matter, there is a very interesting deduction to be made. This is that the force of gravitation must be dependent not upon mass only, as in the Newtonian theory, but upon mass and rapidity of motion combined. For it seems a just conclusion that if the motion of atom groups causes rarefaction of ether within their pores and resultant ether pressure, the rarefaction must increase with increase of speed, and the pressure be correspondingly enhanced. In other words, the force of gravitation bust increase with increase of speed in spheral bodies, and the rapidity with which suns and planets move through space would become an important factor in their force of attraction or gravitative energy.

Thus the gravitative force of each planet of our system toward the sun may be dependent only partly upon its mass and largely upon its speed of motion in its orbit. If this be the case, calculations of the mass of a planet based solely upon its gravitative force may be radically false. The same principle would hold good for the great orbs of space. In the case of the extremely swift star known as 1830 Groombridge, the speed of which may be more than two hundred miles per second, there would, under this hypothesis, be a very marked increase in gravity. It is held that such orbs as these have an energy of motion too great for the restraining power of all the orbs in the universe. But if their force of gravity increases with their rate of speed they may be held captive despite their plunging energy. The effect of speed upon gravity is offered here as a suggestion only. It will appear more in the light of a fact when we return to the consideration of electric and magnetic attraction.

There are some minor suggestions to be offered. While the view has been taken that the gravitation of the planets toward the sun depends upon their orbital speed, it must also be remembered that the sun and its attendant planets are rushing together through space with a speed estimated at twelve or more miles per second. While this may mainly affect the astral relation of the sun, it may also affect the gravitative energy of the planets and of their satellites. There remains to be considered the axial revolution of the planets. In this the atoms move in parallel lines which should have some gravitative effect. This effect we seem to find in their magnetic energy. The earth, in this way, may be converted into a great magnet, with an energy dependent upon its rotatory speed. If such is the case, each orb of the universe is brought into close analogy with the atom magnet, with its circling motions. The sun, whose equatorial speed or rotation is about four times that of the earth, may possess a much more powerful magnetic force, sufficient, perhaps, to affect some of the planets.

The hypothesis here advanced is not in agreement with the view that the ether is a solid mass. Solidity would require that the ether should possess a force of attraction within itself, a condition which would infinitely complicate the problem of gravitation. The doctrine of ether solidity is based on the fact that the irregular agitation of the particles of a gas would destroy or dissipate transverse vibrations like those of light. But may there not be a condition even of a gaseous ether in which this irregular agitation would not exist sufficiently to disturb the vibrations of light? Under the hypothesis here advanced the ether, even if a very rare gas, would seem to be controlled in its movements, these, instead of being irregular and desultory, being in straight lines toward the spheres; its corpuscles, if such exist, being as rigidly controlled in their movements as though it were a solid mass. Being drawn constantly inward toward the spheres by the vacua existing within them, exerting its pressure, and then drifting away as the spheres pass onward, its movements would be, at least in considerable measure, regular in character, and while not a solid in fact, it might be so in effect. Lines of radiant vibration from sphere to sphere would traverse an ether setting steadily toward these spheres, and the ordinary irregular agitation of gaseous atoms be overcome by the directing and controlling agency here indicated. This would be especially the case if the pressure toward the spheres should give rise to a subsidary pressure at right angles to this line of action, the whole body of ether being thus set in motion and controlled in its movements by the one unceasing cause. It is not here intended to maintain that the effect here predicated would be complete, or even large, at a great distance from the sphere, but it may be that even a partial repression of irregular atom motions might suffice to permit the conveyance of light undulations.

Returning to the local forms of attraction, which furnished the clue followed in our study of the spheres, we may seek to apply the hypothesis advanced to these local conditions The movement of the electric current is now thought to be due to the flight of electrons, passing with immense swiftness from atom to atom, the effect being the same as if a multitude of electrons darted side by side through the whole length of the wire. Here we have to do with a motion of material particles in parallel lines, as in the case of the spheres, and while the electrons are far more minute than atoms, their speed is enormously greater, approximating the extraordinary speed of light. They should therefore profoundly disturb the ether through which they move and produce an ether pressure of far greater energy than that of gravitation. This may be the explanation of the fact that the attraction of two wires conveying electric currents so greatly exceeds the attraction of gravitation between these wires.

In magnetic attraction we have much reason to believe that matter in motion is the agency concerned. The helix, with its current, is simply an artificial magnet, and gives abundant warrant for the Amperian theory that each atom is a minute magnet, its energy being due to closed currents circulating around it, as currents of electricity circulate around the circular curves of the helix. In Ampere's day it was impossible to say more. The theory of the atom then entertained gave no clue to the character of these currents. We can now go much farther. In the recent theory of the atom we have to do with a large number of minute particles, which make up the atom-mass and move with an extraordinary rapidity, their speed being brobably many thousands of miles per second. Their mode of motion is unknown, but if we conjecture it to be a circular one, we would have in it the counterpart of Ampere's closed currents. In this case we would possess in each atom an analogue of the rotating sphere, with its axis and poles and its magnetic force, and the movement of the rotating electrons which make up the atom through the ether in which they float, might well produce pressure like that due to the electric current in the wires, or the atom motions in the spheres. Their immense rapidity would serve to explain the great force of magnetic attraction.

In dealing with electric and magnetic forces we have the phenomena of repulsion as well as that of attraction to consider. How shall this force be explained? While electric attraction demands parallel or accordant currents, repulsion follows reversal of currents, and if the former produces ether rarefaction between the conducting wires, may not the latter produce ether compression? The ether whirls produced by the speeding electrons would meet from opposite directions and tend to heap or condense, exerting a pressure outward in their resumption of normal conditions. In magnetic action the same state of affairs may be looked for. When like poles are brought together their currents of atomic motion would sweep round in opposite directions and the ether whirls to which they gave rise meet in opposition, heaping up and forcibly expanding, thus forcing the magnets apart.

Does a similar force of repulsion act between spheres? It well might do so in the case of the requisite conditions between two spheres. If, for instance, two spheres pass each other in apposite directions, repulsion would be likely, under our hypothesis, to result, if they were close enough together to produce conflicting ether whirls between them. This, however, is a matter upon which it is not safe to decide.

Of the hypothesis here presented it may be said in conclusion that it fulfills the conditions of being simple in principle and universal in operation. Its cause is an unceasing one, since matter and motion are co-eternal verities and ether everywhere present and active. All the substantial contents of the universe are concerned in it, matter and ether alike it depending upon a simple interaction between these two constituents of nature. If the action of the ether suggested be an admissible one all else follows. If it is inadmissible, for any sufficient reason, the whole argument falls to the ground.

ORIGIN OF SOUTH AFRICA DIAMONDS.

An interesting paper on "The Diamond Pipes and Fissures of South Africa" was recently read before the British Association by H. S. Harger. The author considers that the age of the Orange River Colony pipes is Triassic (late) or Jurassic, and that the Pretoria pipes are contemporaneous. "They are," he said, "the latest eruptives of South Africa." The origin of the blue ground in the pipes he considers due to the sheltering of the ultra basic rocks, such as eclogite, pyroxenite and lherzolite, all of which are commonly met with and are made up of the minerals which form the bulk of the blue ground. In these rocks garnet occurs plentifully and also olivine and pyroxene. The diamond has frequently been found crystallized in garnet and more rarely in olivine; hence the gem must have had its genesis in the ultra-basic zone in which those minerals originated. The experiments of Crookes and Moissan suggest that the presence of iron was necessary for the formation of the diamond; but to this Mr. Harger objected, owing to the fact that the necessary iron does not exist in the diamond mines and also because Dr. Friedlander's experiments proved that diamonds can be formed in olivine without the enormous pressure and heat aimed at by other experimentalists. In conclusion, the author expresses the opinion that the deep-seated, ultra-basic zone, in which garnet and ferro-magnesian silicates predominate, was the medium in which the crystallization of the diamond occurred.—*Eng. and Min. Jour.*

FIRE TESTS OF WINDOW GLASS.

Windows and skylights of this material were recently tested at the plant of the British Fire-Prevention Committee. In the window test five squares of wire glass, 3 ft. 3 in. by 4 ft. 6 in., were set in a brick wall, one in a brick frame, two in steel frames and three in brick reveals. The glass was ¼ in. thick. The following details are from *Engineering News*, October 5, 1905. The first lasted for forty-five minutes, with temperatures ranging from 650 deg. F. to 1,540 deg. F.; immediately on lighting the gas the glass in all the openings cracked, particularly around the edges, but beyond the cracks increasing no particular change occurred during the firing. The fire did not pass through the glass. On application of water through a ¾-in. nozzle under forty-five pounds per square inch pressure, an irregular patch of small holes was made in the upper portion of the center window, which was also bulged inward; otherwise the glass in all the windows remained in position.

In the skylight tests four squares of glass, about 2 by 2 ft., were placed in the roof of the test house so as to be set horizontally. The fire lasted forty-five minutes and reached a temperature of 1,650 deg. F. Immediately on lighting the gas the glass in all four squares cracked in various directions; beyond this no perceptible change occurred during the fire test. Water was applied to the underside of the glass through hose and also poured on the tops of the glass; its effect was to develop hair cracks all over the glass, but no water passed through.—*Iron Age.*

ELECTRICAL SECTION.

(Stated Meeting, held Thursday, December 21, 1905.)

Electrochemical Calculations.

By Joseph W. Richards.

Electrochemical processes produce chemical changes through the agency of the electric current, and the current is used primarily either for its electrolytic effect or for its thermal effect. The first thing needful, in order to make any calculations connecting electric energy with thermal or chemical effect, is to set forth the values of electric energy as expressed in thermal or mechanical units, in order to get a common energy basis on which to make comparisons, and to postulate the facts concerning the electrolytic effect of the electric current.

ENERGETICS OF THE ELECTRIC CURRENT.

One watt-second of electric energy may be postulated as equivalent to one joule of mechanical work, or 0.2385 gram-calories. This is the energy basis on which electric and chemical phenomena meet on common ground, for the only energy measure of chemical reaction which has as yet been quantitatively investigated is the thermal one. We therefore express the energy of a chemical reaction in the terms in which that energy has been measured, and then have the theoretical basis for calculating the amount of electrical current which can theoretically furnish that amount of thermal energy.

PRINCIPLES OF THERMO-CHEMISTRY.

The measuring of the amount of heat liberated or absorbed in chemical combination or decomposition or reaction is the province of thermo-chemistry. Many enthusiastic scientists have accumulated large numbers of these experimental data, from which, however, very few generalizations have been, up to the present, deduced. There are critical discussions of thermo-chemical data for inorganic compounds in Ostwald's "Allgemeine Chemie;" one can find the most complete collection of thermo-chemical figures yet published, both for inorganic and organic compounds, in Berthelot's "Thermochemie;" in English, Muir's "Elements of Thermal Chemistry" is the only collection of tables on this subject which aims at completeness, and the work is at present nearly two decades old. More or less incomplete or fragmentary tables can be found in works on chemistry or treatises on heat. In giving the data, chemists always give the heat change for molecular weights of the substances concerned; for instance

$$(H^2, O) = 69,000 \text{ Calories}$$

means that when two parts of hydrogen unite with 16 parts of oxygen to form 18 parts of water, 69,000 heat units are evolved. If the weights of the substances be called kilograms, the heat units are kilogram-calories; if called grams, they are gram-calories; if called ounces or pounds, they are ounce- or pound-calories (1°C). The Germans write the above $(H^2, O = 690$ K, in which K stands for 100 ordinary heat units; while the French write it $(H^2, O) = 69.0$ Cal., in which the weights of the substances are supposed to be taken in *grams*, and the heat evolved expressed in *kilogram*-calories. In our opinion, the English style first given, is logically and practically superior to the other two, and is the best to follow.

It must be borne in mind that almost without exception, the thermo-chemical data given in the tables are those which have been determined starting with the constituents at laboratory temperature and ending with the products at the same, or very nearly the same, temperature. The heat quantities therefore apply strictly only to electrochemical processes taking place at ordinary temperature, or very near thereto, and with the constituents or products in the same physical state as was used in the calorimetric determination. If the electro-chemical process is taking place at any different temperature, let us call it t°C., and the ordinary temperature 15°C, the heat of the chemical change at t°, from constituentts at t° to products at t°, or *vice versa*, can be calculated from the rather simple rule, that—

The heat of combination at t° equals the heat of combination at 15° (tabulated heat), plus the heat necessary to raise the constituents from their ordinary state at 15° to their ordinary state at t°, and minus the heat which would be necessary to raise the products from their ordinary state at 15° to their ordinary state at t°.

The above calculation requires a knowledge of the specific heats and latent heats of change of state of the constituents and of the products of the reaction, between 15° and t°, but when these are known, the heat of the reaction at t°, the quantities very often needed in electro-chemical calculations, can be evaluated. These specific heats and latent heats are best obtained from such standard works as Landoldt and Bornstein's "Physikalische-chemische Tabellen," or works of similar scope in English.

FARADAY'S LAWS.

There is another and an entirely different quantitative relation between the amount of electric current used and the quantity of electrochemical action produced, which applies to direct current producing electrolysis. In this case, while the energy relations hold, yet they are conditioned by the fact that every coulomb of electricity passing produces a definite amount of chemical change, equivalent to the setting free of a certain weight of hydrogen, for instance, and the energy required must adjust itself to the bringing about of this amount of chemical change. I am referring, of course, to Faraday's discoveries, which may be briefly summed up in the two statements that

1. The amount of chemical change produced electrolytically by the current is proportional only to the amount of electricity passing, as measured in coulombs, and is independent of the strength or temperature of the electrolyte, or the size or distance apart of the electrodes.

2. The amount of different elements dissolved or set free by the passage of a given amount of electricity proportional to their chemical equivalents.

When it was determined experimentally that 0.00001035 grams of hydrogen is set free by one coulomb, or that 96,540 coulombs set free one chemical equivalent, one gram of hydrogen, the whole scale of relations for all elements whose chemical equivalents were known, became known. The *"Faraday,"* —96,540 coulombs, sets free, or tends to set free, in passing from an anode to a cathode, a chemical equivalent weight in grams of any element. No law of nature has been subjected to more rigorous tests than this law of Faraday, and it appears so far as a law without an exception.

EVOLUTION OF GAS ELECTROLYTICALLY.

A Faraday of current, 96,540 coulombs (representing 26.82 ampere hours) sets free 1 gram equivalent of metal or acid element or radical. If the element or acid radical is gaseous, there is a simple relation between the current flow and the volume of gas evolved, based on the observation that a gram equivalent of a monatomic gaseous element has a volume under standard conditions of temperature and pressure of 22.22 litres, of a diatomic gaseous element, of 11.11 litres, etc. Thus, 96,540 coulombs set free a gram equivalent weight of hydrogen, chlorine, sodium, zinc, etc. The formulæ for these in the gaseous state are Na, Zn, H^2, Cl^2; the molecules of these gases, representing 22.22 litres, contains therefore:

Na molecule—1 gram-equivalents
Zn " —2 " "
H^2 " —2 " "
Cl^2 " —2 " "
O^2 " —4 "

Therefore, 1 Faraday (26.82 ampere hours) sets free 1 gram molecule of sodium vapor (22.22 litres), and ½ a gram molecule (11.11 litres) of vapor of zinc, hydrogen gas or chlorine gas, assumed at normal temperature and pressure. For any other temperature, t, or pressure, p, in m.m. of mercury, the volume would be—

$$\text{Standard Volume} \times \frac{273 + t}{273} \times \frac{760}{p}$$

Example: How much oxygen and hydrogen should be produced per day by 300 amperes passing through 20 cells in series?

Solution: The ampere hours performing electrolysis are
$$300 \times 20 \times 24 = 144{,}000$$
The Faradays' passing are
$$144{,}000 \div 26.82 = 5{,}370$$
The volumes of gas produced will therefore be
$$5{,}370 \times 11.11 = 59{,}670 \text{ litres of hydrogen.}$$
$$5{,}370 \times 5.55 = 29{,}835 \text{ litres of oxygen.}$$

If the gas evolved is a compound gas or acid radical, which is formed at the electrodes, the valence of the acid of basic constituent in a molecule of the gas, is the basis of calculation. For instance, CO, CO_2, CH_4, H_2S all represent molecules of the said gases, of a volume of 22.22 litres, and contain respectively—

CO contains 4 gram-equivalents of oxygen
CO_2 "   4   "   "   "
CH_4 "   4   "   "   " hydrogen.
H_2S "   2   "   "   "   "

One Faraday (26.82 ampere hours) which would produce one gram-equivalent of oxygen or hydrogen, would therefore produce ½ a molecule (11.11 litres) of CO or H_2S or ¼ molecule of CO_2 or CH_4 (5.55 litres).

Another means of calculating the volume of these gases is to note that CO_2 and H_2S are equal in volume to the oxygen or hydrogen contained in them, CO is double the volume of the oxygen contained in it, and CH_4 is double the volume of the hydrogen going into its composition; so that the volumes of the compound gases can be calculated from the volumes of the simple gases going into their formation. The relations alluded to are shown by the Roman numerals indicating molecules or volumes in the following equations:

$$\overset{I}{O_2} + \overset{II}{2C} = 2CO$$
$$\overset{I}{O_2} + \overset{I}{C} = CO_2$$
$$\overset{II}{2H_2} + \overset{I}{C} = CH_4$$
$$\overset{I}{H_2} + \overset{I}{S} = H_2S$$

OHM'S LAWS.

The resistance which the electric current encounters to its flow through the body of a substance is determined by the specific resistivity of the body per unit cube of substance, multiplied by its length and divided by its cross-sectional area. The specific resistivity is given in tables per centimeter cubed, in ohms, and the simple arithmetic described gives us the resistance of any bar or wire of said material. The resistance of the body being known, Ohm's laws furnish us with the relation between the applied voltage and current flow through the body, as follows:

$$\text{Current} = \frac{\text{potential drop}}{\text{resistance}}$$

or

$$\text{Amperes} = \frac{\text{volts drop}}{\text{ohms resistance}}$$

These relations apply to the current *in the body* of an electrolyte just as strictly as to a metallic wire (and tables of specific resistivity of electrolytes are given in almost all books on electrochemistry),* but they do not apply at all to the resistance or drop of potential at the surface of an electrode, that is, at the contact surface where current enters or leaves an electrolyte. That latter potential drop is due to chemical work being performed, and has no connection whatever with ordinary ohmic resistance and Ohm's laws.

RESISTANCE CAPACITY OF VESSELS.

In making electrochemical experiments, as in tubes or in vessels between fixed electrodes, it is often convenient to calculate, as a constant of the apparatus, the "resistance capacity" of the vessel. This is simply its ohmic resistance, between the two fixed electrodes, supposing it to be filled with a liquid whose specific resistance is unity. If the cross section is uniform, and the electrodes of similar area, this resistance capacity

---

*Kohlrausch and Holborn's "Leitungsfähigkeit der Elektrolyte" is the most complete collection of such data.

would be, in ohms, supposing the measurements made in centimeters:

$$\text{resistance capacity} = \frac{\text{length between electrodes}}{\text{cross-sectional area of electrolyte}}$$

Thenceforth, if said vessel is filled between the electrodes with a liquid whose specific resistance is known, the ohmic resistance of the electrolyte will be simply the product of the resistance capacity by the specific resistance of the given liquid used: that is, Ohmic resistance of electrolyte = resistance capacity of the vessel × specific resistance of electrolyte.

DETERMINATION OF OHMIC RESISTANCE.

If the conductor is not electrolysed by the current, its resistance is measured by sending a known current through it and measuring the drop of potential at its terminals.

$$\text{resistance (in ohms)} = \frac{\text{potential drop (in volts)}}{\text{current flowing (in amperes)}}$$

The current used may be either direct or alternating, the former is preferable because of the more accurate measurements possible. Another method, not so often used, but applicable under many circumstances, is to put a delicate thermometer or thermo-couple in contact with the conductor and measure the rate at which its temperature *begins* to rise. Knowing its specific gravity (weight of one cubic centimeter in grams) and its specific heat (per unit of weight), the heat generated per second in one cubic centimeter is known:

Heat generated (gram calories) = specific gravity × specific heat × rise of temperature per second.

But, this quantity is also equal to the heat value of the current used, which is 0.2385 times the watts used, or 0.2385 times the square of the current used into the resistance. We therefore have—

$$\text{resistance (in ohms)} = \frac{\text{heat found calorimetrically}}{0.2385 \times (\text{current used})^2}$$

The above method is particularly applicable to electrolytes

of any kind if they are placed in a long tube of low heat-conducting material, and the rise of temperature thus measured at a point remote from the electrodes. Then, whatever action, chemical or thermal, occur at the electrodes, and whatever potential drop may occur, it is only necessary that the amount of current passing be accurately measured, the specific heat and specific gravity of the electrolyte be known, and the rate of rise of temperature be measured at the first few seconds after turning on the current, to be able to calculate the specific resistance.

The use of alternating current for measurements of ohmic resistance using Kohlrausch's method of Wheatstone bridge and a telephone is applicable with accuracy to non-electrolytic conductors, and has benerally been assumed to be applicable quite as accurately to electrolytes. Recent work on the possibility of electrolysis occurring when alternating current is passed through an electrolyte has rather cast doubts upon the accuracy of these tests. It is possible that most determinations thus made have been free from error, but we can understand now that tests so made might be erroneous in many instances, if made with certain electrodes and with certain frequency of alternation of the measuring current.

For many practical purposes, the ohmic resistance of an electrolytic cell can be determined by the following simple device and calculation: Use electrodes equal to the cross section of the electrolyte, and put in series with a relatively high resistance, so as to keep the strength of current as nearly constant as possible. Measure amperes and voltage drop with the plates as wide apart as possible; draw together till they are exactly half the distance apart, and measure again. If the outside resistance is high enough, the amperes will be constant within the ability of the ammeter to record, while the voltage will decrease. Double the decrease in voltage will be the total voltage drop in overcoming the ohmic resistance of the whole cell. Designating this as the voltage drop due to electrical conductivity of the electrolyte, V^c, we have:

$$\text{Resistance of cell (in ohms)} = \frac{V^c}{\text{amperes passing}}$$

VOLTAGE DROP AT THE ELECTRODE SURFACES.

This is the loss of potential across the electrodes corresponding to the work done at the electrodes. It is practically determinable by measuring the total potential drop, and subtracting from it the drop due to overcoming the ohmic resistance of the electrolyte. Calling V the total drop of potential, and V^d that part of it absorbed in chemical (or physical) work at the surface of the electrodes, then

$$V^d = V - V^c.$$

If V^c has been determined in the manner described in the last paragraph, or has been calculated from the specific resistance of the electrolyte, properly determined, and the resistance capacity of the electrolytic vessel, then V^d represents accurately the voltage drop due to all phenomena occurring *at the surface* of the electrodes, as distinguished from the mere phenomenon of electric conduction, ruled absolutely by Ohm's law, occurring *in the body* of the electrolyte.

It may not be amiss to remark, *en passant*, that the fact that Ohm's law applies absolutely to the conduction of electricity through the body of an electrolyte, in the same manner as in a metallic conductor, combined with Prof. Hopkins' recent determinations that the conducting of the current is practically instantaneous in electrolytes, as it is in solids, and that the body of an electrolytic conductor acts in all respects magnetically, etc., exactly the same as the body of a metallic conductor—all prove the identity of the mechanism of electric conduction through the substance or body of an electrolytic conductor and through solid metallic conductors. The phenomena at the bounding surfaces, the electrodes, are different in the two cases, but there is no experimental evidence of any dissimilarity in the mechanism of the conduction in the body of the conductors in the two cases.

TRANSFER RESISTANCE.

This is supposed to represent resistance to the passage of the current from the electrolyte to the electrode, of the nature of the work done when current is passed across a thermo-electric junction; that is, it is a resistance purely physical in its nature, existing simply because the current passes from one conducting substance to another one, and, finally, a resistance which causes the current to either generate heat, by heating this junction, or to absorb heat, by cooling the junction. It is therefore a reversible phenomenon, either subtracting potential from the current in such quantity as that the heat thus generated represents the heat equivalent of the watts thus lost, or else contributing potential to the circuit in such quantity that the potential thus furnished represents, when multiplied by the amperes flowing, the watt equivalent of the heat energy absorbed.

In your lecturer's opinion, this transfer resistance must be very small. Since it would be of different signs at the two electrodes, absorbing voltage at one and generating nearly an equal amount at the other, the difference between two quantities in themselves small, must be of a low order of magnitude. Further, no reliable determinations are at hand concerning these + and − thermo-electrical potentials, because of the inevitable complication of the measurements by purely chemical changes. For instance, one investigator kept two zinc electrodes in zinc-chloride solution, but at 20° C. difference of temperature, and measured the difference of voltage, calling it thermo-electric difference of potential; but aside from the fact that this ignores any thermo-electric difference of potential between the hot and the cold solutions or in any part of the external circuit, it is certain that it ignores the difference between the heat of formation of zinc chloride in aqueous solution at two temperatures 20° apart, which might easily be equal to the whole potential difference noted. Until, therefore, physicists have cleared up satisfactorily this whole subject of thermo-electric potential between electrodes and solutions, we are making a less error in leaving out its consideration than in trying to account and allow for it—particularly since we know that some of the so-called allowances are certainly erroneous.

I have left out of the definition of transfer resistance that produced by a change in the electrode whereby a film of insoluble salt or gas is produced and so chokes off the current. Such action is polarization, and such change in the original conditions, practically introducing modified or even new electrode surfaces, is not transfer resistance, properly speaking.

VOLTAGE REQUIRED FOR CHEMICAL WORK.

We arrive here at the kernel of electrolytic calculation, the sole and sufficient basis being that the amount of electrical energy expended in doing chemical work must equal the energy equivalent of the chemical work done. If that position does not hold in this question, then energy could be created or lost, and the principle of the conservation of energy violated. We must admit, however, that if an electrolytic cell cools off while the current is passing, that external heat energy is being supplied which will diminish, by the amount so supplied, the work being done by the current. However, that quantity is in the nature of a possible correction, while the heat of the chemical reaction produced is the principal factor.

We must remark at the outset that the chemical work done means the whole change from the system before electrolysis to the system after electrolysis. There is no division of the chemical work of the current into that required for assumed primary reactions and that for assumed secondary reactions. Only changes taking place at the surface of the electrodes, however, affect the energy requirements; chemical reactions taking place away from immediate contact with the surface of the electrodes neither absorb energy from the circuit nor deliver energy to it,—they are purely incidental and independent chemical phenomena.

Knowing that 96,540 coulombs set free one chemical equivalent weight in grams of an element, or decompose one chemically equivalent weight in grams of a substance, or, in more general terms, reduce by one valency atomic weight of any chemically basic element, or increase by one valency atomic or molecular weight of any chemically acid element or radical,— we can soon foot up the chemical energy of the change produced. The thermochemical heat absorbed in the separation of a chemical equivalent weight of an element is the energy required to isolate it; and since the passage of 96,540 coulombs produce that change, the voltage drop must be such that the calculated heat equivalent of the electric energy expended—

$$96{,}540 \times \text{voltage drop} \times 0.2385$$

is equal to the thermochemical heat absorbed (Q for 1 equivalent). The voltage drop for the decomposition (V^d) must therefore be—

$$\text{Voltage drop} = \frac{Q \text{ for 1 equivalent}}{23{,}040}$$

These calculations have been made and verified by experiment on many chemical compounds. There are some few exceptions, but many such have subsequently been shown to be only apparent exceptions, the real reaction occurring during electrolysis not having been exactly understood, or else the thermochemical data not applying strictly to the conditions under which electrolysis took place.

(*To be concluded.*)

Book Notices.

Eléments de Chimie Inorganique. Prof. Dr. W. Ostwald. Traduit de l'allemande par L. Lazard. II Partie. Metaux. Paris: Gauthiers-Villars, 1905. (Price, fcs. 15.)

This work constitutes a large 8-vo. of some 450 pages giving to French readers a satisfactory translation of the inorganic portions of Ostwald's standard treatise on the Elements of Chemistry.   W.

The Insulation of Electric Machines. By Harry Winthrop Turner and Henry Metcalf Hobart, with 162 illustrations. Large 8-vo., pp. xvi 297. Whittaker & Co., London. 1905. (10s. 6d. net.)

The effective insulation of electric machines has been one of the most difficult technical problems with which the electrical specialist in this field has had to contend, and much still remains to be done. The present work gives in satisfactory form a summary of the present state of the art, and may be consulted profitably by those directly interested in the subject.

W.

Cement and Concrete. By Louis Carlton Sabin, B. S., C. E., Assistant Engineer, Engineering Department U. S. Army, &c. Large 8-vo., pp. x + 507. New York: McGraw Publishing Co. 1905. (Price, $5.00.)

The use of cement in construction has of late made such important progress that a new treatise on the subject, giving the results of the application of the material in its most advanced phases, from the hands of an expert, will be received with singular interest; more especially since the advance in this branch of the constructive arts has been so rapid of late as to have outstripped the literature in book form, and inquirers have been compelled to seek their information in the scattered pages of the publications of the various engineering societies. To such the present volume will prove very helpful.   W.

Elektrolytische Verzinkung von Sherard Cowper-Coles, London. Ins Deutsche Uebertragen von Dr. Emil Abel. Chemiker der Siemens & Halske A. G., Wien. Mit 36 Figuren and 9 Tabellen im Text. Halle a. S. Verlag von Wilhelm Knapp, 1905. (Price, 2 marks.)

Dié elektrolytische Chloratindustrie. Von John B. C. Kershaw, F. I. C. London. Ins Deutsche Uebertragen von Dr. Max Hutte, Chemiker der Siemens & Halske A. G., Berlin. Mit 39 Figuren and 3 Tabellen in Text and einem Anhang welche die Wörtliche Wiedergabe der wichtigsten Patente enthält. Halle, a. S. Verlag von Wilhelm Knapp, 1905. (Price, 6 marks.)

The two works above named constitute volumes xviii and xix of the highly-useful Electrochemical Monographs that have been appearing during the past few years from the Knapp press.

The first discusses the subject of zincing by electrolytic methods, which, within a recent period have been so substantially developed that they have been able to compete in the commercial way with the old process of galvanizing by dipping. The second treats of the electrolytic methods for the production of chlorates, perchlorates, bromates and iodates, which have achieved considerable commercial success. A useful compendium of the patent literature of various countries bearing on the subject is presented in an appendix. W.

Alternating Current Engineering practically treated. By E. B. Raymond, Testing Department General Electric Co., with 112 illustrations. 8-vo. pp. vii + 225. New York: Van Nostrand Co., London. Kegan Paul, Trench, Trübner & Co. 1904. (Price, $2.50 net.)

This subject is treated in two parts; the first treating of the general laws of magnetism and alternating currents; and the second, to modern alternating current apparatus, methods of operation and testing. The work is intended specially for the use of those who are inexperienced in the management of this class of electrical machinery, and should be found very useful. W.

Le Abitazioni Popolari (Case operaie), dell' Ing. Effren Magrini. Con 151 incisioni. Milano: Ulrico Hoepli. 1905. 12mo. pp. xvi × 309.

This work treats in a concise and masterly way the subject of the habitations of workingmen. Part first treats of the legislation of various countries on the subject, and the character and influence of various local institutions, such as building and other public and private coöperative societies. The second part treats specially of the plans and methods of a number of the great industrial establishments, such as Krupp's, in Germàny; Schneider & Creusot, in France; the Gemeinnützige Bau-Gesllschaft, in Manheim; the Willemantic Linen Co., in this country, and numerous other corporations for the housing of their operatives. The illustrations are numerous and instructive. W.

The Proceedings of the Society for the Promotion of Engineering Education. Twelfth annual meeting, held in St. Louis, Mo., September 1-3, 1904. Edited by C. Frank Allen, Fred. W. McNair, Milo S. Ketchum, Committee. New York: Engineering News Publishing Co. 1905.

Recent Advances in the Metallurgy of Iron and Steel. R. S. Hutton. Reprint from the Journal of the Society of Chemical Industry, 15 June, 1905.

Notes on Heat Insulation, particularly with regard to materials used in furnace construction. R. S. Hutton and T. R. Beard. Reprint from the transactions of the Foundry Society, August, 1905.

Correspondence.

HIGH-GRADE SILICON FOR PURIFYING CAST-IRON.

To the Editor:—Since the publication of my paper on "Recent Progress in Metallurgy," in the December issue of the *Journal* of the Franklin Institute, I have received the enclosed interesting letter from Dr. Moldenke, which he has kindly authorized me to send to you as a contribution to the discussion of the subject.

As Dr. Moldenke is a metallurgical chemist of international reputation, who has had wide experience, especially in cast-iron industries, his views are valuable and it is especially gratifying to find that he states that the process alluded to is "entirely new."

His theory regarding the seeming paradox of obtaining "increase in strength, where the reverse would be expected when softening," agrees exactly with my own views, as will appear from the following quotation taken from the *Journal* of the Franklin Institute, March, 1888, in which I announced the result of experiments made in adding ferro manganese to car-wheel iron:

"A remarkable effect is produced upon the character of hard iron by adding to the molten metal, a moment before pouring it into a mould, a very small quantity of ferro-manganese, say one pound of ferro-manganese in 600 pounds of iron, and thoroughly diffusing it through the molten mass by stirring with an iron rod. The result of several hundred carefully-conducted experiments which I have made, enables me to say that the transverse strength of the metal is increased from thirty to forty per cent., the shrinkage is decreased from twenty to thirty per cent. and the depth of the chill is decreased about twenty-five per cent., while nearly one-half of the combined carbon is changed into free carbon; the percentage of manganese in the iron is not sensibly increased by this dose, the small proportion of manganese which was added being found in the form of oxide in the scoria. The philosophical explanation of this extraordinary effect is, in my opinion, to be found in the fact that the ferro-manganese acts simply as a deoxidizing agent, the manganese seizing any oxygen which has combined with the iron, forming manganic-oxide, which being lighter than the molten metal, rises to the surface and floats off with the scoria. When a casting which has been artificially softened by this novel treatment is remelted, the effect of the ferro-manganese disappears and hard iron results as a consequence."

The high-grade ferro-silicon (50 per cent. silicon) acts in a similar manner not only increasing strength but softening the metal.

Small pulleys cast from iron treated in the way I described are now being turned at 40 feet per minute, whereas the usual speed of turning the same class of pulleys from the same grade of soft iron untreated is 25 feet per minute.

I stated in my paper that as a result of a large number of tests I had found an average increase in strength of about 15 per cent. accompanied by marked increase in softness. I now have records showing much larger gain in strength and resistance (over 25 per cent.) accompanied by remarkable increase in softness. These tests were cast from iron containing 15 per cent. of steel in the mixture.

Very truly,

A. E. OUTERBRIDGE, JR.

PHILADELPHIA, PA., Dec. 23d, 1905.

AMERICAN FOUNDRYMEN'S ASSOCIATION,

Secretary-Treasurer Richard Moldenke.
President, Thos. D. West.   Natchung, N. J.
Thos. D. West Foundry Co., Sharpsville, Pa.

Office of the Secretary,
WATCHUNG, N. J., Dec. 17th, 1905.

MY DEAR MR. OUTERBRIDGE:—

Thank you very much for your valued article. I have read it with great interest. The portion anent the addition of high-grade silicon is especially valuable and entirely new. You will no doubt find that the increase in strength, where the reverse would be expected when softening, is due to the purifying effect of the silicon addition when not made in the cupola. This is similar to that produced by ferro-manganese when added to the car-wheel mixtures. I hold that these alloys when they are effective at all, do take out dissolved oxygen or perhaps other gases, bringing out silicon or manganese as the case may be, and passing into the slag. The result is a cleaning up and better adherence of the crystals, raising the resilience and general strength.

I think you will find further that where steel additions have been made, you will get still better results, as the melting point being raised, these alloys get in their work better.

Your paper shows that we have not yet exhausted our resources in the foundry. May you discover many new ways and means yet.

Sincerely yours,
(Signed)   RICHARD MOLDENKE.

Sections

SECTION OF PHYSICS AND CHEMISTRY. *Stated Meeting*, held Thursday, December 7, 8 P.M. Dr. Edward Goldsmith in the chair. Present, thirty-seven members and visitors.

The communication of the evening was presented by Dr. Henry Leffmann. The speaker exhibited and discussed a number of new laboratory appliances and analytical methods. Dr. W. J. Williams and the speaker discussed the subject. The chairman expressed the thanks of the meeting to the speaker. Adjourned.

WM. H. WAHL, *Sec'y pro tem,*

SECTION OF PHOTOGRAPHY AND MICROSCOPY. Thirty-fifth Stated Meeting, held Thursday, December 14th, 8 P.M. Dr. Henry Leffmann in the chair. Present, seventeen members and visitors.

Mr. J. W. Ridpath exhibited and described an inexpensive dark-room lantern. He said in substance, "About five years ago I made for my own use an inexpensive dark-room lantern, by which the unpleasant effects of heat and smell were avoided.

"The first lantern, which is still in use, was experimental, and made from a cigar box 3½x5½x6½ inches.

"The lid was removed and the orange-colored paper lining left in place. A small hole was bored in one end, intended for the bottom, to allow the expanding air to escape. A larger hole was bored in the other end, intended for the top; into which was fitted a receptacle for an Edison incandescent electric light; this was held in place by a few drops of glue. Into this receptacle was screwed a low c. p. incandescent lamp of the proper voltage. A sheet of orange-colored glass, the same size as the original lid, was put in place where the lid formerly belonged. This glass was secured in place by binding strips. A sheet of ruby glass as long and wide as the outside dimensions of the box was then placed over the orange glass and secured in place by binding strips. A suitable plug to screw into an Edison receptacle was placed on the outer end of the cord, and the lamp was ready for use.

"For developing plates particularly sensitive to colored light, a few curtains made of cherry-colored tissue paper were hung in front of the glass, being pasted to the top of the box. After development has proceeded sufficiently for safety, one or more of these paper curtains can be raised and placed on the top of the lantern. This lantern is quite convenient, cool and comfortable to work by, especially in warm weather."

Mr. Hugo Bilgram referred to an interesting specimen of quartz which exhibited the so-called Brownian movements, and exhibited a specimen of the same mounted as a microscopic slide, which the members were given the opportunity of examining. Dr. Edward Goldsmith made some comments on the frequency with which such and similar enclosures are found in quartz.

Dr. Leffmann showed with the lantern a picture of the mosquito, which is believed to be the direct source of yellow fever infection and another showing the parasite which is the cause of the characteristic anemia so commonly found in semi-tropical countries. Adjourned.

M. I. WILBERT, *Sec'y.*

ELECTRICAL SECTION. *Stated Meeting*, held Thursday, December 21, 8 P.M. Mr. Thomas Spencer in the chair.

Present, forty members and visitors.

The paper of the evening was read by Prof. Joseph W. Richards, of Lehigh University, on "Electro-Chemical Calculations."

The subject was discussed by Prof. Carl Hering, Mr. C. J. Reed, Mr. A. Simonini, Mr. Thomas Spencer, Mr. S. S. Sadtler, Dr. E. Goldsmith and the speaker.

The thanks of the meeting were voted to the speaker of the evening and the session was adjourned.

WM. H. WAHL, *Sec'y pro tem*

MECHANICAL AND ENGINEERING SECTION.—*Stated Meeting*, held Thursday, January 11th, 8 P.M. Present, twenty members and visitors.

President Charles Day in the chair.

The annual election was then held, and resulted in the choice of the following officers to serve for the year 1906:

President, Mr. Chas. Day.
Vice-Presidents, Mr. Kern Dodge, Prof. H. W. Spangler.
Secretary, Mr. Francis Head.
Conservator, Dr. Wm. H. Wahl.

The paper of the evening was read by Prof. Wm. H. Marks, of Philadelphia, on "The Finances of Engineering Enterprises," which was discussed by Mr. James Christie, Dr. Edward Goldsmith, the President of the Section, and the speaker.

Adjourned.

FRANCIS HEAD.

SECTION OF PHOTOGRAPHY AND MICROSCOPY.—Thirty-sixth Stated Meeting, held Thursday, January 18th, 8 o'clock P.M. Present, sixty-five members and visitors. Dr. Henry Leffmann in the chair.

The annual election of officers resulted in the choice of the following, to serve for the year 1906, viz.:

President, Dr. Henry Leffmann.
Vice-Presidents, Mr. J. W. Ridpath. Mr. U. C. Wanner.
Secretary, Mr. Martin I. Wilbert.
Conservator, Dr. Wm. H. Wahl.

The communication of the evening was presented by Mr. U. C. Wanner on Floral Photography, and the coloring of lantern slides.

Mr. Wanner gave an interesting description of his method of procedure, and proceeded thereupon to exhibit a considerable collection of colored lantern photographs of floral subjects, which in respect both of photographic technique and exquisite coloring, represented the highest order of excellence. The communication was freely discussed by a number of members and the author.

Dr. Leffmann exhibited an interesting lantern photograph, showing the ancient Indian Reservation near Second and Walnut streets, set apart in Colonial times for use in occasional conferences between the authorities of the Province of Pennsylvania and the neighboring Indian tribes.

Some additional slides were shown, after which the meeting was declared adjourned.

M. I. WILBERT, *Secretary*.

Franklin Institute.

Proceedings of the annual meeting held Wednesday, January 17, 1906.

HALL OF THE FRANKLIN INSTITUTE,
PHILADELPHIA, January 17, 1906.

PRESIDENT JOHN BIRKINBINE in the chair.

Present, 123 members and visitors.

Additions to membership since last report, 9.

The annual reports of the Board of Managers with appendices embracing the annual reports of the several committees of the Institute and the Board, and of the Trustees of the Elliott Cresson Medal Fund were presented and accepted.

President Birkinbine made an address referring to the present occasion of the 200th anniversary of the birth of Dr. Benjamin Franklin and outlined briefly the plans contemplated by the Institute for celebrating the anniversary. (The address appears as an appendix to these minutes.)

The Secretary presented and read a communication received from The American Philosophical Society, inviting the Institute to be represented at the Franklin Bi-Centennial Anniversary to be held by the Society, April 17th to 20th, 1906. The communication was referred to the Board of Managers.

The President then introduced Prof. Albert H. Smyth, Professor of English Literature in the Central High School of Philadelphia, who gave a most eloquent and interesting address on "Franklin as a Man of Letters."

The meeting passed a vote of thanks to the speaker, with the request that he prepare his remarks for publication in the Franklin Memorial issue of the *Journal*, which it is contemplated to publish in due course.

Mr. W. N. Jennings followed with an exhibition of lantern photographs relating to Franklin and his works.

The tellers of the annual election reported the election of all the candidates nominated at the stated meeting of Wednesday, December 20, 1905.

who were thereupon declared elected to the offices for which they were respectively named. The tellers were given a vote of thanks.

Adjourned.

WM. H. WAHL, *Secretary.*

ADDRESS OF PRESIDENT BIRKINBINE.

Two hundred years ago Benjamin Franklin was born in Boston, but as so much of his time was spent in Philadelphia, where he developed most of his admirable qualities, and where he attained prominence as a printer, philosopher, author and statesman, it is appropriate that the anniversary of his birth be celebrated in Philadelphia, and that this celebration should start at the Franklin Institute.

The Board of Managers have decided to recognize this anniversary by a series of commemorative addresses and papers, the first of which you will listen to to-night. "Franklin as a Man of Letters" will be presented to you, and at the next monthly meeting "Franklin as a Man of Science" will be discussed.

The Board of Managers had also taken the initiative steps for an industrial exposition the coming Fall, the only drawback to assuring this enterprise being the uncertainty of obtaining accommodations for the exhibits. Tentative arrangements had been made for the use of the main exhibition hall and one of the permanent pavilions, and also the outside ground connected with the Commercial Museum in West Philadelphia. These plans may be interfered with by the avowed intention of the City authorities to use the pavilion as a temporary addition to the Almshouse. Hence, definite announcement of the Exhibition cannot be made at the present time.

In April the American Philosophical Society of this city, which joins the University of Pennsylvania in crediting Franklin as its founder in 1743, will give expression of its appreciation of Franklin's worth by an assemblage of international scientists of world-wide reputation, an occasion which all Philadelphians may look forward to with satisfaction.

It is not my function to discuss Franklin and his achievements, but I think that we can appreciate what he has accomplished better if we picture to ourselves the conditions in and about Philadelphia in the time of Franklin.

Can you see a runaway apprentice, not yet eighteen years of age, completing a journey of three hundred miles, part of it on foot, and part of which he literally worked his passage by rowing a boat, arriving in Philadelphia wearied and hungry, with a dollar of available funds in his pockets? And can you see the same man twenty-five years later, practically retired from business with a competence sufficient to sustain him, and with fame which spread far and wide, a truly self-made man?

The neighborhood of the Institute home is replete with reminiscences of Franklin. Dying at the age of eighty-four years, his remains are interred within four blocks of this hall, and somewhere in the open fields, between where you sit and the present post office site, it is claimed he made his famous experiment of conducting lightning from the clouds, in 1752. Most of his work was done within half a mile of this locality, for Philadelphia was then a moderate-sized town. Its navigation was confined entirely to sailing crafts and row boats, for there were no steamships; in fact no steam engines. Transportation was on horse-back, in carriages, or by stage coach. The wood pump and old oaken bucket were the source of water supply, and consequently there were no filtration scandals, and there was no excitement over gas works lease. There were no railroads, and none to sorrow over the withdrawal of passes. The abundance of clay, supplemented by liberal drafts of cobbles from the river, precluded any asphalt complications.

With wood for fuel, there was no demand for ordinances abating smoke nuisances, or fear of coal strikes. There were no department stores, unless we credit Franklin with initiating them by his modest enterprise in which it is reported he disposed of a variety of goods, from books to human chattels.

The absence of sky-scrapers eliminated one danger from elevator accidents, and Franklin could pursue his studies free from interruption from a telephone call, or from insurance or book agents.

To-night we see our City Hall outlined with thousands of lights to honor the memory of the man who first controlled electricity.

Hundreds of thousands of our town people have to-day ridden in cars propelled by electricity or steam.

Tens of thousands of horse power are hourly developed by coal mined far below the earth's surface.

Over fifteen hundred miles of streets cover water and gas pipe, sewers, light and telephone conduits, etc., probably sufficient to cover one-half the earth's circumference, and messages flash, or power is transmitted by overhead wires aggregating a greater distance.

Report of the Board of Managers for the Year 1905, with Appendices Embracing the Annual Reports of the Various Committees and Sections.

Presented and Accepted at the Annual Meeting of the Institute, held Wednesday, January 17th, 1906.

To the Members of the Franklin Institute:

In presenting this report your Board of Managers expresses the hope that the 200th Anniversary of the birth of Benjamin Franklin may be recognized in such manner as to materially advance the Institute which bears the name of this illustrious American.

Eighty years of sustained effort in advancing the Mechanic Arts has honored the name of Franklin, and won for the Franklin Institute worldwide fame, to which it is justly entitled. But this effort has failed to enlist the financial support necessary to permit the self-imposed work being so prosecuted as to keep the Institute as well in advance as it should be.

Our unequalled Technical Library demands each year large expenditures to maintain it as a complete reference library, and the opportunities for scientific investigation need liberal expenditures.

The Committee on Exhibitions is seriously considering a celebration of the Bi-Centennial of Franklin's birth by an exhibition in Philadelphia, believing that the time is opportune. What more fitting recognition than an exhibition of the Graphic Arts and a display which would demonstrate the marvellous advance, since close to our Institute home, Franklin drew electricity from the clouds?

The reports of the various committees of the Board and of the Institute, and records of the schools, which are important features of the Institute's work, appended hereto, give details of what has been accomplished during the year, but the results may be summarized as follows:

For eight decades the Night Schools of the Franklin Institute have attracted men, who, in the endeavor to help themselves, received in these practical aid, and many who later achieved prominence, made their initial start in these schools. During the past year 601 students were in attendance in the night schools; 447 studied drawing; 96 machine design, and 28 naval architecture.

It would be interesting, were it possible, to trace the advance in education, and in science, due directly to the conscientious work of preceptors and scholars in the Franklin Institute schools. The gratuitous assistance given students by the Institute's Professors and Lecturers deserve cordial recognition.

The Committee on Science and Arts reports the recommendation of the award of 21 medals, 2 certificates of merit, and making five advisory reports, with 41 cases pending at the close of the year.

The work of this Committee has, as in the past, honored the Institute, and through it given public recognition to meritorious discoveries and inventions.

Our Library, with 59,089 volumes, and pamphlets, maps, drawings, etc., making an aggregate of 107,470 titles, is a feature of the Institute which must be preserved and maintained.

The Board regrets that the income limits the appropriation to the Library Committee and forces it to use, not only economy but parsimony, in increasing and caring for the unique collection of scientific works.

The Activity of the Sections drew to the 32 meetings many contributions which otherwise would be presented at the Institute meetings, but at the latter, subjects of general interest have been considered, and in addition two courses of popular, well-attended lectures were given in the Young Men's Christian Association Auditorium.

The subjects, and discussions thereon, have supplied ample material for the *Journal* of the Franklin Institute, which carries to all parts of the world the record of work accomplished.

The Roll of Members, now 1,554, is far below what it should be to properly sustain the work above outlined, for much of the expense of this, as the financial statement indicates, is met by membership dues.

The Board has accepted grateful contributions of books and other documents, which have been added to the library. It also acknowledges donations which have somewhat augmented the available funds of the Institute. But the limited endowment, and the decreased interest obtainable, demand that every expenditure be reduced and seriously limits the good which the Institute could accomplish. The total invested funds of the Institute, including all special funds, amounts to $141,325.10, and the value of building site and contents will fully double this.

The Board wishes to express its thanks to the lecturers, most of whom have rendered gratuitous services, and to the various committees of the Institute and Board which have done so well with restricted facilities.

By order of the Board,

JOHN BIRKINBINE,
President.

FINANCIAL STATEMENT FOR 1905.

RECEIPTS.

Balance on hand January 1, 1905			$2,255 08
Committee on Publications		$2,074 74	
Committee on Library		13 64	
Committee on Instruction		1,952 25	
Committee on Science and Arts		119 39	
Curators (a reimbursement)		20	
Contributions of members	1904-5	$1,056 25	
" " "	1905-6	4,448 00	
" from Stock	1901-2	6 00	
" " "	1902-3	12 00	
" " "	1904-5	102 00	
" " "	1905-6	340 00	
" from non-res. mem.,	1903-4	12 00	
" " "	1904-5	647 00	
" " "	1905-6	928 00	
Entrance fees non-resident members		140 00	7,691 25
Fees from life memberships			425 00
Interest Phila. & Reading Gen. Mtg. Bonds			200 00
" C. R. R. of N. J. 5 per cent. Bonds			100 00
Dividends on Penn'a R. R. Stock			66 00
Income, Estate of Robert Wright			2,212 26
Income from General Endowment Fund			2,162 52
Income from Bloomfield Moore Fund			622 82
Income from M. Carey Lea Fund			192 80
Income from Jas. T. Morris Memorial Fund			180 00
Donation of Mrs. Alice Gibson Brock			1,000 00
Donation of Entertainment Committee Iron and Steel Institute (for library purposes)			942 50

Certificates of Membership..........................	10 00	
Membership Badges	22 00	
Scott Legacy Premiums..............................	220 00	
Bills Payable	16,320 98	
Index to *Journal of Franklin Institute*................	4 00	
Membership Directory	329 60	36,861 95

$39,117 03

PAYMENTS.

Committee on Publications..........................	$7,340 34
Committee on Library...............................	1,795 02
Committee on Instruction...........................	2,223 76
Committee on Meetings.............................	399 52
Committee on Science and Arts......................	357 50
Com. on Elections and Resignations of Members.....	735 50
Curators	1,980 98
Incidental Expenses	765 19
Salaries and Wages.................................	5,464 33
Joint Section Account..............................	520 31
Insurance	466 31
Income, Gen. End. Fund (an interest payment)......	200 00
Income Bloomfield Moore Fund.....................	116 62
Income M. Carey Lea Fund.........................	13 80
Publicity Fund	125 00
Expenditures from Donation, Iron and Steel Inst. Committee	53 72
Certificates of Membership..........................	3 75
Membership Badges	10 50
Scott Legacy Premiums.............................	320 00
Bills Payable	14,000 00
Bloomfield Moore Fund, Expenditures:.............	176 23
M. Carey Lea Fund, Expenditures...................	122 97
Jas. T. Morris Mem. Ed., Expenditures..............	116 67
Memo. Library Fund, Expenditures.................	76 86
Interest on Temporary Loan........................	106 46
Membership Directory..............................	485 12
Fees for Life-Memberships (turned over to Trustees).	425 00 38,463 33

Balance January 1st, 1906..................... $653 70

ENDOWMENT FUNDS.

(In the hands of the Institute.)

Bloomfield H. Moore Memorial Fund...............	$15,000 00	
Memorial Library Fund............................	1,000 00	
H. H. Bartol Fund.................................	1,000 00	$17,000 00

(In the hands of Elliott-Cresson Trustees.)
The Elliott-Cresson Medal Fund.................... 4,811 66

(In the hands of the Board of Trustees of the Franklin Institute.)

The Legacy of George S. Pepper.....................	$38,937 50
The Legacy of Eugene Nugent.......................	1,000 00
The Legacy of Emily B. Nicholson..................	1,520 00
The Edward Longstreth Medal Fund.................	1,000 00
The Jas. T. Morris Memorial Fund..................	4,000 00
Donation of unknown friend........................	5 00
Donation of a friend who desires his name withheld..	5,000 00
Life Membership Fund since October 1st, 1894......	3,803 00
Journal Endowment Fund...........................	138 00
Donation of E. G. Acheson.........................	20 00
Donation of E. M. Walsh...........................	10 00
Special Endowment Fund............................	11,940 00
By the will of John Turner, deceased, one-fourth of net income on 2 per cent. of his residuary estate, yielding about $100 or more per year, equivalent to a principal of.....................................	2,000 00
By the will of Robert Wright, deceased, the institute is now in receipt of the income of a principal sum of about	50,000 00 119,373 50

$141,185 16

(An increase in 1905 of $569 15.)

REPORT OF THE LIBRARY COMMITTEE FOR THE YEAR 1905.

To the President and Members of the Franklin Institute:

The Committee on Library respectfully submits its annual report as follows:

The accretions to the library during the year 1905 comprised 2,132 titles, as against 2,397 in the preceding year.

The new acquisitions include:—

    1,059 bound volumes,
    292 unbound volumes,
    768 pamphlets,
    10 charts,
    3 photographs.

These additions were derived from the following sources:

    83 from the Bloomfield Moore Fund,
    44 " " M. Carey Lea "
    31 " " Jas. T. Morris "
    9 " " Memorial Library "

    81 through the *Journal* of the Institute,
    12 "   "   exchange of duplicates,
   272 by binding of periodicals,
  1800 through bureaus of Federal and State governments, from various scientific institutions and societies at home and abroad, and through gifts from individual donors. Especially notable have been the contributions received from Messrs. George Westinghouse, Louis Teal, Caleb Milne, Carl Hering, Richard L. Humphrey, Arthur Falkenau, Cyrus Chambers, Edward D. Adams, Dr. Henry Leffmann, Dr. Solomon Solis-Cohen and Prof. Lewis M. Haupt. Some 45 volumes relating to the Paris Exposition of 1900 have been conditionally deposited in the library by Mr. Carl Hering.

The diminution of the number of volumes obtained by purchase—167 this year against 263 in 1904—while in some measures resulting from the fact that no part of the general library appropriation has this year been available for other than current necessities, is partly accounted for by the Committee's recent policy of utilizing the several special library funds for the purchase of larger and more important publications, such as are generally not within the compass of private libraries and which, on the other hand, are naturally to be sought in the collections of a scientific institute. Even this limitation has had to be further limited for lack of means adequate to the occasion, but, as far as possible, purchases have been made with the view of meeting the most general requirements of the library.

A number of imperfect volumes of general literary magazines, 218 in all, and a lot of duplicates, have been disposed of to make room for more strictly scientific publications. Duplicate public documents were returned to the various Government bureaus for re-distribution and some material was sold. The net result of the year in the library brings its contents to
    59,089 volumes,
    43,347 pamphlets,
       690 drawings,
     1,247 photos,
       192 newspaper clippings,
        31 manuscripts.

A total of..107,470 titles.

The library is greatly indebted to the efforts of Mr. Walter Wood, for a fund of $942.45, which was an unexpected balance of the money collected for the entertainment of the members of the Iron and Steel Institute at the recent meeting of that body in this city, and which, through Mr. Wood's initiative, was turned over to the Franklin Institute for the uses of its library. This fund has enabled the committee to make some much-needed improvements in the book stack and in the lower book rooms, and also to undertake the binding of a large number of volumes of periodicals which have hitherto had to be stored in parts, as well as to forward the arrangement and classification of the library's valuable collection of pamphlets.

A total of 336 volumes were bound up during the past year, including 309 volumes of serial publications, in addition to this some necessary rebindings and repairs were effected.

The library is in receipt of 558 different serial publications, an increase of 12 over the preceding year. As heretofore, nearly all of these are obtained through exchange for the *Journal* of the Franklin Institute, and they comprise issues in various languages from all quarters of the globe.

The library has been kept open until 10 o'clock on Thursday evenings, when the Section meetings occur, and on the occasion of the monthly meeting of the Institute on the third Wednesday evening. The number of visitors after night-fall tends rather to diminish than otherwise, and appears to be limited almost entirely to members attending the meetings. The daily attendance, on the other hand, is on the increase, indicating the growing importance of the library as a factor in the city's intellectual life.

LOUIS E. LEVY,
PHILADA., January 1, 1906. *Chairman.*

REPORT OF THE COMMITTEE ON SCIENCE AND THE ARTS FOR THE YEAR 1905.

To the President and Members of the Franklin Institute:

The Committee on Science and the Arts has the honor to submit the following account of its operations during the year 1905:

The total number of cases pending at the close of the year 1904 was 43, and the new cases proposed in 1905 numbered 26.

The total number of cases disposed of during 1905 was 28, leaving 41 cases pending at the present time.

There were granted, 1 "Elliott-Cresson Medal;" 14 "John Scott Medals;" 6 "Edward Longstreth Medals;" (21 medals in all). Two "Certificates of Merit" were awarded, and in five cases advisory reports were made.

The detailed statistics of the Committee will be found hereto appended.

The Chairman of the Committee has much pleasure in reporting that the efficiency of the Committee has been well maintained.

Respectfully sumbitted,

E. GOLDSMITH,
Chairman for 1905.

PHILADELPHIA, PA., January 3, 1906.

APPENDIX.

DETAILS OF THE COMMITTEE'S WORK IN 1905.

Number of cases pending on December 31, 1904..............43
Number of cases proposed in 1905............................26
  Total number of cases before the Committee in 1905........—   69
Total number of cases finally disposed of in 1905..............28
Total number of cases pending on December 31, 1905..........41
                                                              —   69

The 28 cases disposed of were determined as follows:
  Award of the Elliott Cresson Medal....................... 1
  Award of the John Scott Legacy Premium and Medal........14

Award of the Edward Longstreth Medal of Merit........... 6
Award of the Certificate of Merit........................... 2
Reports made advisory..................................... 5

Total 28

DETAILS OF AWARDS, ETC.

AWARDS OF THE ELLIOTT CRESSON MEDAL.

2281. Prof. M. I. Pupin, for his "Art of reducing the Attenuation of Electrical Waves and Apparatus."

AWARDS OF THE JOHN SCOTT LEGACY PREMIUM AND MEDAL.

2216. The Nernst Lamp Co., for their "Nernst Lamp."
2310. Dr. Persifor Frazer, for his "System of Quantitative Colorimetry."
2312. David J. Kurtz, for the "Cleveland Cap Screws and Bolts."
2313. John Bonner Semple, for his "Shell Torch or Tracer."
2322. Henry T. Hallowell, for his "Stardand Pressed Steel Products."
2329. Edward Parkinson, for his "Knitting Machine."
2338. Dr. A. Wehnelt, for his "Interrupter for Induction Coils."
2339. Dr. B. Walter, for his "Schaltung."
2342. Walter A. Rosenbaum, for his "Letter-Copying Press."
2357. Prof. Lino F. Rondinella, for his "Photo-Printing Machine."
2358. John M. Browning, for his "Automatic Pistols."
2360. Frederick E. Ives, for his "Replicas of Rowland Diffraction Gratings."
2368. Byron E. Eldred, for his "Process of Flame Regulation."

AWARDS OF THE EDWARD LONGSTRETH MEDAL.

2340. Miley & Son, for their "Color Photographs."
2348. Standard Elevator Interlock Co., for their "'Elevator Safety Devices."
2352. Phonosphere Mfg. Co., for their "Phonosphere."
2358. Colt Patent Firearms Mfg. Co., for their "Colt Automatic Pistols."
2364. Folmer & Schwing Mfg. Co., for their "Graflex Camera."
2367. Theodore Alteneder & Sons, for their "Drawing Pen."
2372. Henry Emerson Wetherill, for his "Physical Diagnostic Instruments."

AWARDS OF THE CERTIFICATE OF MERIT.-

2331. International Burglar Immunity Co., for their "Electric Protective Devices."
2335. Samuel Eastman & Co., for their "Improved Fire Nozzle System."

REPORTS MADE ADVISORY.

2354. Emanuel Metzger, for his "Railway Car Cuspidor."
2355. Richard Raby, for his "Metallic Railway Tie."
2359. Mitford C. Massie, for his "Alternating Current Motor."
2377. William L. Barker, for his "System of Car Braking."

DISMISSED OR WITHDRAWN.

2344. "The Warren Rotary Engine."
2346. "Morris's Portable Roof."

REPORT OF THE COMMITTEE ON MEETINGS FOR THE YEAR 1905.

To the President and Members of the Franklin Institute:

The Committee on Meetings has arranged the programs of the usual number of meetings, with the assistance of the Secretary of the Institute. The Committee has nothing of special interest to report, and can only reiterate the statement contained in its report for the previous year, namely, that the growing activities of the Sections render it increasingly more difficult to retain for the monthly meetings the conspicuous place they occupied for many years.

The policy of preserving for presentation and discussion at the monthly meetings, subjects of general popular interest will be continued.

Respectfully submitted,
WASHINGTON JONES.
Chairman Committee on Meetings.

PHILADELPHIA, PA., January 3, 1906.

REPORT OF THE COMMITTEE ON PUBLICATIONS FOR THE YEAR 1905.

To the Board of Managers:

GENTLEMEN:—The Committee on Publications respectfully reports that the quantity of material offered for publication has been ample for its requirements. The Committee feels that the managers and members have cause for congratulation in the fact that the contributions to its pages through the Sections, meetings of the Institute and the several committees have made the *Journal* in recent years, what its name implies, the *Journal* of the Franklin Institute, indicating as it does, the continued scientific activity of the Institute. By reason of its contract entered into early in the year, the Committee is pleased to be able to say that the cost of publication has been greatly reduced and that it hopes to continue this economy for the future.

H. W. JAYNE,
Chairman Committee on Publication.

PHILADELPHIA, PA., January 4, 1906.

REPORT OF THE COMMITTEE ON INSTRUCTION FOR THE YEAR 1905.

To the Board of Managers:

The Committee on Instruction has been much gratified by the interest shown by members in the annual courses of popular scientific lectures, which have been given in coöperation with the Central Branch of the Young Men's Christian Association. The experience of the past year was quite satisfactory. The subjects presented were all attractive and the participation of members large. The lecturers, with an occasional exception, have given their services gratuitously to the Institute, and due acknowledgement therefor has been made.

The Schools of Drawing, Machine Design and Naval Architecture have been fairly well patronized, as the accompanying comparative figures will exhibit:

	1904.	1905.
Drawing School	349	260
" " (Branch School)	201	217
School of Machine Design	87	96
School of Naval Architecture	38	28
Total	675	601

The Committee takes pleasure in acknowledging the effective coöperation of the recently-appointed professors and lecturers which has been of substantial value.

Respectfully submitted, WM. H. WAHL,
Chairman of Committee on Instruction.
PHILADELPHIA, PA., January 3, 1906.

ANNUAL REPORT OF THE COMMITTEE ON ELECTIONS AND RESIGNATIONS OF MEMBERS FOR THE YEAR 1905.

To the Board of Managers:

GENTLEMEN:—The Committee on Elections and Resignations of Members respectfully submits the following membership figures for the year 1905:

The systematic efforts to obtain new members has been continued, and the Committee's work, supplemented by that of a number of members, and the considerable assistance rendered by Mr. R. C. H. Brock, member of the Board, has resulted in placing on the rolls the names of 85 new members. The losses by death, resignation and non-payment of dues are noted in the accompanying schedule, showing a small loss numerically. The Committee again urges the members in general to participate in the work of increasing the numerical strength by inducing new persons to join. It is certainly possible for each member to obtain one new member in a year. The result of such action is easy to estimate.

The membership at the close of the year 1904 was............1597
New members secured in 1905................................. 85
                                                            ────
                                                            1682
Loss by death, resignation and non-payment of dues.......... 128
                                                            ────
Members at the close of 1905................................1554

Respectfully,
ALEX. KRUMBHAAR,
PHILA., January 1, 1906.   *Chairman.*

REPORT OF THE COMMITTEE ON SECTIONAL ARRANGEMENTS FOR THE YEAR 1905.

To the Board of Managers:

During the past year there were held no less than 32 meetings of the Sections, all of which were given up to the presentation of papers on subjects of scientific and technical interest, or to the discussion of live topics of general utility.

Much of the material printed in the *Journal* is derived from this source, and the uniform excellence of these articles affords the best testimony that could be desired to the value of the work of the Sections.

In accordance with resolutions passed by both Sections, the Chemical Section and the Physical Section have been united into a single Section, under the title of the Section of Physics and Chemistry. This union has received the Board's sanction.

The report of the Secretaries follows.

JAMES CHRISTIE,
Chairman Com. on Sectional Arrangements.

PHILADELPHIA, PA., January 3, 1906.

APPENDIX.

REPORT OF THE SECTIONS FOR THE YEAR 1905.

To the Committee on Sectional Arrangements:

The Sections of the Institute have just completed a year of gratifying activity. There were held 32 joint meetings, which were generally well attended, while the subjects presented and discussed were, as a rule, highly interesting and instructive. The *Journal* has benefitted greatly by the abundant supply of valuable material derived from this source.

The Sections in the past year, returned to their original plan of conducting their respective business affairs directly, instead of through an Executive Committee. It is thought that this will have the effect of keeping alive the idea of separate organizations, which was being gradually lost under the Executive Committee plan.

WM. H. WAHL,
For the Secretaries.

PHILADELPHIA, PA., January 3, 1906.

JOURNAL

OF THE

FRANKLIN INSTITUTE

OF THE STATE OF PENNSYLVANIA

FOR THE PROMOTION OF THE MECHANIC ARTS

VOL. CLXI, No. 3   81ST YEAR   MARCH, 1906

The Franklin Institute is not responsible for the statements and opinions advanced by contributors to the *Journal*.

ELECTRICAL SECTION.

(*Concluded from vol. clxi, page 142*)

Electrochemical Calculations.

BY JOSEPH W. RICHARDS.

POTENTIAL FOR MIXED ELECTROLYSIS.

The above data are easily applied when the current is doing only one thing, *i. e.*, is separating out or dissolving only one element. But in many cases the current is causing a compound anode to dissolve, as when copper and silver both dissolve from a silver bullion anode; or two or more metals may be simultaneously deposited, as when copper and zinc are deposited together as brass. In all such cases, the simplest procedure is to foot up the thermochemical energy of the total electrochemical change for any given time or number of coulombs passing. Then express this per 96,540 coulombs passing, and divide by 23,040; the quotient will be V^d

$$\text{Voltage drop } (V^d) = \frac{Q \text{ for 96,540 coulombs passing}}{23,040}$$

Example: An electrolyte contains zinc sulphate with some copper sulphate, and is electrolyzed with a copper anode. There is deposited upon the cathode in one hour 15 grams of brass, containing one-third zinc and two thirds copper. What voltage drop will occur in the cell in addition to that necessary to overcome its ohmic resistance?

Solution: The coulombs passing are those necessary to deposit 5 grams of zinc (chemical equivalent 65÷2) and 10 grams of copper (chemical equivalent 63.6÷2)

$$\text{Coulombs for zinc} = \frac{5.0000}{0.00001035 \times (65 \div 2)} = 14{,}864$$

$$\text{\textquotedbl \qquad \textquotedbl \ copper} = \frac{10.0000}{0.00001035 \times (63.6 \div 2)} = 30{,}383$$

$$\text{Sum} = 45{,}247$$

The weight of copper dissolved will be, on the same principles—

$$45{,}247 \times 0.00001035 \times (63.6 \div 2) = 14.892 \text{ grams.}$$

The solution of 14.892 grams of copper and deposition of 10 grams of the same element, shows that the only thermochemical energy required is that absorbed for the solution of 4.892 grams of copper and deposition therefor of 5.0 grams of zinc. The thermochemical equation is—

$$\text{Zn SO}^4\text{aq} + \text{Cu} = \text{Cu SO}^4\text{aq} + \text{Zn}$$
$$-248{,}000 \qquad +197{,}500 \qquad = -50{,}500 \text{ calories,}$$

Showing that 50,500 calories is absorbed for every 65 grams of zinc (its atomic weight) thus deposited. This equals an absorption of 3,885 calories per 5 grams of zinc, or for the whole period, during which 45,247 coulombs passed through the cell. For every 96,540 coulombs passing, the thermochemical heat absorption is

$$\frac{3,885}{45,247 \times 96,540} = 8,289 \text{ calories.}$$

and the voltage drop is

$$V^d = \frac{8,289}{23,040} = 0.36 \text{ volt.}$$

The above calculation is on the assumption that the copper and zinc deposit separately, as mixed crystals. If they really deposit as a chemical combination, the heat of formation of the alloy would need to be considered also in the above calculation.

The main point sought to be emphasized by the foregoing statements and illustration is that the current may, and very often does, do two kinds of chemical work simultaneously, requiring, if each alone took place separately, different drops of potential to accomplish the chemical work; in such cases, the only safe ground is to say that the coulombs passing must furnish the sum total of all the energy for all the chemical changes induced at the electrode surface, and therefore must drop sufficiently in potential to furnish the energy required (in addition, of course, to the ordinary drop of potential caused by the ohmic resistance of the bath).

CURRENT FOR MIXED ELECTROLYSIS.

When the current is dissolving or depositing only one element, the calculation of the quantity of material concerned with a given transfer of electricity is a simple question: The Faraday, 96,540 coulombs, dissolves or deposits one chemical equivalent weight in grams.

If, however, two or more elements are simultaneously dissolved or deposited, as occurs frequently in metal refining or plating, the calculation is not so simple. The proper procedure in that case is to find how many coulombs are required for the separate weights of each element dissolved or deposited, and add these together to obtain the total current needed. Or, if the current used is given, and the question is the amounts of mixed metals dissolved or deposited, obtain the coulombs necessary to dissolve or deposit one kilogram of the alloy or mixed metals, and divide this into the total coulombs used.

Example: In the Wohlwill process of refining gold bullion, the anodes consist of 80 per cent. gold, 8 per cent. silver, 10 per cent. copper and 2 per cent. platinum. One hundred and fifty amperes pass through each cell, attacking or corroding the anodes unifornily, while the silver deposits at the bottom of the bath as Ag Cl, gold is deposited pure on the cathodes, and platinum and copper accumulate in the solution at Pt Cl^4 are— Cu Cl^2 respectively.

Required: (1) How much weight of anode is corroded per twenty-four hours?

(2) How much more gold is deposited than is dissolved?

(3) What voltage of decomposition, to represent chemical work, must be furnished?

Solution: (1) One kilogram of anode contains—
  Gold.........800 grams
  Silver......... 80 "
  Copper........100 "
  Platinum...... 20 "

The coulombs necessary to dissolve these weights, to to convert these metals into Au Cl^3, Ag Cl, Cu Cl^2 and Pt Cl^4 are—

Gold.....800÷(0.00001035×197 ÷3)=1,177,076 coulombs.
Silver.... 80÷(0.00001035×108 ÷1)=   71,570   "
Copper...100÷(0.00001035× 63.6÷2)= 303,831   "
Platinum.. 20÷(0.00001035×195 ÷4)=   39,638   "
                                  ─────────
                                  1,592,115   "

Coulombs available, per 24 hours
  150×60×60×24=12,970,000

Anode corroded away in 24 hours, per cell
  12,970,000÷1,592,115=8.146 kilograms
            =8,146 grams   (1)

(2) The weight of gold deposited, by the total current, is
  12,970,000×(0.00001035×197÷3=)=8,815 grams.
Amount dissolved,
  8,146×0.80   =6,717   "
                     ─────
Shortage in each cell, per 24 hours =2,098 " (2)

This shortage, it will be recalled, has to be made up by dissolving some of the bullion in acid, and adding to the cells. There would be required 2,098÷0.80=2,747 grams of bullion thus dissolved, making the total bullion treated per cell per day 8,146+2,747=10,893 grams, of which 8,146 grams or 74.8 per cent. would be treated electrolytically in the cell, and the rest, 25.2 per cent., would be dissolved chemically, and yet when added to the electrolyte, its gold deposited electrically.

(3) Basing calculations on the solution of a kilogram of anode, requiring the passage of 1,592,115 coulombs (16.49 Faradays), the 800 grams of gold dissolved has a corresponding 800 grams deposited, so that no chemical energy needs to be expended for it. For the other elements we have the following equations and heat absorptions or evolutions:

$$Au\ Cl_3 aq + 3Ag = Au + 3\ Ag\ Cl$$
$$-27,200 \qquad\qquad +3(29,000) = +59,800 \text{ calories.}$$
$$2\ Au\ Cl_3 aq + 3Cu = 2Au + 3\ Cu\ Cl\ 2aq.$$
$$-2(27,000) \qquad +3(62,500) = +133,500 \quad\text{"}$$
$$4\ Au\ Cl_3 aq + 3Pt = 4Au + 3\ Pt\ Cl_4 aq.$$
$$-4(27,00) \qquad\quad +3(79,800 = +131,400 \quad\text{"}$$

The above heat evolutions are for the solution of, respectively,

    3 gram atomic weights of silver   =324 grams.
    3   "   "   "   " copper =190.8 "
    3   "   "   "   " platinum=585.0 "

The heat evolution per kilogram of anode corroded will therefore be for:

80 grains silver $= \dfrac{80}{324} \times 59,800 = 14,765$ calories.

100 " copper $= \dfrac{100}{190.8} \times 133,500 = 69,980$ "

20 " platinum $= \dfrac{20}{585} \times 131,400 = 450$ "

$$\text{Sum} = \overline{85,195} \quad\text{"}$$

This heat evolution takes place for the passage of 1,592,115 coulombs (16.46 Faradays). For one Faraday (96,540 coulombs) it will be,

$$85{,}195 \div 16.49 = 5{,}167 \text{ calories.}$$

and since one Faraday falling in potential one volt represents 23,040 calories, the heat *evolution* will *generate*

$$5{,}167 \div 23{,}040 = 0.224 \text{ volts}$$

which means that the fall of potential for chemical work is negative, *i. e.*,

$$V^d = -0.224 \text{ volts} \qquad (3)$$

It is interesting to note that, the observed total drop of potential across the baths being 0.7 volt, the voltage drop to overcome ohmic resistance must be *greater* than this, *i. e.*,

$$V = V^d + V^c$$
$$0.7 = -0.224 + V^c$$
$$V^c = 0.7 + 0.224 = 0.924 \text{ volt.}$$

making the ohmic resistance of the cell

$$R = \frac{0.924}{150} = 0.00616 \text{ ohms}$$
$$= 6.16 \text{ milli-ohms}$$

Further, if no external source of current were used, the cell would have available 0.224 volt electromotive force, which should, if short circuited, send

$$\frac{0.224}{0.00616} = 36.3 \text{ amperes}$$

through the bath, that is, run it at 24 per cent. of the present rate, without any external generator of current.

RISE IN TEMPERATURE OF BATHS.

A subject of much importance is that of the temperature of an electrolytic cell being maintained above the room temperature by the heating effect of the current. It is difficult to calculate ahead, before the apparatus is constructed, how high its temperature will rise above its surroundings, but it may be determined with considerable approach to accuracy how high it will be heated as soon as it has been constructed, and before electric current is applied, by a simple test.

The tank is filled with electrolyte, the electrodes put in place, and note made of the weight of electrodes and electrolyte. The tank is heated by a steam pipe until at desired temperature above the room, and then allowed to cool by radiation, with thermometer immersed in the electrolyte so as to determine the rate of fall of temperature per minute. An ordinary tank will cool approximately twice as fast if uncovered as if covered. Tests on a small tank showed, for example, the following results:

At 90° C. falling 3°.2 per minute.
" 80° "   "   1°.7   "   "
" 70° "   "   1°.1   "   "
" 60° "   "   1°.0   "   "

Knowing the weight and heat capacity of the electrodes and electrolyte (to which may be added half the heat value of the tank), it can be calculated how much heat is being lost per minute at any of the above temperatures. To maintain the bath at that temperature would therefore require just that much heat to be generated within it electrically in overcoming the ohmic resistance, and if we know or have calculated the electric resistance of the bath we have,

Heat generated in bath = (current used)$^2 \times$ resistance $\times$ 0.2385 calories per second,

and therefore the current required to supply any required quantity of gram-calories per minute will be,

$$\text{Current} = \sqrt{\frac{\text{Heat to be generated in bath per minute}}{\text{resistance of bath in ohms} \times 0.2385 \times 60}}$$

This same method of investigation can be applied to apparatus working with fused salts, or electric furnaces, providing that means of measuring the temperature satisfactorily are at hand.

ENERGY REQUIRED FOR CHEMICAL WORK.

This energy must be equal in amount to the chemical work done, and the only measure of this that we have are the thermochemical heats of combination of the separated-out products to re-form the original material. A most striking generalization has been arrived at by the discussion of thermochemical data, viz., that taking the heats of formation of salts plus their heat of solution in excess of water (usually called "heat of formation to dilute solution"), these quantities are found to be *additive* in their nature, such being composed of the sum of two quantities, one characteristic once for all of the base, in all its combinations, and the other being characteristic once for all of the acid radical, in all of its combinations. There is thus for each basic element a *thermochemical constant*, which represents the amount of heat it contributes to the formation-heat of a salt, the latter taken in dilute solution; and for each acid radical a similar *thermochemical constant*, representing in a similar manner the part which it contributes; the sum of these two thermochemical constants is the formation heat of the salt, from its elements, to dilute solution. The thermochemical constant is nothing more nor less, in the case of a base, than *energy drop* which represents the decrease of free energy in the element as it passes from the free, uncombined state into the combined state in dilute solution; in the case of an acid element, the statement is entirely similar; in the case of an acid (or basic) radical, it is the total energy drop from free elements to the state of combination as a radical.

The arbitrary basis selected to which to refer these thermochemical constants is best that of hydrogen constant equal to zero. This would make the thermochemical constant of every basic element the heat evolved when it displaces hydrogen from a dilute solution of acid; and that of an acid element or radical the heat of formation of the corresponding acid in dilute solution. Practically, the thermochemical constants have been evaluated by the comparison of the heats of formation of many acids and salts, and the selection of the most probable experimental values. The following tables have been thus derived by the writer from an extensive discussion of the subject matter. The constants are given in every case *per one chemical equivalent* for the base or acid, element or radical, and *not* for the normal number of equivalents which are designated by the symbol and dots (·) or accents (').

I have also given the voltage drop corresponding to an energy drop of the designated number of calories per chemical equivalent, by the simple procedure of dividing the latter by 23,040.

THERMOCHEMICAL CONSTANTS OF BASIC ELEMENTS.

	Per Chemical Equivalent.	Corresponding Voltage.
Li ·	+62,900 calories	+2.73
Rb ·	62,000 "	2.69
K ·	61,900 "	2.69
Ba · ·	59,950 "	2.60
Sr · ·	58,700 "	2.55
Na ·	57,200 "	2.48
Ca · ·	54,400 "	2.36
Mg ·	54,300 "	2.36
Al · · ·	40,100 "	1.74
(N+H$_4$) ·	33,400 "	1.45
Mn · ·	24,900 "	1.08
Zn · ·	17,200 "	0.75
Fe · ·	10,900 "	0.47
Cd ·	9,000 "	0.39
Co ·	8,200 "	0.36
Ni ·	7,700 "	0.33
Fe · ·	3,230 "	0.14
Sn ·	1,900 "	0.08
Pb ·	400	0.02
H ·	0	0
Tl ·	— 900	—0.04
Cu ·	— 7,900 "	—0.34
Hg ·	—14,250 "	—0.62
Pt · · · ·	—19,450 "	—0.84
Ag ·	—25,200 "	—1.10
Au ·	—30,300 "	—1.32

THERMOCHEMICAL CONSTANTS OF ACID ELEMENTS.

	Per Chemical Equivalent.	Corresponding Voltage.	Salt.
F_2''(gas)	+52,900	+2.30	Fluoride
Cl_2'' "	39,400	1.71	Chloride
Br_2'' "	32,300	1.40	Bromide
Br'(liquid)	28,600	1.20	"
Br'(solid)	27,300	1.18	"
I_2''(gas)	20,000	0.87	Iodide
I'(liquid)	14,600	0.63	"
I'(solid)	13,200	0.57	"
S'' "	— 5,100	—0.22	Sulphide
Se''(met.)	—17,900	—0.78	Selenide

THERMOCHEMICAL CONSTANTS OF ACID RADICALS.

(From their elements.)

Constituents.	Radical.	Per Chemical Equivalent.	Corresponding Voltage.	Salts.
O_2(gas),H_2(gas)	$(OH)'$	+55,200	+2.40	Hydrate
S(solid),H_2 "	$(SH)'$	3,400	0.15	Sulphydrate
Se(met),H_2 "	$(SeH)'$	19,100	0.83	Selenhydrate
Cl_2(gas),O_2 "	$(ClO)'$	27,500	1.19	Hypochlorite
" " " "	$(ClO_3)'$	21,900	0.95	Chlorate
" " " "	$(ClO_4)'$	39,400	1.71	Per-chlorate
Br_2 " O_2 "	$(BrO)'$	28,600	1.24	Hypo-bromite
" " " "	$(BrO_3)'$	12,500	0.54	Bromate
I " O_2 "	$(IO_3)'$	65,000	2.82	Iodate
" " " "	$(IO_4)'$	52,600	2.28	Per-iodate
S(solid) O_2 "	$(S_2O_3)''$	71,750	3.11	Hypo-sulphite
" " " "	$(SO_3)''$	75,100	3.26	Sulphite
H_2(gas)" " " "	$(HSO_3)'$	149,400	6.48	Bi-sulphite
" " " "	$(S_2O_5)''$	115.200	5.00	Pyro-sulphite
" " " "	$(SO_4)''$	107,000	4.64	Sulphate
H_2(gas)" " " "	$(HSO_4)'$	211,100	9.16	Bi-sulphate
" " " "	$(S_2O_8)''$	158,100	6.86	Per-sulphate
" " " "	$(S_2O_6)''$	138,500	6.01	Di-thionate
" " " "	$(S_3O_6)''$	136,500	5.92	Tri-thionate
" " " "	$(S_4O_6)''$	130,600	5.67	Tetra-thionate
" " " "	$(S_5O_6)''$	133,100	5.78	Penta-thionate
Se(solid)O_2	$(SeO_3)''$	60,050	2.61	Selenite

"	"	"	"	(SeO$_4$)"	72,800	3.16	Selenate
	P(solid)O$_2$	"	(PO$_4$)'''	99,300	4.31	Phosphate
H$_2$(gas)	"	"	"	(HPO$_4$)"	152,750	6.63	Mono-H-Phosphate
"	"	"	"	"(H$_2$PO$_4$)'	307,700	13.35	Di-H-Phosphate
	As(solid)O$_2$	"	(AsO$_2$)"	102,150	4.43	Arsenite
"	"	"	"	(AsO$_4$)'''	70,200	3.05	Arsenate
H$_2$(gas)	"	"	"	"(HAsO$_4$)"	71,700	3.11	Mono-H-Arsenate
"	"	"	"	"(H$_2$AsO$_4$)'	217,200	9.43	Di-H-Arsenate
	N$_2$(gas) O$_2$	"	(NO$_3$)'	48,800	2.12	Nitrate
"	"	"	"	(NO$_2$)'	27,000	1.17	Nitrite
"	"	"	"	(NO)'	—3,800	—0.16	Hypo-nitrite
	C(amor.)O$_2$	"	(C$_2$O$_4$)"	99,800	4.33	Oxalate
"	"	"	"	(CO$_3$)"	82,450	3.58	Carbonate
H$_2$(gas)	"	"	"	(HCO$_3$)'	169,100	7.34	Bi-Carbonate
"	"	"	"	(CHO$_2$)"	104,600	4.54	Formate
"	"	"	"	"(C$_2$H$_3$O$_2$)"	120,500	5.23	Acetate
N$_2$(gas)	"	"	"	(CNO)'	37,100	1.61	Cyanate
"	"	N$_2$	"	(CN)'	—34,900	—1.51	Cyanide
S(solid)	"	"	"	(CNS)'	—18,100	—0.79	Sulpho-cyanide
Fe(solid)	"	"	"	"(FeC$_6$N$_6$)""	—25,600	—1.11	Ferro-cyanide
"	"	"	"	"(FeC$_6$N$_6$)'''	-52,800	—2.29	Ferric-cyanide

In using above tables, it must be borne in mind that the sum of the thermochemical constants for the basic and acid constituents of any salt, gives the heat of formation of the salt from the constituent chemical *elements*, in dilute solution. Conversely, the heat of formation thus obtained, is the energy necessary to separate the salt in dilute solution into its constituent chemical elements. The sum of the voltages, corresponding to this amount of chemical energy, is the voltage necessary to decompose the salt in solution into its constituent, free chemical elements. If these elements re-combine in the process of decomposition, to form other chemical compounds, then the whole chemical reaction must be taken into account,—best on the total energy basis; and when this is expressed in calories per chemical equivalent concerned, the necessary voltage for producing the change is obtained by dividing this by 23,040.

Lehigh University, December, 1905.

ALUMINUM FOR PATTERNS.

Aluminum is now being used to some extent as a pattern metal, the former difficulties connected with soldering it having been solved. The most satisfactory alloy for soldering consists of 1 part of aluminum, 1 of phosphor tin, 11 of zinc and 29 of tin. To avoid loss of the more easily volatile of these metals the aluminum is melted first, then zinc is added in small pieces, then tin in small pieces and lastly the phosphor-tin. For the soldering no acid is used, but the surfaces to be joined are first covered with a thin coat of the solder in the usual way and then brought together and heated with the soldering copper or a blow pipe or torch until the solder already upon them is melted, when pressure is applied and the joint is made. Aluminum must be heated to about 660 degrees F. before it can be soldered.—*Iron Age.*

UNUSUALLY HEAVY RAILS.

The rails on the belt line railroad around Philadelphia are the heaviest used anywhere in the world, weighing 142 pounds per yard, or 17 pounds more than any previously fitted. They are ballasted in concrete, with 9-inch girders to bind them. All the curves and spurs have the same heavy rails, which were made especially for the Pennsylvania Railroad by the Pennsylvania Steel Company. An officer of the railroad states that this section of road ought to last for 25 years without requiring repairs. The section is considered superior to any other in existence.

INCREASING SPEED OF WARSHIPS.

The average speed of warships is rapidly increasing. Excluding torpedo craft, the warships built and building for the several powers July 1, 1899, had a mean speed of 16.92 knots. This speed has become successively 17.24 in 1900, 17.49 in 1901, 17.89 in 1902, 18.17 in 1903, 18.39 in 1904 and 18.71 in 1905. This increase of nearly two knots is due largely to the construction of large numbers of large armored cruisers, most of which are capable of steaming 22 knots or over, while the building of smaller and slow ships has heavily decreased. The mean speed of the warships of the eight greatest naval powers in 1905 are: England, 19.82; Japan, 19.41; Italy, 18.79; Austria, 18.65; Unied States, 18.64; France, 18.56; Germany, 18.18, and Russia, 17.29 knots. The fastest navy is that possessed by Chili which with only eleven ships, largely very swift cruisers, has a mean speed of 20.71 knots.

A NEWLY INVENTED GERMAN TYPEWRITER,

Not yet upon the market, prints syllables and short words instead of only single letters. It is said to be remarkably rapid. With a language as complex as English it seems doubtful if such a device could be made practicable, unless it were to have a wholly prohibitive number of characters.

Mechanical and Engineering Section.

(Stated Meeting, held Thursday, November 16, 1905.)

The Selection of Material for the Construction of Hydraulic Machinery.

By Arthur Falkenau.

Member of the Institute.

In the designing of hydraulic machinery one of the most important considerations is the determination of the material to be used. The determining factors are, first: the function of the machine or appliance; second: The space available for placing the apparatus; third: The hydraulic pressure involved; fourth: The velocity of the flow of water or other liquid; fifth: In the case of pumping machinery, the kind of fluid to be pumped. Possibly I should have mentioned the last consideration first, but in speaking of hydraulic machinery I had in mind mainly hydraulic presses working under the higher pressures. As to the functions of hydraulic machines or appliances they can be divided into four great classes: 1st, Presses and Intensifiers; 2nd, Pipes; 3rd, Valves; and 4th, Pumps.

Of course the first consideration in designing a press is the total pressure to be exerted, and the second is what hydraulic pressure is to be used or is available. This is at times governed by the space available for the machine, and I have known of instances where in order not to have too cumbersome a machine, and the space being limited, it became necessary to use a factor of safety of three or four referring to the ultimate strength of the materials of construction. Of course, in every construction the expense is to be considered, and this frequently determines the choice of cast iron or steel castings for hydraulic cylinders. At times extreme high pressures indicate the desirability of using steel forgings for cylinders. If the cylinder is to receive a plunger which can be packed externally, cast iron or steel castings prove very satisfactory, but where a piston is to operate in the cylinder, on account of the wear of the leather packing a forged cylinder, bored and ground, gives the best results. With castings a copper lining is frequently resorted to in order to present a smooth surface to the leather.

A most important consideration is the density of the material entirely aside from the tensile strength. Water under 3000 or 4000 pounds pressure will ooze through cylinder walls made of ordinary gray iron 3″ to 4″ thick, that is, if it is open-grained iron. For these high pressures it is usual to use air-furnace iron, which, besides giving a tensile strength of 30,000 to 32,000 pounds, furnishes a very dense material, and now that steel castings are so much more reliable than formerly, steel castings are supplanting the air-furnace iron.

In this connection I would say that in my experience, however, I have known air-furnace iron to fail where a good ordinary casting was successful. This was largely due to the manner of the casting and local shrinkages from gates and risers. The failure of cylinders or valve body castings to be thoroughly impervious to the water is frequently the cause of great annoyance and expense in the construction of hydraulic machinery. I have had cases where on account of the anxiety of the customer to obtain the machine we have had to pass cylinder or valve bodies which when first tested failed to hold the pressure, the water oozing through the walls quite rapidly. We remedied the defect by pumping starchy fluid prepared from potatoes into the cylinders, and after half an hour to an hours work, the cylinders were bottle tight. These cylinders were put to work under water pressure of 1500 pounds, and they have remained permanently tight. I designedly use the word "water-tight," as a later experience proved to me that the starch caulking method is not oil tight. In building some 200 ton pressure with 10 inch cylinders we used cupola iron castings. As they proved defective we next ordered some of the air-furnace iron. After trying three of these, all of which proved defective, in order to meet the importunities of our customers, we concluded to try the starct caulking method. Within a half hour the cylinders were perfectly tight, and after having a hydraulic pressure of 400 pounds per square inch applied to them, and locking this pressure in the cylinder, a drop of only ten pounds was recorded after twelve hours. We thought that our annoying problem was solved, and expedited the machine to the customer's work. The next morning we were informed that the cylinder was leaking badly, and on inspection found that our customer was using oil, and that the oil oozed through the cylinder at an apparently greater rate than the water had done originally. I suppose that the oil must have had some dissolving effect upon the starch. As we had had such unsatisfactory results with the air-furnace iron I concluded that the only rapid solution of our trouble would be some other way of sealing up the pores of the cylinder. As the leak indicated, the porisity was mostly at the bottom of the cylinder. We therefore had the inside of the cylinder towards the bottom brazed by the ferrofix process. This proved entirely successful and the cylinder has remained sound ever since. I understand that in the case of steel cylinders the sealing by means of the thermit process has been successfully used.

It may be interesting to note here that in the year 1849 the first cylinders used in the hydraulic presses constructed for raising the tubes of the Brittania bridge into position proved porous and leaky, and were made tight by pumping oat meal gruel and sal ammoniac into them. These cast-iron cylinders were 20 inches in diameter and had walls $8\frac{3}{4}$ inches thick. When they failed, they were replaced by cylinders made of wrought-iron with 8 inch walls. When the wrought-iron cylinders were first put to work, the engineers were discouraged due to the fact that the cylinders expanded under the great pressure causing the pistons to leak. New pistons were made but the expansion continued. The outer diameter, however, remained constant, and this encouraged the engineers to persist in making new pistons until the inner portion of the cylinders had taken a permanent set.

In valves and pumps where water under high pressure attains a high velocity it has been a general experience that cast-iron and steel are frequently subjected to a peculiar cutting action. According to my own observation this cutting action has been decidedly more rapid and marked when two dissimilar metals were in contact; thus in the Loss valves which we built we originally used steel valves and bronze bodies. In several instances after a year or two we found the steel valve apparently eaten out as if by an acid. In one particular instance, believing that acid or grit in the water was the cause of the trouble, a water filter was put in place and only pure filtered water was used throughout the system. The new steel valves were soon eaten as badly as the former ones. It may be that some tannic acid washing out of the leather packing had something to do with this action, or the action may be of an electric nature. We replaced the steel valves by bronze ones so that two like metals were in contact and no further trouble was experienced. I have examined samples showing this peculiar pitting action which, as the location showed could not have been caused by the impact of the water, due to high velocity in passing out of the valve. Still the fact that this action occurs near the point of efflux, and not so much elsewhere, might lead one to discontinuance the electric couple theory. This peculiar action, I believe, has been observed by a great many engineers, but does not seem to have been satisfactorily explained. For small structures under high hydraulic pressures, say from 3000 to 8000 pounds, forgings are far more satisfactory than castings, and I have found bronze under these high pressures unsatisfactory solely owing to the low or uncertain elastic limit of the same. The castings seem to gradually expand and get leaky, although figured with the factor of safety of 6 to 10 based on the ultimate strength.

The material to be used for packing is also an important consideration. Where U or Cup leathers are used a close-grained flexible leather is desirable; of course, such leathers should not be taken except from the middle of the back of the animal. Leather treated with paraffin has given good results. There is no doubt that the method of preparation of the leather is an important factor in its imperviousness to water, and I have within recent years tried the Vim leather, which has given better results than any I had heretofore used. The manufacturers of the Vim leather claim that their peculiar process of tanning preserves the fibre and brings the fibres into closer contact. The process of tannage is one of oxidation by the use of a mineral, and for this reason the leather is not affected by oxide of iron as are oak- and hemlock-tanned leather. For light pressures the leather is furnished without any filler, but for high pressure the leather is filled with a lubricant which primarily hardens the leather and renders it more impervious. It is claimed that owing to the process of tanning the Vim leather will absorb 45 per cent. of lubricant as compared with 15 per cent. absorbed by oak-taned leather. Furthermore, in moulding the leather no water is used, the leather being heated and thus sufficiently softened. The leather is not affected by hot water.

The blame for the failure of leathers, however, is frequently not chargeable to the material, but to the construction of the metal against which the leather rests. The U leather should, as far as posisble, be backed by the metal over its curved portion, and should have either a metallic ring or hemp or other material inserted between the flaps. Furthermore, the surface over which the leather runs should be as smooth as possible. Some constructors in building cylinders in which pistons travel, line them with brass, aiming at the double purpose of furnishing a smooth surface and covering any porous structure which may appear in the cast-iorn or cast-steel cylinders.

Where leathers are used in valves in such a way that they cross ports the construction should be such as to avoid blowing the leather out into the ports. I desire, however, to confine myself to the question of the kind and quality of the material to be used, and only call attention to these points of construction which, if neglected, may cause damage which should not be charged to the material.

In choosing the material for pumps used in removing fluids, such as mine pumps, ammonia pumps, etc., the question of chemical action on material is of great importance. In many of the mines in the anthracite coal regions the water is so permeated by sulphuric acid that cast-iron cannot be used with any satisfaction. For several years I had preserved a specimen, which I had broken from a pump that had been in use some four years. The metal had been three-fourths of an inch thick. This was no longer cast-iron, as the iron had been removed and only the graphite remained, and I could write with it as with a pencil.

This subject, "The selection of a suitable material for the construction of hydraulic machinery," of course, is a very large one, and can not be adequately treated in a short paper. My object was simply to present a few notes that might bring out interesting discussion.

THE COURSE OF THE MINING INDUSTRIES DURING 1905.

Dr. David T. Day, of the United States Geological Survey, who brings out each year a volume entitled "Mineral Resources of the United States," is authority for the statement that the general prosperity of the year 1905, due to good crops, excellent foreign trade, and business confidence, was emphatically expressed in the greatest production of minerals ever known in the history of our country. Good crops stimulated the railroads to increase their rolling stock. The direct consequence of that increase was a heavy demand on the iron and steel trades, which in turn extended the expansion movement to coal and coke.

The preliminary estimate made by the Director of the Mint of the production of gold and silver in the United States during the calendar year 1905 shows a gain approximately $6,000,000 in gold and 1,000,000 ounces of silver over the output of 1904. The gain in gold is represented almost entirely by the increased output of Alaska, which is placed at $14,650,100 as against $9,160,500 in 1904. The Alaskan gain is nearly all in the Tanana or Fairbanks district, the returns for which are $5,107,000. California shows a loss in gold product of about $1,500,000, due to prolonged draught, which not only brought hydraulic operations to a standstill but interfered, to some extent, with quartz mills. Nevada shows a gain of about $400,000, and Utah an equal increase. Colorado's gains amounted to about $1,000,000.

The ill wind of war and revolution which brought disaster to the Russian petroleum industry carried prosperity to the productive oil areas of the United States. To supply the increased demand from abroad, which resulted from Russia's deficit, strained the petroleum resources of this country to their utmost.

The demand for copper during 1905 was intense and prices were high. This demand for copper affected other metals sympathetically, especially aluminum, which floats on the copper market as a substitute for copper conductors, and is more or less sought, according to the price of copper.

General prosperity was sufficiently pronounced to make itself felt in the building trades, is shown by the increased production of structural steel, brick, terra-cotta, and stone. Of all the structural materials, cement profited most by the forward movement, which came, fortunately, at a time when a favorable market was necessary to take care of existing stocks.

Prosperous conditions were reflected even in the sale of precious stones, although a large majority of the business consisted of sales of diamonds from abroad, which brought unusually high prices.

IMPROVEMENT IN MAKING ARMOR PLATE.

Charles C. Davis, of Germantown, Philadelphia, has patented a process of cementing, or carbonizing, armor plate. Carbonaceous material is packed between two armor plates and a direct electric current is passed through the carbonaceous material and the plates, facilitating the absorption of carbon. The plates are maintained at a temperature of from 800 to 850 degrees C.

(Stated Meeting, held Thursday, November 16, 1905.)

Some Data Relating to the Heating of the Edgar F. Smith House, Dormitories, University Pennsylvania.

BY H. W. SPANGLER.

The University Dormitories are a group of twenty-one buildings under a common roof, divided into separate houses, each with its own entrance from an interior quadrangle. These buildings would extend about 1325 feet if in one row, and when the entire system is completed they will have a combined length of 2475 feet. At present they will house about 589 students, there being seventy-nine three-room suites, consisting of a study and two bed-chambers, two four-room suites, sixty-three double rooms and 291 single rooms. These buildings are of brick with stone trimmings, terra-cotta floors and ceilings and have a slate roof on wooden rafters. The rooms are used for living rooms only, and toilets and baths are provided on each of the upper floors. These buildings are all heated and lighted from the Central Station, about 1200 feet away. High-pressure steam is carried underground for heating, the main steam supply extending about 2000 feet from the boiler house to the extreme building. The cost of heat furnished to the occupants is included in the rental, and the Dormitories pay the Light and Heat Station their pro rata of the cost of operation. In that portion of the Dormitories originally built, the charge for light is included in the rental, but in all later houses the occupants pay by meter.

That portion of the Dormitories originally built, housing 361 students, is heated indirectly, stack of indirect radiators being located in the basement. An independent mixing damper, controlled from each room, is intended to be used for regulating the heat received, the entire basement acting as a cold air chamber. As several rooms are heated from one stack of in-

Fig. 1—University Dormitories—Edgar F. Smith House. Southern Exposure.

directs, the control of the heat is not entirely satisfactory; first, because hand-dampers in the heat pipes must be regulated from time to time during the winter, and secondly, it is possible, as for instance when the wind is blowing on the north side of the house, that an open window on one floor will permit cold air to pass down into the top of the heat stack in the basement and thence up into a room on another floor, supplied from the same stack, and thus entirely cut off the heat supply.

In building the second lot of houses, consisting of 146 apartments, it was decided to use direct radiators for heating, with an auxiliary fan system for very cold weather. The real service obtained from an auxiliary system was so small, that, after a few of the direct radiators had been increased in size, the fan system was no longer necessary.

The last house built is shown in Figs. 1 and 2, Fig. 1 being the southern exposure, the blank wall on the eastern end being

Fig. 2—Northern Exposure.

the separation between this house and the next one, now being erected. In building this house, called the Edgar F. Smith house, which contains thirty-two apartments, it was determined to heat all the rooms by direct radiation, the steam main being carried in the attic and a two-pipe system being employed. Figs. 3 and 4 show the working drawings from which the heating system of this building was laid out. The floor plans show the distribution of the radiating surface together with the general

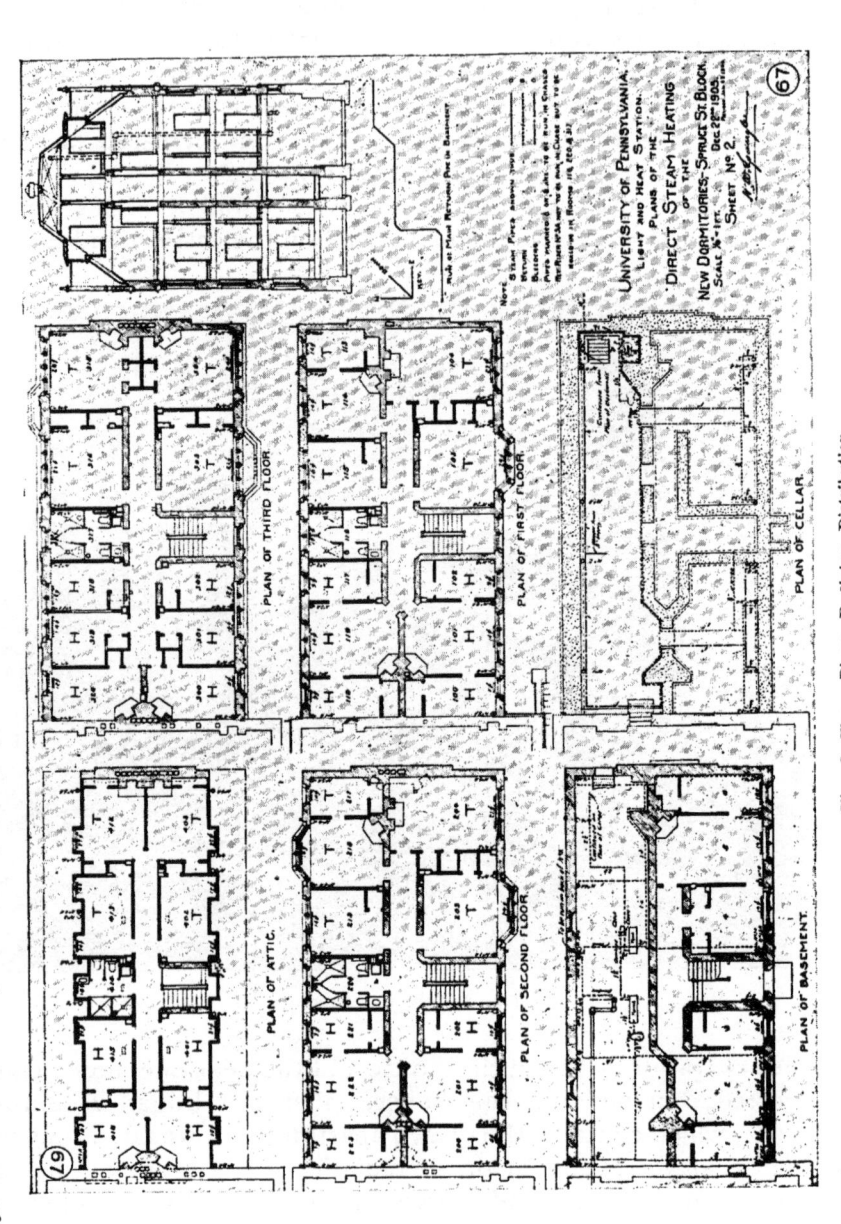

Fig. 3—Floor Plans—Radiator Distribution.

location of the risers and returns. Most of the piping is concealed in chases, the pipes being covered with one inch of cork sectional covering before being closed in.

The amount of radiation required was calculated carefully from the exposed surface. A form is used for making the calculation, one being used for each room. The one reproduced in Fig, 5 is for room 219 in this house. In the lower left-hand corner will be seen a sketch of the room, with dimensions. The desired temperatures are as shown, zero on the north side, 70 in the room, 70 above and 70 below the room, 60 on the south side, 70 on the east and 70 on the west.

The table shows, in the first column, the location of the wall, in the second and third the kind of wall and the next column the area. ΔT is the temperature difference on the two sides of this wall, C is the constant for the heat transmitted per square foot of surface per degree difference of temperature. $E \times C$. is a constant depending on the exposure of the building, nearness of surrounding buildings, etc. The last column shows the expected heat loss from the room. On a basis of 300 B.T.U. per hour, the number of feet of radiating surface is determined.

The constants used are practically those developed by Kinealy with some modifications and have been tested by us in other of our buildings and found to satisfactorily determine the amount of radiator surface required to heat our rooms under varying conditions. They are as follows:

Glass, S.		1.09	Ceiling, F. P. 12 in.	.10
Brick	9 in.	.35	9 in.	.15
	14 in.	.28	Door, outside 2 in.	.42
	18 in.	.24	inside 1 in.	.40
Brick and Stone	42 in.	.24	Lath and Plaster	.34

The following table shows the actual volume and the square feet of radiator surface provided in the various rooms. The radiators marked "T" are controlled by Thermograde valves and those marked "H" by ordinary hand valves on the supply, the return being controlled by an automatic check. It may be of interest to state that no trouble is experienced with these check-valves if put in as shown in Fig. 4. that is, if the check is dropped to about four inches below the outlet on the radiator.

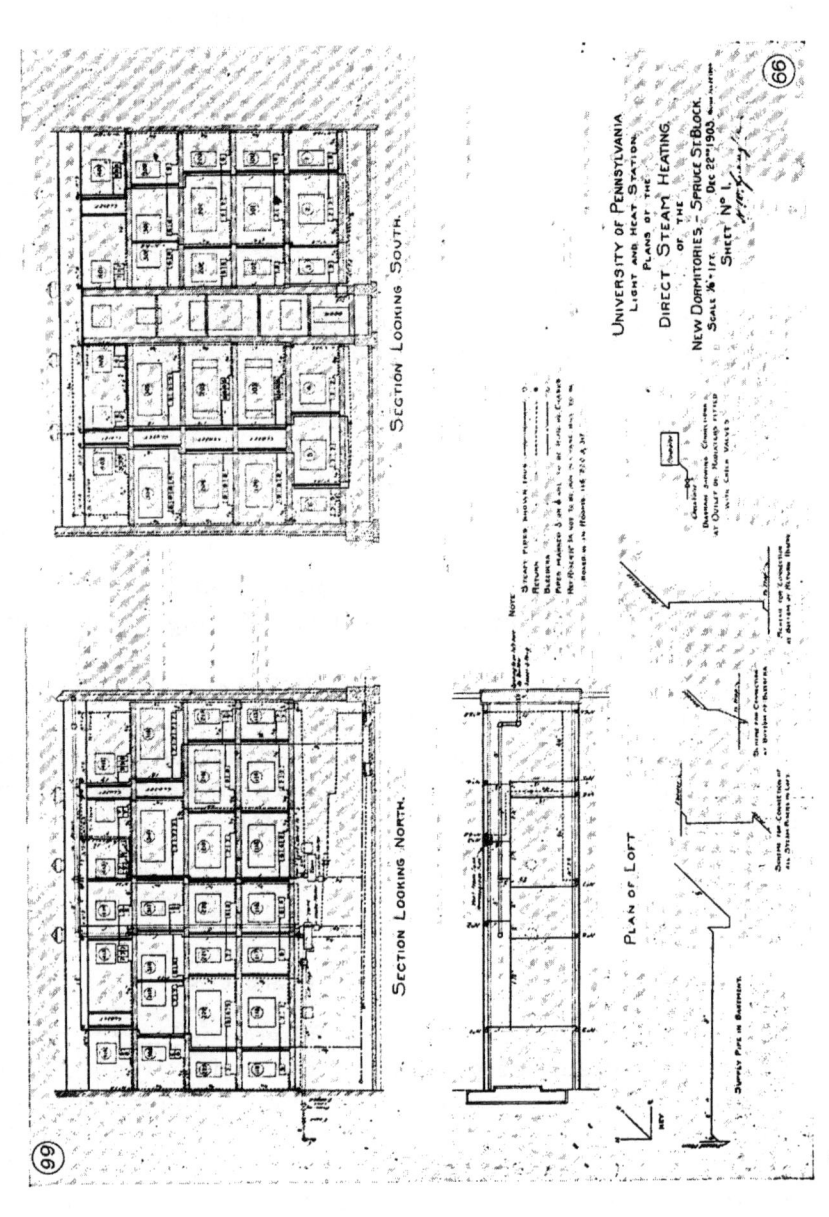

Fig. 4—Riser Diagram.

VOLUME AND RADIATOR SURFACE.

Room	Vol. cu. ft.	Rad. surface sq. ft.	Ratio	Room	Vol. cu. ft.	Rad. surface sq. ft.	Ratio
1	1080	9	120	219	1575	14 T.	113
2	2160	14	154	220	1215	10	122
3	900	9	100	221	945	7 H.	135
4	1575	14	113	222	2160	16 H.	135
5	2385	14	70	223	1125	7 H.	161
6	1260	18	70	Hall	2790		
1st Hall	1390			300	1620	9 H.	180
Vesti'le	1035	21		301	1260	12 H.	105
100	1125	6 H.	188	302	1125	12 H.	94
101	2160	14 H.	154	303	2385	21 T.	114
102	945	9 H.	105	304	2115	24 T.	88
103	2340	24 T.	97	315	2070	28 T.	74
104	2700	27 T.	100	316	2250	21 T.	107
113	945	14 T.	68	317	1215	10	122
114	1665	14 T.	119	318	1125	12 H.	94
115	1575	18 T.	88	319	1215	14 H.	87
116	1215	12	101	320	1620	12 H.	135
117	945	9 H.	105	Hall	3465		
118	2160	14 H.	154	400	1230	15 H.	82
119	1125	9 H.	125	401	1200	21 H.	57
Hall	2790			402	1310	24 T.	55
200	1125	6 H.	188	403	1555	27 T.	57
201	2160	14 H.	154	412	1555	27 T.	57
202	945	10 H.	95	413	1300	28 T.	46
203	2340	24 T.	97	414	800	18	44
204	2700	27 T.	100	415	1155	21 H.	55
217	945	14 T.	68	416	1250	18 H.	69
218	1800	18 T.	100	Hall	2665		
				Entire Bld'g	90790	810	112

No radiators were put in the halls, except in the vestibule on the first floor. The number of cubic feet of space per square foot of radiator surface is large, and the large variation of this ratio is very notable, reaching a maximum of 188 cubic feet of space per square foot of radiator surface in well-protected rooms with small glass area, and a maximum of 44 cubic feet in exposed portions of the building.

In designing the plant it was the desire to obtain a temperature of 70° in the living rooms, 60° in the halls, and it was expected that a temperature of 50° would be obtained in the attic and cellar.

After the heating had been planned a proposal was made to us to use thermograde valves, and as the building was one in which such an installation could well be tested, an agreement was made to install such valves in part of the building, on a guaran-

No. 29								
BDG. Dorm.	ROOM 219	LOCATION House 21	CEILING 9 FT.

WALL	KIND	THICK	AREA	ΔT	C	Ex. C	HEAT LOSS
N	Brick	18"	95	+70	.24	1.2	1900
	Glass	S.	20	+70	1.09	1.2	1800
E	Wood stud	5"	125	0			
S	Brick	14"	115	+10	.28		300
W	Wood stud		125	0			
CEIL.	F. P.	12"	175	0			
FLOOR	F. P.	12"	175	0			

TOTAL 4000

SKETCH: N 0°, 12½; W 70°, 14, 70, 70, 70 E 70°; 60° S; VOLUME 1575 CU. FT.

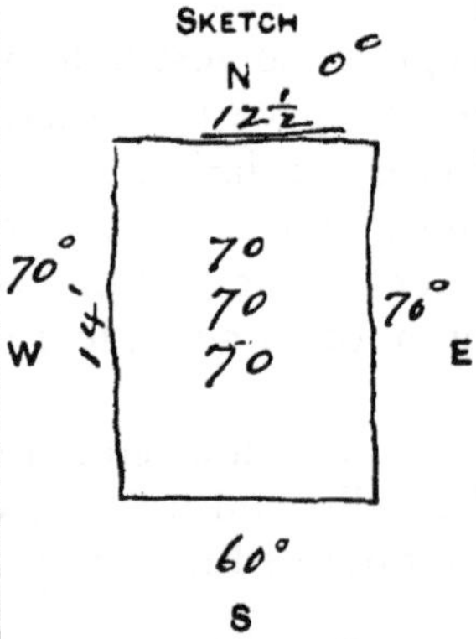

VENTILATION
OCCUPANTS No. .... HT. ....
AIR SUPPLY .... CU. FT.
T. CHANGE .... HT. ....
FLUES ....

HEATING SURFACE

KIND	MAKE	SQ. FT.	ST.	D
1	Fr W. Wall	14	3/4	3/4

Fig. 5

tee to remove them at the end of the first winter if they were not in every way satisfactory to the writer.

As a result of this agreement, all the radiators in the rooms marked "T" on the heating drawings were supplied with thermograde valves and the returns were connected together and led through a meter for measurement.

All the radiators in the rooms marked "H" were connected to a second meter.

The following table gives some of the data relating to these rooms:

	On hand control.	On Thermograde control.
Number of rooms	22	18
Cubic contents	29,720 cu. ft.	33,125 cu. ft.
Square feet glass	340 sq. ft.	455 sq. ft.
Outside wall	1,935 sq. ft.	2,915 sq. ft.
Equivalent glass area (about)	824 sq. ft.	1,183 sq. ft.
Radiating surface including runouts	282 sq. ft.	406 sq. ft.
Outside equivalent glass per cu. ft. under the two systems is as	1 is to	1.26
Radiator surface per cu. ft. of space is as	1 is to	1.29
Pounds steam required to heat 1 cu. ft. of space is as	1 is to	.556

As it is necessary that the entire return from the thermograde system should be at atmospheric pressure the return risers were connected at the top in the attic and vented there, and at the bottom they discharged into a tank which also was vented. The supply risers are all connected at the bottom by a bleeder system, which takes care of the condensation in the risers. The returns from all the rooms marked "H" discharge through a trap and then into a tank and meter. The remaining rooms in the building are on an independent hand-control system.

The discharge from each of the two systems was taken to a tank and from this tank discharged through a tilting meter into the general return pipe leading back to the Central Station. Fig. 6 is a view of the basement showing the two tanks near the ceiling, the nearer one being that of the hand-controlled system, the trap on this system showing clearly above this tank. The meters are on the opposite wall.

The pressure on the heating system is regulated, in the adjacent house, by a plug reducing valve, of which a large number are in use on the general system. These valves require careful attention during the first season and but little thereafter. Figs. 7 and 8 are charts taken from the heating system of this building on the dates shown. Fig. 7 was taken when the valve was working entirely satisfactorily, having been cleaned two weeks before and practically having received no attention until the date at which this chart was taken. The chart, Fig. 8, shows the variation in pressure when the re-

Fig. 6—Basement—Showing Position of Return Tanks.

ducing valve was not working smoothly. The diagram hereafter reproduced shows when this valve was overhauled during the season.

To determine the accuracy of the meters, tests were made of them by discharging the return water after it had passed through the meter into a tank on scales, in which the discharge was weighed, Fig. 9 showing the connections for testing. The temperature of the discharge water was also taken.

Test of meter 510—Hand.				Test of meter 511—Thermograde		
Dial reading Gals.	Wt. in tank Lbs.	Temp °F	Dial reading Gals.	Wt. in tank Lbs.	Temp. °F
0.0	209.0	146	1⅛	216.5	
1.0	216.0	146	6¼	255.0	
2.0	223.5	146	10⅛	285.5	
3.0	231.5	146	11⅛	293.25	
4.0	239.0	147	16¼	232.0	
5.0	246.3	147	20⅛	362.5	
6.0	254.0	146	25⅛	400.5	
7.0	261.5	146	26¼	408.0	
8.0	269.0	147	30⅛	438.5	
9.0	275.3	146	31⅛	446.25	
10.0	283.8	145	35⅛	476.5	
16.0	331.5	164	36¼	484.5	
20.0	361.3	148			
27.0	416.3	163			
30.0	438.0	163			
35.0	476.8	149			
40.0	514.8	151			
40.0	305.8	150	35⅛	268.	89°F.

Apparent Wt. per gallon, 7.65 lbs.   Apparent Wt. per gallon, 7.63 lbs.

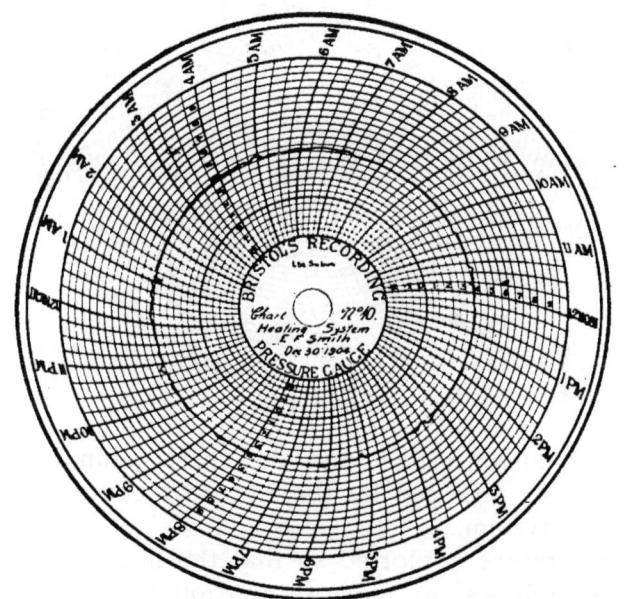

Fig. 7—Pressure Recording Gauge. Dec. 30, 1904.

The temperature of the return water averaged 89°F. for the thermograde system and 150°F. for that under hand control. The operation of the thermograde system is an intermittent one allowing part of the radiators to act as a hot water radiator until the discharge valve opens, discharging the partly cooled water.

During the entire winter the readings of these meters were taken at noon each day and the quantity of return water was plotted on the chart, Fig. 10, showing the values obtained. At the beginning of the chart will be seen the scale to which it has been drawn, one of the larger divisions vertically representing .1 gallon of return water per square foot of radiating surface per twenty-four hours. Using the second column of figures one can read the total gallons returned per day by the radiators

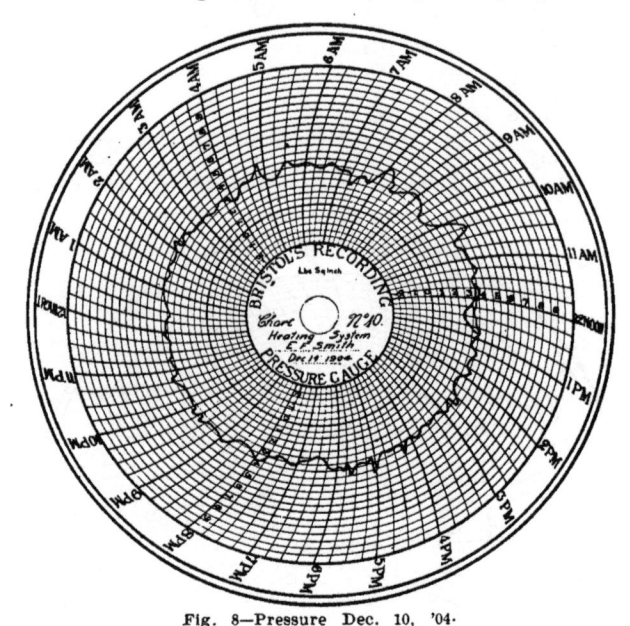

Fig. 8—Pressure Dec. 10, '04.

that were hand-controlled, and the third column gives the gallons per day returned by the radiators controlled by the thermograde system.

The temperature is plotted so that the line of variation in temperature follows the line of variation in water returned. That is, the zero of termepature is at the top of the diagram and of water returned is at the bottom. Four irregular lines are plotted on the diagram. The upper one shows the variation in temperature at 8 A.M. each day. Next below this, and crossing it generally on stormy days is the mean temperature

Fig. 10 A.

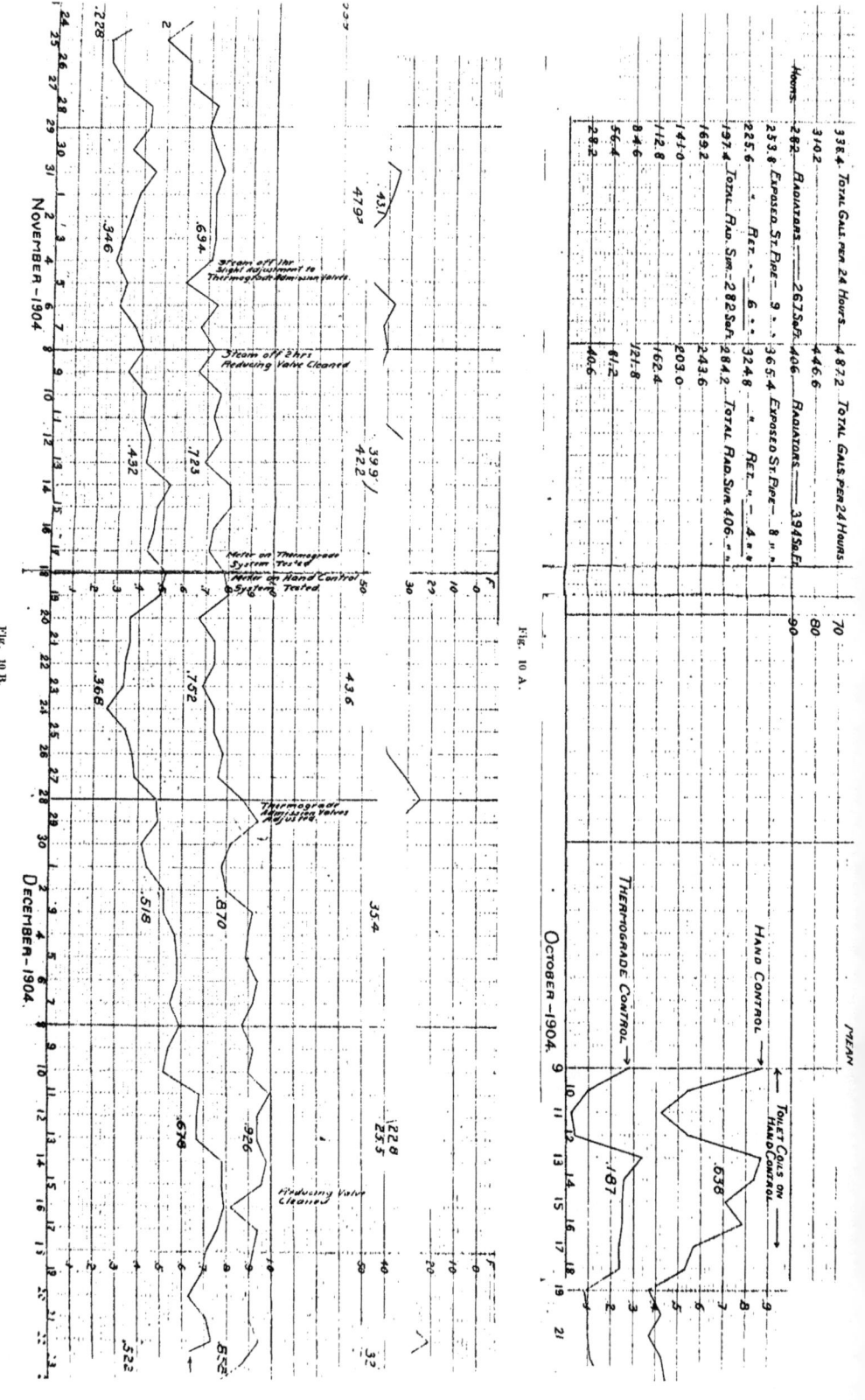

Fig. 10 B.

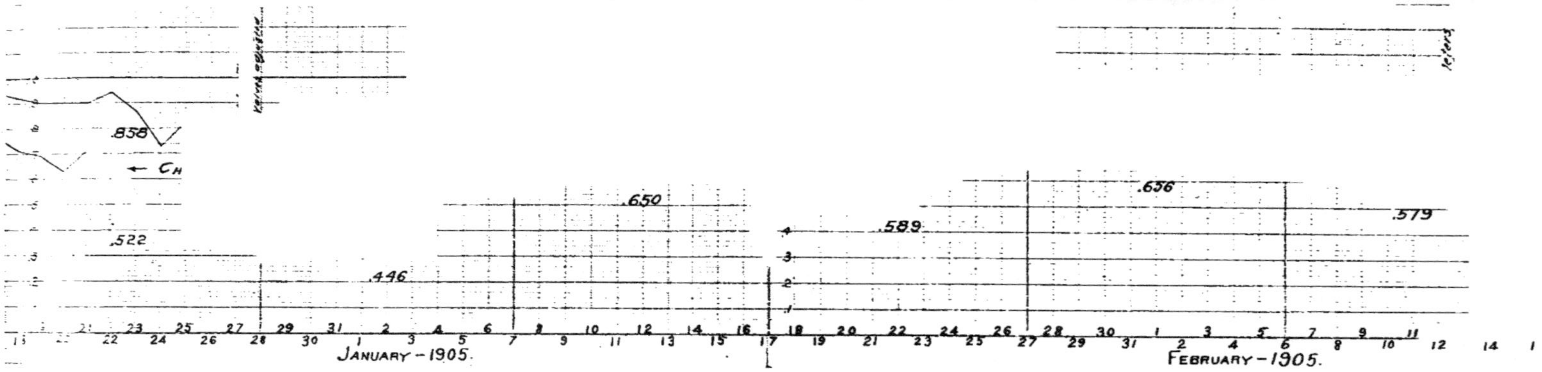

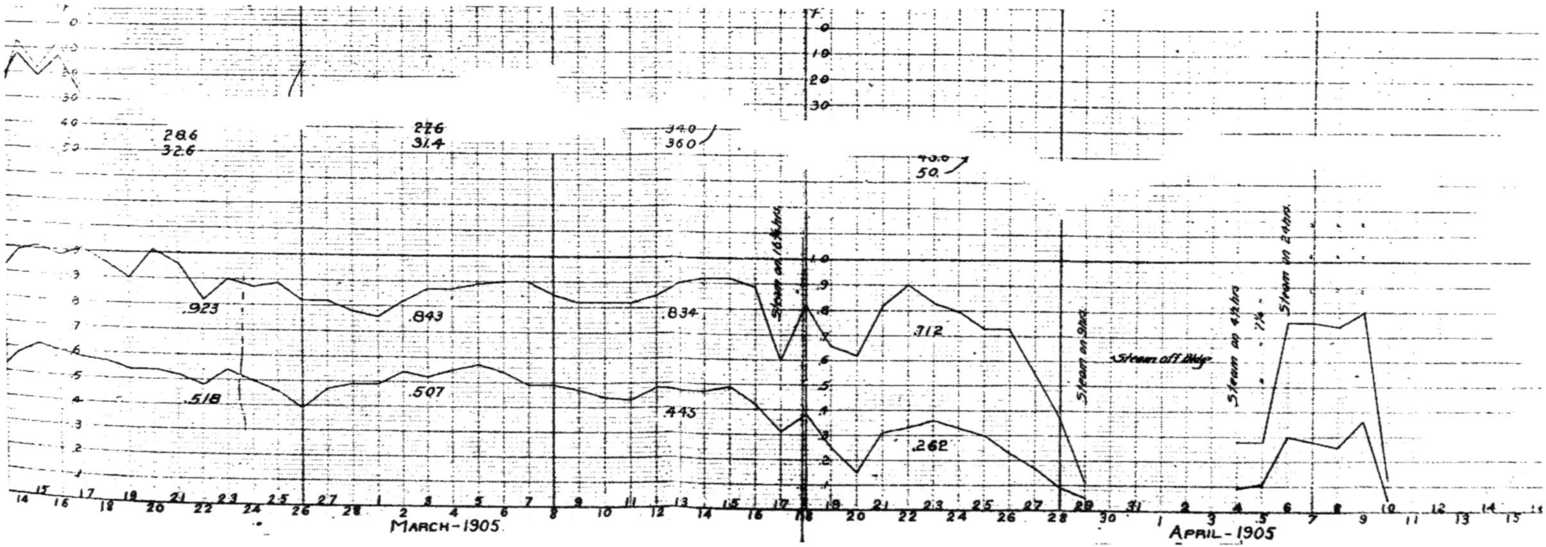

Fig. 10 D.

for each twenty-four hours. The next line shows the amount of steam condensed per square foot of radiating surface per day by those radiators which are under hand control. The lowest line shows the amount of steam condensed by the radiators under thermograde control. It will be noticed that all these lines follow the same general course, rising and falling in nearly the same way. The decided difference between the lower two curves for December 22nd to January 4th (Christmas Holidays) is interesting.

Fig. 9—Testing Meters.

It will be noticed that the reducing valve was cleaned on November 9th, December 15th and on January 24th. At the end of the first week in January all the windows facing the north were weather-stripped in the belief that the quantity of condensed steam returned from the hand control side would be greatly reduced.

From this diagramatic record the following table has been made showing the average value of the various quantities for each ten days, from the time steam was turned on in the Fall until it was turned off at the end of March, during the first warm weather. The values in this table are shown also on Fig. 10.

			Gallons of steam Cond. per sq. ft. per 24 hours	
1904	Mean Temp.	8 A. M. Temp.	Hand control	Thermograde
Oct. 9 to 19.......	59.6°F.	52.4°F.	.638	.187 Toil's on hand con-
Oct. 19 to 29......	53.9	49.2	.520	.228 Toilets off. [trol-
Oct. 29 to Nov. 8.	47.4	43.1	.694	.346
Nov. 8 to 18.......	42.2	39.9	.723	.432
Nov. 18 to 28......	43.6	40.1	.752	.368
Nov. 28 to Dec. 8.	35.4	33.0	.870	.518
Dec. 8 to 18.......	25.5	22.8	.926	.678
Dec. 18 to 28......	31.3	32.2	.858	.522 Holidays.
Dec. 28 to Jan. 7...	36.1	32.2	.885	.446 Holidays.
1905		WEATHERSTRIPPED NORTH WINDOWS.		
Jan. 7 to 17........	30.3	29.2	.926	.650
Jan. 17 to 27......	30.0	27.3	.886	.589
Jan. 27 to Feb. 6:.	20.4	17.4	.995	.656
Feb. 6 to 16.......	23.6	22.2	.939	.579
Feb. 16 to 26......	32.6	28.6	.923	.518
Feb. 26 to Mar. 8..	31.4	27.6	.843	.507
Mar. 8 to 18.......	36.0	34.0	.834	.445
Mar. 18 to 28.....	50.0	45.6	.712	.262

The diagram shown in Fig. 11 gives all the above data except the 8 A.M. temperature. Horizontal distances are average outside temperatures for ten days. Vertical distances are the number of gallons (7.64 lbs.) of steam condensed per twenty-four hours per square foot of radiating surface.

The points fall into two groups, the upper ones relating to hand-control and the lower ones to thermograde-control. Certain of the points are those determined after the windows were weather-stripped.

On the diagram lines are drawn representing the most probable relation between the temperature and steam condensed. Using these average lines for calculation purposes will enable us to determine the value of the thermograde valve and also of the weather-stripping from the standpoint of the amount of steam required to heat the building under various conditions.

It will be noticed first that the thermograde lines cross the base line at nearly 70° and the amount of steam used varies directly at the temperature falls below 70°, that is, with these valves there is real heat regulation. Again it will be noted that, with the windows weather-stripped, the amount of steam

March, 1906.]   *Some Data Relating to Heating.*   193

required or the amount of heat used is less by about 10% for the season than if the weather-stripping is not used.

With hand-control, while the amount of heat used decreases as the temperature increases the value does not tend to disappear at 70°, in other words, moderately mild weather means a consumption of much more steam than would be required to do the *necessary* heating. The line showing the average consump-

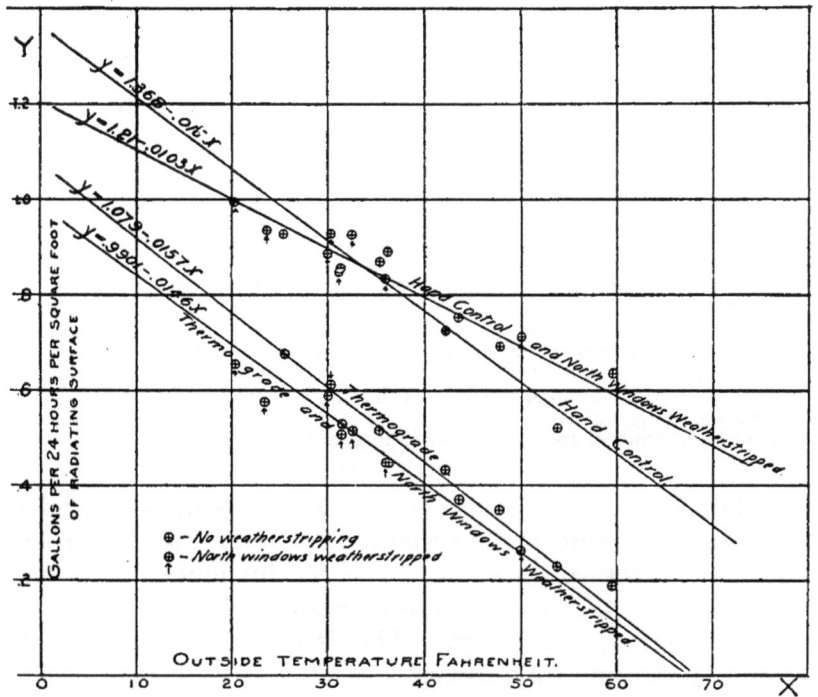

GALLONS OF WATER CONDENSED PER SQUARE FOOT OF RADIATING SURFACE PER 24 HOURS FOR VARYING OUTSIDE TEMPERATURES (MEAN OF 10 DAYS READINGS)

Fig. 11—Average Results.

tion under hand-control with the windows weather-stripped differs from what might have been expected. Above 33° the heat required is more with the weather-stripping than without, while below 33° the reverse is true.

The formulæ connecting the outside temperature and the amount of steam condensed are as follows, in which y is the gallons of condensed steam per square foot of radiator surface per twenty-four hours and x is the outside temperature Fahrenheit.

With thermograde-control and without weather-stripping:
$$y = 1.0792 - .0157 x$$
With thermograde-control and weather-stripping:
$$y = .9901 - .0146 x$$
Hand-control with no weather-stripping:
$$y = 1.369 - .015 x$$
Hand-control and weather-stripping:
$$y = 1.21 - .0103 x$$

The application of the formulæ given on this diagram results in practically the same steam consumption for the total time over which the experiments extended, and they have been applied to the season's work to determine the money value of the four methods of heating this part of the building:—

(A) Had the thermograde control system been on all these rooms throughout the season, there would have been return in condensed steam....................................57,650 gals.
(B) Had the same conditions as the above existed and in addition had all the windows been weatherstripped............51,800 gals.
(C) Had all these radiators been on hand control with no weatherstripping ....93,200 gals.
(D) Had hand control and weatherstripping been used throughout ....98,200 gals.

As the return water was cooler with the thermograde-control, the real value of the system is less, than that shown by the amount of return water, by 5.9%, as this much additional heat must be supplied at the Station before the returned water is again converted into steam. This is taken into account in determining the following figures:

The charge made against the Dormitories for one year averaged .2015 cents per gallon of return water, which covered cost of fuel ($3.00 per ton), labor, repairs and all items except fixed charges which are charged in bulk. on this basis the charges under condition (A) would be $123.44 per year
" " (B) " " 110.92 " "
" " (C) " " 187.80 " "
" " (D) " " 197.87 " "

As the thermograde-valves for all the rooms cost $473.34 more than hand-valves, the advantage to the user is clearly shown from a comparison of the figures just given.

The depreciation and repairs are likely to be slightly more with thermograde-valves, but allowing for these items and interest on extra cost there is still a reasonable advantage in using them.

The cost of weather-stripping the north windows was $90.00 and the saving per year, using it with the thermograde-valves, amounts to $12.52. It must be remembered that these buildings are particularly well built and the outer walls are better built than the average dwelling.

It must be remembered also that the above figures are not the cost of heating the building, as the bleeder system is not

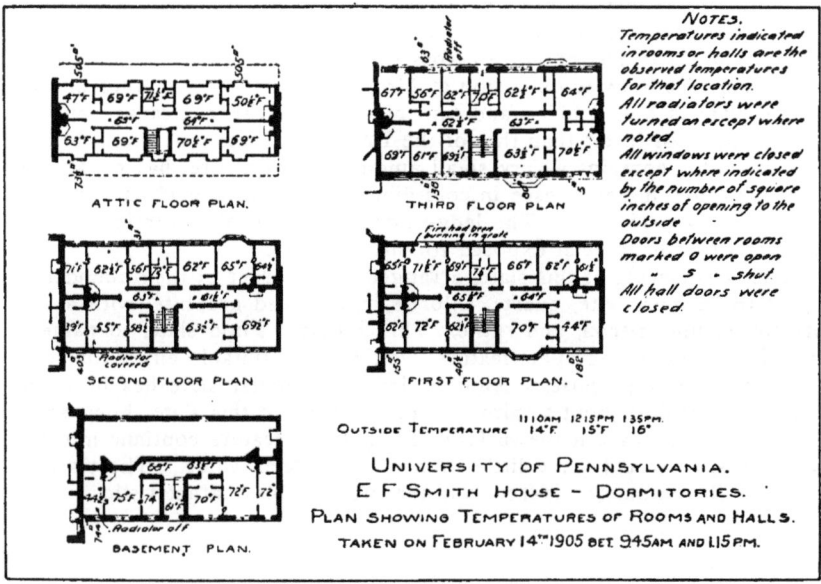

Fig. 12—Actual Temperatures.

connected to either system, but the data above given is strictly comparable as it covers radiators and returns only in each case.

It may be of some interest to know the condition of these various rooms during the coldest weather of the winter. The rooms are almost invariably aired during the time of cleaning in the morning and, in many cases, windows are kept open in the most severe weather. The diagram herewith, Fig. 12, shows the actual temperature at the time of taking together with such data as may be of interest relating to the occupancy of the rooms.

The average temperature in the attic on December 14, 1904, was 64°, with an outside temperature of 19°F., and in the cellar 69°; and on January 26, 1905, with an outside temperaturt of 14°F., the temperature of the attic averaged 54.7°.

University of Pennsylvania, May 15, 1905.

A NEW VARIETY OF MAINE SLATE.

A new variety of slate has been discovered by Prof. T. Nelson Dale of the United States Geological Survey, in the town of Forks. Somerset County, in Central Maine, between the Kennebec and Piscataquis rivers.

The slate crops out in the bed of Holly Brook, where it is exposed for a thickness of thirty feet or more across the cleavage. The nearest railroad is the Somerset Railway extension at Mosquito Narrows, six miles distant.

The slate is bluish black and fine of texture, with a cleavage surface which shows less luster than that of the Brownville slate, but is still bright. It is graphitic, contains a very small amount of magnetite, has no argillaceous odor, does not effervesce in cold dilute hydrochloric acid, is sonorous, and is readily perforated. The ledge does not show discoloration nor do fragments that have been exposed for fifteen years.

The constituents of this slate, arranged in the order of their abundance, appear to be muscovite, quartz, chlorite, pyrite, and graphite, with accessory tourmaline, zircon, and rutile. This Pleasant Pond slate, to name it after the nearest topographic feature, would prove suitable either for roofing or mill stock purposes. Another ledge of similar slate has been exposed by trenching about a third of a mile away, but this slate shows some false cleavage, at least at the surface. Should that feature continue into the mass the slate would have little or no commercial value. The slate of the Holly Brook outcrop is free from that undesirable characteristic.

AMERICAN SKYSCRAPERS IN ENGLAND.

Some discussion has gone on in England recently over the proposal to introduce the skysraper steel structure in London. B. H. Thwaite, the well-known metallurgical engineer, writes very favorably in London *Public Works* of the American tall building. He says there are in the United States many examples of artistic excellence in the lofty buildings of New York, Chicago and other cities, pointing out also that steel frame buildings are less costly and can be erected more rapidly than any other type of permanent buildings. The London County Council, however, interprets the Building acts as opposing the erection of steel frame buildings in the way that would render them most advantageous. An effort is being made to secure some relaxation of the rules in respect to the use of steel.

(Stated Meeting, held Thursday, January 11, 1906.)

The Finances of Engineering.

BY WM. D. MARKS, PH.B. C. E.

Honorary Member of the Institute.

The brain of the engineer is the dormant intellectual force of to-day, and it should be the dominant one.

For more than 140 years it has been busy in bending the greater forces of nature to the service of man.

I speak of 140 years ago, for about that time Watt told us how to harness the force of heat, and so to-day man's feeble strength is multiplied a million fold by countless steam engines and internal combustion engines.

Faraday has shown us how electricity could be generated by the movement of a copper disc in a magnetic field, and to-day thousands of engineers are busy designing, erecting and operating electric dynamos and motors for the purpose of generating, conveying and utilizing the unseen force of electricity in thousands of places with millions of horse-power. Our railway cars are moved by it, our shops driven by it, and our houses and streets are lighted by it.

In 1792, William Murdock lit his shops with gas distilled from coal, and now we can almost say that the world is lit with coal gas, from vast works designed and operated by gas engineers.

The invention of Watt has been so multifariously adapted by engineers to the necessities of man that we now see vast engines driving enormous ships one-eighth of a mile long at a speed of twenty-five miles an hour across the oceans, or on land find locomotives dragging trains at an average speed of fifty miles per hour half way across our continent: or again moving slower freights on 297,000 miles of track.

The intricate and perfect machinery that drives our ships or drags our trains; the strong, yet wonderfully light bridges which span abysses; the dark tunnels, driven true to a hair through miles of mountains, or under rivers, and the level, smooth paths of iron rails laid upon them are the works of our engineers.

On the grade, in the drawing room, in our shops and in great rolling mills, the engineer's brain is ever thinking and directing vast enterprises.

Our tall iron buildings, rising to dizzy heights, or our deep mines, our vast power and irrigation dams—all are constructed after being thought out by an engineer.

Our water works, our sewer systems, our public improvements on a large scale, all require his thought and his skill, before all else.

His tasks are beset with many difficulties, for nature's obstacles are often hidden, and nature's laws are immutable and merciless.

His knowledge of facts must be large and of the laws of nature exact and profound. His work is on a large scale and his responsibilities are enormous.

Incorrect reasoning, ignorance, or superficial theories on his part are sure to result in great losses of money and often also of life.

Could we present the claims of our engineers upon their fellow men to an inhabitant of another planet unaware of the social organization that binds us, he would decree to them princely rewards for their beneficent labors.

Unfortunately, there is no such impartial tribunal, and instead of being princes of the earth engineers are usually very poorly-paid toilers, who are so much interested in the nature of things with which they deal that they have forgotten the ways of men, particularly those of the employing financiers who find their profit in promoting our great enterprises of which the engineer is the real brain power.

Without a mercenary thought, regardless of his own value, buried in the delight of his work, he toils for a meagre salary, confiding the question of the future success of the enterprise to his employers. If it should not succeed he is usually forced to share the blame of failure, although his engineering work may have been perfect. If it succeeds his is the jackal's share, the lion's share of the profit going to the promoter of the enterprise.

Very often impossibilities are asked of him, and the whole burden of censure due to financial failure is unjustly thrown upon him.

Engineering in its various branches has nevertheless grown into an honored profession, and to successful engineers is awarded an abundant meed of praise and public recognition.

Why then is it that financially the engineer is so inadequately rewarded for his beneficent services?

May it not be that engrossed in the study of the laws of nature he too often neglects to study the laws of business? Will not his pecuniary rewards be greater if in addition to his scientific studies of them he gives a portion of his thought to a financial investigation of the projects with which he may be connected?

By the word business is meant the various avocations by which men earn or obtain the money required by all to support them and their dependents.

We can divide business into two classes, legitimate and illegitimate. Whenever a man, for a fair profit or wage, furnishes his fellow men with the necessities or luxuries of life he is doing a legitimate business. If a man takes cunning advantage of the weakness, the ignorance, the misfortunes or the vices of his fellow men to obtain more than a just profit or a fair wage from them, then he is doing an illegitimate business.

To this latter class belong most of our very wealthy men, but not our engineers, for their's is a distinctly legitimate business, and for that reason it cannot result in sudden and vast accumulations of money. The engineer must earn his money honestly and slowly.

The honesty of thought and action demanded by his professional work moulds his character so that it is well nigh impossible for him to practice crimes of cunning in illegitimate business enterprises. A conscientious engineer always becomes an honest man.

Every commercial enterprise, great or small, may for the purpose of consideration be divided into three factors:

First—The investment, on which we must reckon the depreciation, the interest and the profit.

Second—The fixed expense, such as clerk hire, stationery, taxes, insurance, and many other items of expense which go on regardless of the volume of business transacted, and with very little variation.

Third—The operating expense, which includes the raw material and the productive labor utilized, and which varies in proportion to the volume of business done.

After having said so much about the characteristics of the engineer, I should first consider him as a commercial or business proposition. Until he is graduated from his college he certainly is only an investment. And when at twenty-one or twenty-two years of age he is thrown on his own resources he may, with good luck, hope for forty years of earning capacity.

Perhaps, however, thirty years is nearer his average working life, and we will use this figure.

The investment in an educated engineer may be put down about as follows:

Living expenses to eighteen years of age	$12,000
College expenses, four years	5,000
Total	$17,000
Depreciation (for thirty years)	$3\frac{1}{3}$ % annually
Interest	5 " "
Profit	8 " "
Total	$16\frac{1}{3}$ " "

If we assume this engineer to hold a salaried position his fixed expense is the cost of his living and his operating expense is nothing.

We then have as an adequate average salary for a technically-educated engineer—

$16\frac{1}{3}$ % of $17,000	$2777
Cost of living (for a single man)	1200
	$3977

Of course he will at first be obliged to begin on a very small salary, but he ought to hope to average $4,000 salary per year for thirty years, and earn $120,000 in his probable years of usefulness.

If an engineer opens an independent office and becomes a consulting engineer, his fixed and also operating expenses both exist, and his risk of loss and chance of profit both increase.

However, if an engineer prefers to live upon a salary and give only the occasional thought required of him by the commercial side of his various enterprises, $4,000 per year may be regarded as his probable average income for thirty years from purely scientific work.

In America a storm wind of prosperity has been driving all before it for years, and our ever-increasing schools of technology are pouring out civil engineers, mechanical engineers, hydraulic engineers, electrical engineers, marine engineers, sanitary engineers, mining engineers and chemical and gas engineers by the hundreds to join the ranks of our industrial army as real captains of industry, trained to direct it in its honorable toil, but not to manipulate it for personal profit.

Above our line of average earnings rise as brilliant exceptions a Morse, a Corliss, a Roebling, a Brush, an Edison, a Thomson, a Welsbach, and a score of less largely-rewarded technical workers, who, possessing the power to discover in nature new relations between old things are called geniuses, and having conferred vast benefits upon their fellow men by their discoveries (whose value it is impossible to overestimate), have had doled out to them by reason of our patent laws relatively beggarly gifts of a few millions from those who have profited many millions from the exploitation of their inventions in a commercial way.

These financiers, promoters, or exploiters of the public and its needs, own and control our steam and electric railways, our electric light stations and gas works, our fleets of steamships and our great hydro-electric power stations, and hire and pay our engineers to build and operate them. Those of our engineers not having the genius to force a well-earned fortune from nature by a profound and practical study of its laws, must, if they aspire to more than our average modest stipend of $4,000, study the laws of business and the methods and mental processes of our financiers, so frequently miscalled "captains of industry," for in truth the great majority of them are merely parasites upon the industries which they control, knowing nothing of their practical operations and relying upon engineers to do the producing.

Throwing aside simply lucky men, such as the inheritors of fortunes or the finders of mines as sporadic instances of wealth, and counting as entirely legitimate financiers the organizers and owners of great industries, and which, beginning at the bottom, men of force and genius have built up a great business while directing its details from practical knowledge acquired in working their way from grade to grade until one has reached the top, we have left as yet a great majority of parasitic financiers whose great wealth is the result of methods which should be clearly understood but not practiced by the aspiring engineer. An absolutely cold-blooded and scientific direction of this parasitic type is necessary to our self-protection.

This type is wise because it knows that average men have neither reasoning power nor moral courage. They have only memory and fear of pain. He knows that for the average man the law of the herd is supreme, and he prefers to be wrong with the herd to being right and alone.

This type of financier is not dissipated or unconventional. It scrupulously respects the religious beliefs and the social prejudices of the community in which it exists.

It is economical, because in the pursuit of money it is careful to practice all the economical virtues and to have no expensive or unconventional vices.

It is seldom dishonest, because it usually keeps within the laws of the land, only venturing on legal crimes in times of desperate need.

It is quite as devoid of any sense of justice as it is of practical knowledge of the business from which it reaps a profit.

With truly diabolical cunning it arranges to close the door of opportunity and adequate reward to the real workers and producers, upon whom it relies for service.

With an acuteness amounting to genius it takes advantage of the weakness, the vice, the ignorance and the misfortunes of men to legally rob them of their money. It gains control of the sources of supply of coal, of oil, of transportation, of electricity, of gas, of water, and of the commodities of life, usually through the agency of a corporation, and getting State and country under its control proceeds by myriads of petty thefts of unjust profits to make counterfeit certificates and bonds (representing no real property) profitable and saleable.

It is self-denying, for it denies itself the pleasure of generous friendship or kindly assistance to men.

This is the type of the conventionally-respectable and parasitic financier of whose doings our daily papers delight to tell us.

Neglecting our criminal financiers, such as Fisk and Gould of the Erie 1872, as too obvious and despicable to require discussion with intelligent men, we will take up our problem from the point of view of the legitimate financier, of the engineer who, having practically and theoretically learned his business by working up from grade to grade, aspires to become a real captain of industry, and by the greatness of his work in organization reap for himself a just and generous reward in money power.

The possession of money creates a commander of an industrial army, and if this commander is wise, just, unselfish and temperate in his acts, his well-earned wealth makes for the good of all men who serve him, and for the advancement of his community.

It may be a work of supererogation in a discussion of finances to devote so much time to a discussion of the types and methods of financiers before taking up the statistical side, but stripping off all cant, it is a fact that the man who is scrupulously honorable and fair in his dealings with all men imposes upon his abilities a burden which the merely legally honest and yet unfair man does not carry in the struggle for money.

The really honorable man must therefore work harder and be satisfied with less than prominent financiers, even though his energies may possibly make him a millionaire.

It is, too, exceedingly important to recognize the facts that all wealth is not the result of dishonor, and that it is impossible, since all men are not equals in ability or opportunities, for all men to have equal possessions, or positions of authority, or leisure for recreation.

The laws of nature prohibit such utopian dreams, for they demand work of all us for the development of man's better faculties, and it is by work alone that they can grow and expand.

Then, too, we all of us have a limited susceptibility of education. We must divide our vocations and by iteration of work in our own specialty become adepts in it of greater skill than otherwise would be possible, and then in turn rely upon adepts in other vocations for the needs we ourselves do not supply.

It appears as if nothing could be more foolish than to fail to recognize and accept the relentless laws of nature against which the puny efforts of man avail nothing.

However good, however delightful, however humane our theories may be, unless nature consents and assists they can result in nothing but failure.

To succeed in any enterprise we must organize, we must recognize and obey well defined laws and principles and we must have superiors and inferiors in our industrial armies—and may I add our superiors should only be honest veterans, whose diligence and ability shown in every department and grade of their commands has proved them worthy officials.

It will be rather difficult to discuss the cost of construction and operation of the various forms of engineering enterprises in our limited space, but a few selections will open the gates to you and perhaps interest you sufficiently to lead you to a reference to the authorities quoted which will give you details impossible to go into here.

STEAM RAILWAYS.

The report of the Interstate Commerce Commission for June 30, 1904, gives the total mileage of railway tracks in the United States, wherever located or however used, as 297,073 miles, and their length of road bed as 212,243 miles.

Total capital stock issued...................$6,339,899,329
" funded debt........................ 6,873,225,350

" railway capital.....................$13,213,124,679

The total length of road bed of a steam railway is not a criterion of its cost, for it may have four tracks and vast freight yards and stations and miles of siding.

It may be built regardless of cost of construction, as is the Pennsylvania low-grade freight line from Harrisburg to Atglen, where 4,000,000 cubic yards of rock excavation sometimes are required per mile of road bed, or it may be built like the many tens of thousands of miles of single track stretching across the prairies of Mississippi Valley, which can be best described as two streaks of rust on a mud bank, traversed by few trains.

But the total length of single track, wherever located or however used, will give us a rough index to the difference in construction cost of a four-track, a double- track, and a single-track road, and also amply cover the extra cost of sidings and freight yards.

From our figures, we find that every mile of railway track in this United States is capitalized at about $44,480, whatever its real cost may have been.

To find this real cost is somewhat difficult.

Suppose we take the oaths of the railway officials to the tax gatherer as to taxes and reported by Interstate Commerce Commission June 30, 1904:

AD VALOREM TAXES JUNE 30, 1904.

On values of real estate and personal property....$43,410,020
"   "   stocks, bonds, earnings, etc...........   6,305,807
"   "   property not used in operation........   1,324,808

Total (omitting special taxes)................$51,040,635

Delaware, Maine, Maryland, Minnesota and Vermont are the only States not accepting the basis of ad valorem taxation as their principal factor in levying taxes.

One per cent. is certainly as low an average tax rate for State or county taxes as can be reckoned upon and will fix a maximum valuation of $5,104,063,500, or about $17,520 average value of total construction property and equipment per mile of single track wherever built in this United States.

Perhaps this is not convincing. Perhaps it will prove more convincing to select an average State, such as Indiana, and give the tax valuation of its State Commissioners for 1904 upon all its steam railways.

We find returns of trackage and values as follows:

Main track .... 6,730.55 miles
Second main track....................... 615.69 "
Side tracks............................. 2,846.17 "

 Total of all tracks...................... 10,192.41 "

And we have as their tax valuation upon all tracks, rolling stock and improvements upon right of way $165,863,367, or $16,274 per mile of track.

A careful study of the details of the Report of the Tax Commissioner of Indiana for 1904 will prove convincing as to its fairness in stating the truth concerning the true cost of Indiana railways.

If we take Indiana as an average railway State we obtain for 297,073 miles of single track $4,834,666,002 as the total cost of all the railways of the United States.

Without going into the details here a most liberal and careful analysis of the excellent Report of the Massachusetts State Railroad Commissioners for 1903 would appear to prove the average cost of construction and equipment of a mile of single track to be less than $25,000. In Massachusetts the car equipment is very large ($10,000 per mile), because the manufacturers are many and the population dense, so we can regard this figure as a maximum average cost per mile of single track for every State.

Perhaps a detailed estimate of cost of a single-track railway will make its cost clearer and still more convincing.

We will at first omit the extra costs of urban right of way, of long bridges, of heavy rock cuts, and of extraordinary tunnels and excavations. We will, however, allow an average of 10,000 cubic yards excavation at 35c, or $3500 per mile. If excavation and embankment are balanced in each mile we have 102 cubic feet per foot of track or an average cut or fill sixteen feet wide and six and a-half feet deep, which of course you will recognize as an excessive allowance.

ESTIMATE OF COST OF ONE MILE SINGLE TRACK TO THE TOP OF THE RAIL.

Spikes, four per tie, $5\tfrac{1}{2} \times \tfrac{9}{16}$ 5.280 lbs., @ 4c.................... 211
Preliminary legal expenses and right of way..................... $700
Surveyors ....$ 100

Grading	3500	
Ties, 2640, @ 50c	1320	
Rails, 70 lbs., @ $30 (118 tons)	3540	
Joints and bolts, 352 @ $2	704	
Switch irons	150	
Road ballast, 10 ft. x 2 in. (50c per cu. yd.)	652	
Fencing wire with posts	480	
Crossings	20	
Labor laying track	1000	11,577
Architectural work—stations, shops and houses		1,500
Extraordinary expenses—bridges, tunnels, etc		2,000
Total for construction of one mile		$14,877
Equipment from State tax Comm. Indiana		1,781
Total cash cost of construction and equipment		$16,658

The item $1,781 is obtained by dividing the assessed value of all rolling stock on Indiana railways by their total of single track 10,192 miles. The items of $1,500 and $2,000 cannot be verified as completely as others, but they are ample, and the majority of our nearly 300,000 miles of single track has never benefitted by the expenditure of this $3,500 item.

We have endeavored to get at the real cost of construction and equipment of average railways, and reached the following results:

(1) From the taxes sworn to by railway officials ....$17,250 per mile
(2) From the Report of the Indiana Tax Com. 16,274 " "
(3) From a careful and overloaded estimate.. 16,658 " "

Before this I have stated to you that Massachusetts with its hilly and rocky topography might average a cost of some where near $25,000 per mile of single track, and now you also see that $17,500 per mile cost is a high average estimate for the vast net-work of railways that covers the Southern States and the Mississippi Valley. $44,480 per mile is the average capitalization found. This is divided up as follows:

Funded debt per mile $23,137
Capital stock " " 21,341

$44,478

Comparing these figures with the average cost of construction and equipment, $17,500, we find that about 76% of the par value of our railways' funded debt has been utilized for their creation and that the sale of stock has not been required at all.

Remember, I am dealing with averages. Fortunately there are exceptional railways in which both bonds and stocks have been paid for and the money legitimately used for the purpose of creating them, but they are few.

On the average the capital stock of our railways has no more real basis as a token of labor or of existing property than a counterfeit bank note, and the par value of our bonds is greater than the cash cost of our railways by about 33%.

You ask how can this have come about? I answer: There is no more facile instrument for the cunning man than a corporation allowed to work in secret.

When an average railway is built, a construction company is generally formed, which agrees to build it for its securities, and doing so receives them, and placing the bonds, through dealers in securities, reserves all or a portion of the stock for itself.

The actual cost spent on a railway appears to average 76% of the value of the bonds, the remaining 24% goes as profit to the bond broker and his allies in the construction company, unless the bonds are sold below par, say at 90%, and then the profit is reduced to 14% cash and the speculative profit which may be obtained from the sale of as much stock as they may be able to retain for themselves.

If we are extremely liberal with these railways we might allow $20,000 per mile of single track, and then it would appear as if 86% of the par of the bond issue, $23,137, has been appropriated for construction and equipment.

Careful appraisements of the cost of reproduction of existing railways have been made by the State of Texas, in 1895. The best and most careful engineers, valuing the Texas railways mile by mile, obtained an average of $18,000 per mile of road bed for all construction, equipments and investment.

It would have been clearer and fairer if they had given the cost per mile of single track, wherever laid or however used, for their railways, which is much less.

Usually after a few years of operation a railway is consolidated with or purchased by some larger system of railways and on this account a new issue of securities is usually born and sold.

Of course there are many other causes of the existing status of our railways, but I have given you the usual method of inflation and the main cause of their enormous over-capitalization.

These methods have also been applied to almost every class of industry operated by corporations in this country. Electric railways differ from steam railways because of their more costly construction of power stations and transmission systems, and also because being sometimes built in the busy thoroughfares of a metropolis, or again over the fields and country roads as an interurban railway, they differ widely in first cost. From the special reports of the Census Office, 1902, I obtain the following capitalizations of electric railways:

FOR URBAN CENTERS PER MILE OF TRACK.

Population	500,000 and over	$209,162
"	100,000 " "	113,026
"	25,000 " "	63,197
"	25,000 " less	38,445
Interurban railways, fast		52,371
" " ordinary		50,313

From practical experience one-half of their capitalization in cash would appear to me amply sufficient to build any of these different classes of electric railways.

Ninety-six thousand two hundred and eighty-seven dollars per mile was the average capitalization of all the electric railways in the United States in 1902, but this figure is of no value, for the demands of locality vary too widely to permit of arranging electric railways as we do steam railways.

The special report of the Census Office, 1902, upon central electric light and power stations is instructive.

The actual cash cost per installed horse-power of the 10,000 H. P. Philadelphia Edison Station did not exceed $250, and in many cases of smaller and less massively-built stations it goes as low or lower than $150 per H.P., including the distribution system.

The total capital stock and funded debt of all these stations in 1902 was $627,515,875, and the total horse-power installed in dynamo capacity was 1,743,000, giving us an average capitalization of $426 per horse-power. The cost of construction per horse-power of electric light and power stations should vary widely, and it does.

Perhaps $150 per horse-power can be called an average for the smaller stations of the Mississippi Valley and $450 for the remote localities of the far West; $250 per H.P. a proper average of all. A large investment per horse-power can be afforded for hydraulic over steam power, and that has contributed to the occasional larger first cost of stations in the far West.

There prevails however the same general inflation of capitalization due to "water" in the stock in both electric light and gas companies. Of course my figures are averages only, and averages are often widely departed from in individual instances, but still I trust I have made my picture clear and you can see the patient engineer striving to reduce the cost of construction and struggling with operating costs, while the industrious promoter overcharges the public as much as it will stand and sells as many watered securities to it as he can get it to take.

The receipt for making a modern parasitic millionaire can be inferred from this. Create a corporation, preferably under the obliging laws of New Jersey, and be careful to work it in secret. Hire as many engineers as are necessary for all of its legitimate business functions and the direction of productive labor. Arrange that the finances shall be controlled by a board of cunning financiers, skilled in illegitimate business. Get out as many bonds and as much stock in excess of real cost as the public can be pursuaded to buy. Arrange if possible for a monopoly, and overcharge the public as much as it will bear for the product, whether it be ton miles, passenger miles, kilowatt hours, gas, or iron or oil. In a short time your parasitic millionaire will emerge. You hear of a new one in New York every day,

I am not preaching a sermon to you on the ethics of this procedure. I am only giving you, to the best of my knowledge, a cold-blooded dissection of this modern "Chevalier d' Industrie" of the United States.

Bear in mind that there is a limit to the producing capacity of most men and that great geniuses (such as an Edison) only can honestly and rightfully grow colossally rich by honest work. Bear in mind that if you wish to keep, and to have others keep, what they have honestly earned you must not be satisfied with your own rectitude, but also insist that all with whom you deal act honestly, however disagreeable the task may be.

The recent awakening of the public conscience bids us all hope for better conditions than have existed for many years, for if our present methods of corporate legal theft are not stopped all the property in the United States will soon be in the hands of gilded thieves.

THE ZAMBESI POWER SCHEME.

Prof. W. E. Ayrton in the London *Times* criticizes severely the Zambesi power scheme, the details of which have recently been given in these columns. He denies the possibility of such a scheme being commercially successful, and condemns the comparisons made between the power schemes at Victoria and Niagara, as the latter only delivers electrical energy over a distance of 30 miles while the former proposes to deliver it over 745 miles. At present the Rand consumes about 150,000 hp at a cost of something like $100 per year per horse power. Niagara sends 24,000 hp to Buffalo, 30 miles away, and sells it at 14 cents per horse-power-year, an eight-hour customer, which is equal to $125 per horse-power-year. There is a plentiful supply of coal in the neighborhood of Johannesburg, and estimates have been obtained and accurate analysis made of about 20 different kinds of coal, giving the specific gravity and the percentage of carbon, ash, volatile matter, coke, as well as the evaporation factor, with the result that at a price of $2.60 to $3.00 a ton excellent coal can be delivered on the Rand itself. How is it possible, therefore, asks Prof. Ayrton, to "transmit electric energy 745 miles and sell it more cheaply than it can be obtained from coal on the spot?" He also points out the dangers from native interference with the line, and summarizes his objection to the scheme in the sentence that "such lengthy transmissions, more than three times as long as have ever yet been made, because more than three times as long as even the enterprising Americans have ever yet considered commercially possible, could cretainly never succeed in such an unsettled country as South America."

ELECTRIC FURNACE PROCESSES.

Under the heading "Recent Electrochemical Development," the following interesting account of the latest experiments with electric furnaces appeared in the *Electrical World and Engineer*, December 16, 1905:

The revelation in workshop practice, brought about by the arrival of high speed tool steels, has been intimately connected with the revolution in the ferroalloy industry due to the introduction of the electric furnace. It is only in the electric furnace that most high percentage ferroalloys, low in carbon, can be made. The only process which the electric furnace industry has to compete with is the aluminothermic method, using aluminum as the reducing agent (instead of carbon in ordinary electric furnace practice). By the aluminothermic method several carbon-free metals, like chromium, &c., are produced. The feature which distinguishes both the electric furnace and the aluminothermic methods from the blast furnace process is that very high temperatures may be easily obtained in the former two methods. Nevertheless, it appears that so far it has been impossible to reduce some extremely refractory oxides, like that of titanium, to the metallic state by either of the two methods. It is not strange that to reach this goal a combination of both methods is now being tried. The recent American Electrochemical Society paper of O. P. Watts described experiments made in this direction by adding the heat of the electric arc to that of the aluminothermic reaction.

A. J. Rossi, the pioneer of ferrotitanium, is also working in a similar direction, as shown by a patent recently granted to him. The feature which distinguishes his method from the aluminothermic process is that he does not use finely powdered aluminum, but scrap or pieces of aluminum in any form, which he introduces first into the electric furnace. When the current is turned on the aluminum is fused. The refractory oxide to be reduced is then introduced into the fused bath of alumnium. Thus not only the electrically generated heat is utilized, but also the oxidation heat of the aluminum (of course minus the reduction heat of the oxide).

A new form of electric furnace for the production of calcium carbide has recently been patented by Appleby. It is chiefly characterized by a movable bottom, so that constantly fresh portions of the charge are introduced into the zone of the stationary elestrodes.

ARTIFICIAL GRAPHITE.

The International Acheson Graphite Company, Niagara Falls, N. Y., with the recent competition of a large extension, equipped with the most modern electrical and mechanical appliances and machinery, has doubled the capacity of its plant for creating graphite in the electric furnace and has closed a contract with Niagara Falls Power Company for another 1000 horse-power of electrical energy in addition to the 1000 horse-power previously used. The commercial importance of this artificial graphite may be estimated from the fact that the United States geological report for 1904 states that for that year the value of Acheson graphite made in this country amounted merely to $341,372. New York State, which is considered rich in deposits, according to the official report of 1904, produced natural graphite worth $119,509. Acheson graphite is largely used in the manufacture of metal protective paints, dry batteries, stove polish, packing and as a lubricant. The electrochemical processes also consume an extensive supply.

Section of Physics and Chemistry.

Read at the Stated Meeting held Thursday, November 2, 1905.

The Administration of the Imported Food Law.

By W. D. Bigelow.

Chief, Division of Foods, Bureau of Chemistry, United States Department of Agriculture.

I believe that posterity will refer to the present time as a period when the public conscience was awakened to an unprecedented extent—a period when the people at large took an active interest in the establishment of right, and the overthrow of wrong. Illustrations of this tendency occur within our everyday life. We have had numerous instances in the political world during recent years of party lines giving way to vital questions of right and wrong. The mass of the population takes a deeper and more intelligent interest in the affairs of the nation than ever before, and the individual probity of public officers has attained a new and greatly increased importance. In the elections of a year ago the people took occasion to express in an unmistakable manner their confidence or lack of confidence in the integrity of individual candidates, and the executives of several States and of the nation were elected by majorities so great as to represent the confidence, not of a dominant party, but of the entire people, in the personal integrity of certain individual citizens.

This disposition is not characteristic of any municipality or State but extends throughout this country and throughout the world. I believe the effort that is being made towards national, state, and municipal purity in administrative measures is characteristic of the age, and must be productive of important far-reaching results. I am aware that the opinion is frequently expressed that we have fallen upon evil times, and that the corruption that has at times been found to exist in State and municipal organizations, and in corporations entrusted with the control of public utilities, as an evidence of the inability of our form of government to meet the situation that confronts it. It is my opinion, however, that such is not the case, but that the publicity now given to corrupt conditions, wherever they are found—whether general or local, whether in public affairs or in the affairs of corporations controling public utilities or administering funds held in trust—is due to the determination of the people at large that such evils shall be corrected. The public has been awakened and interested to an unprecedented extent, and is in a better position than ever before to remedy conditions of which it does not approve.

If this be true, it is but natural that greater attention should now be devoted to our food supply than was formerly the case. It is but natural that the public at large should take an interest in the wholesomeness of our foods as well as in the larger moral question of fairness and honesty in our commercial relations.

As a matter of fact, we are more interested and better informed with respect to the nature and quality of the food we consume than ever before. We are studying the methods for the preparation of foods and the quality, digestibility, nutritive value and wholesomeness of the finished article. We are investigating methods for their transportation and for their preservation, both in their natural state and in an advanced state of manufacture. We are acquiring information regarding the relative digestibility of various foods and the relative wholesomeness of substances used in their preparation. We are establishing standards of composition and studying methods for the determination of the purity of foods and for the detection of adulterants. We are studying these questions in Federal, State, and municipal laboratories; in food and dairy commissions, experiment stations and boards of health; in schools and colleges; in popular lecture courses; in farmers institutes and womens clubs; in factories and grocery stores; and in the home. We are assisted by information circulated by the public press, though hindered by sensational articles and misleading advertisements in the same agency. We shall not master it in a year—we may not in a generation—but we are making progress.

Until less than twenty-five years ago no attempt was made in the United States to control in any way the character or quality of our foods. A general food inspection law was enacted by the thirteen original States, but from the character of that legislation it is apparent that its enforcement was not contemplated. In 1883, the first practicable food inspection law of the United States was enacted by the State of Massachusetts. This law was immediately put into effect and has been enforced regularly and uniformly from the date of its enactment. Since then other States have enacted and enforced food laws, until at the present time twenty-five of our States and Territories are seriously attempting to regulate the character and quality of the foods sold in their markets.

In 1881, the Bureau of Chemistry—then the Division of Chemistry—of the United States Department of Agriculture began the study of the question of food adulteration and since that time has given much attention to the subject. Until 1898 we studied foods in general and no particular attention was given to the origin of these foods. Until that time the samples examined in the laboratory were secured in the open market and were intended to represent the foods sold in the retail trade, no attention being given to the locality of their production. During the summer of 1898, however, the study of imported foods was begun. Samples of a number of classes of foods were secured from the Customs officers at the various ports of entry and examined. Additional samples of foods, alleged to be imported, were secured in the open market and the results of their examination were compared with those obtained at the ports of entry. On March 3, 1903, Congress enacted as a feature of the appropriation act of the Bureau of Chemistry a law relative to imported foods. This law has been slightly amended in succeeding appropriation bills. As now enforced the portion of the law relating to imported foods authorizes the Secretary of Agriculture—

"To investigate the adulteration, false labeling, or false branding of foods, drugs, beverages, condiments, and ingredients of such articles, when deemed by the Secretary of Agriculture advisable, and report the result in the bulletins of the Department; and the Secretary of Agriculture, whenever he has reason to believe that such articles are being imported from foreign countries which are dangerous to the health of the people of the United States, or which shall be falsely labeled or branded, either as to their contents or as to the place of their manufacture or production, shall make a request upon the Secretary of the Treasury for samples from original packages of such articles for inspection and analysis, and the Secretary of the Treasury is hereby authorized to open such original packages and deliver specimens to the Secretary of Agriculture for the purpose mentioned, giving notice to the owner or consignee of such articles, who may be present and have the right to introduce testimony; and the Secretary of the Treasury shall refuse delivery to the consignee of any such goods which the Secretary of Agriculture reports to him have been inspected and analyzed and found to be dangerous to health or falsely labeled or branded either as to their contents or as to the place of their manufacture or production, or which are forbidden entry or to be sold, or are restricted in sale in the countries in which they are made or from which they are exported. * * * (Section of Appropriation Act of March 3, 1905.)"

This law became effective on July 1, 1903, and was immediately enforced. The investigations of imported foods that had already been made by the Bureau of Chemistry made it possible for us to immediately begin intelligent work upon this subject. It was desired to enforce the law as completely as possible, and for this purpose it became necessary to secure complete information respecting shipments of foods to the United States. It is a matter of general information that all shipments of merchandise sent to this country, valued at $100 or more, must be accompanied by an invoice certified to before a United States Consular officer. Arrangements were made with the State Department that copies of all such invoices relating to food products should be sent from the various consulates directly to the Bureau of Chemistry and should be accompanied by a declaration of the country in which the products were grown, from which they were exported, and of the presence or absence of added preservatives and coloring matter. The following is a copy of the declaration which is required to accompany all invoices of food products sent to the United States:

DECLARATION OF SHIPPER OF FOOD PRODUCTS.

1. Seller or owner, or agent of seller or owner.

2. Country.

3. Town and Country.

I, the undersigner, am the 1—— of the merchandise mentioned and described in the accompanying consular invoice. It consists of food products which contain no added substances injurious to health. These food products were grown in 2—— and manufactured in 3——. They bear no false labels or

4. State coloring matter used, if any.

5. State preservatives or other articles used, if any.

6. Place.

7. Date.

marks, contain no added coloring matter except 4——, or preservative (salt, sugar, vinegar, or wood smoke excepted) except 5——, and are not of a character to cause prohibition or restriction in sale in the country where made or from which exported.

I do solemnly and truly declare the foregoing statements to be true, to the best of my knowledge and belief.

Dated at 6—— this 7—— day of———, 190 .

(Signature)———

Arrangements were also made with the State Department by which special food invoices were required even for consignments of foods valued at less than $100. By means of these invoices the Bureau of Chemistry was kept informed of all shipments of food products sent to the United States. During the first year of the enforcement of the law all work relating thereto was conducted in Washington. The invoices referred to above were examined daily, and when an invoice was found which it was desired to inspect, a request was made on the Secretary of the Treasury that the Collector of Customs of the port of entry to which that shipment was consigned be instructed to send a sample of stated size to the Bureau of Chemistry. This sample was then examined and when found to conform to the law a statement of that fact, with the request for the release of the shipment to the consignee, was forwarded to the Secretary of the Treasury, and by him transmitted to the Collector of Customs. When a shipment was found to be contrary to the law, considerable leniency was shown at the beginning in order that foreign shippers and American importers might have time and opportunity to become fully acquainted with the law. It was desired to interefere with commerce as little as possible. The earlier shipments which failed to conform to the law only in the fact that they were not correctly labeled were admitted without prejudice to future decisions and the importers were notified that they were in violation of the law. Later, importers were permitted to re-label shipments with pasters, on which was given a declaration correcting any omission or misstatement by reason of which they failed to conform to the law.

During the first year of the enforcement of the law it was found that the delay occasioned by transmission of samples to Washington was so great as to cause considerable inconvenience to importers. Accordingly, in September, 1904, a laboratory was established in connection with the Appraiser's Stores in New York for the inspection of the foods imported at that port. The organization of this laboratory was found to greatly expedite the law in a much more complete and thorough manner. As a result of our experience in that port, laboratories have been established during the last few months in five additional ports of entry—San Francisco, Boston, Philadelphia, Chicago, and New Orleans. By far the majority of foods imported into the United States came in at these six ports. Shipments arriving at ports at which no laboratories have been established are subjected to the old system of inspection, that is, the invoices are received at the Bureau of Chemistry and samples are sent directly to the Bureau, or to the nearest laboratory, from the ports of entry.

By far the greatest part of the shipments of food products are now inspected at the port laboratories referred to above. It is believed that in this connection a description in some detail of the methods employed in inspection will be of some general interest.

At the time of, or before, the entry of the vessel bearing a consignment of goods, the importer or his broker calls at the Customs House, delivers his copy of the consular invoice, on which the shipment or shipments in which he is interested are described, and advances the required amount in payment of the estimated charge for duty. After passing through the necessary channels in the Customs House, the invoice is sent to the Appraiser's Stores, where it is finally referred to the examiner, who is charged with the valuation of the shipment. While in the hands of this examiner, the representative of the Department of Agriculture inspects the invoice and decides whether or not it will be advisable for him to inspect the shipment it represents.

All invoices of food products going through the port must pass through the hands of the food inspector. It may be that he will immediately recognize the manufacturer as one whose goods have frequently been inspected and always found to conform to the law. It may be that the goods represented on the invoice are such as he knows by experience usually, and perhaps always, conform to the law. If, for any reason, the inspector decides that it is not advisable to examine the goods described on the invoice, he stamps the invoice with the words "No Sample Desired by the Department of Agriculture." If he considers that the products are of such a nature that they should be examined in the laboratory, he attaches to the invoice a sample tag like the following:

U. S. DEPARTMENT OF AGRICULTURE,
    BUREAU OF CHEMISTRY.

FOOD PRODUCTS.

Food Laboratory No._____   Sample Requested.

Marks and Nos.	Article	Amount

Sample {ordered / forwarded} _____, 190

Examiner.

Chief, Food Laboratory, Port of_____

On this is indicated the particular case from which a sample is to be taken and the amount of sample that is desired. If he is undecided and cannot determine without seeing the goods themselves whether it will be advisable to examine a sample in the laboratory, a detention tag like the following is attached to the invoice:

FOOD PRODUCTS.

INVOICE DETENTION SLIP.

Inspection deferred. Invoice to be detained by, or returned to, examiner of food porduct item upon completion of examination of other items therein.

....
Chief, Food Laboratory.

For the convenience of those handling the invoices, the "detention tag" is yellow, while the "sample tag" is white. The examiner is instructed that when a shipment of foods described by an invoice bearing a yellow detention tag reaches him, he shall notify the Chief of the Food Laboratory of the Department of Agriculture, and give him an opportunity to inspect the shipment on the floor and decide whether a sample is desired or not. The "detention tag" is especially useful in the case of foods on whose labels a declaration of some ingredient or ingredients is required. Frequently the inspector only requires to see the label to determine whether the declaration has been made. If it has not an analysis is necessary to determine whether the declaration should have been made or the nature of the goods has been changed since previous inspections. After seeing the shipment the inspector decides whether or not a sample should be taken for the food laboratory. He then tears off the yellow detention tag, and either stamps the invoice with a statement that no sample is desired, or affixes a sample tag, requesting the amount of sample he desires.

It is thus seen that eventually the inspector marks each invoice of food products, either by stamping on it the statement that no sample is desired, or by affixing to it a tag requesting the amount of sample that shall be sent to the laboratory. On the arrival of the goods represented by an invoice that is marked with a sample tag, the examiner forwards a sample to the laboratory. When the shipment is appraised on the docks, as is done with certain articles, the examiner sends an order for a sample to the proper officer. He then cancels the word "ordered" or the word "forwarded" and signs the form. When the examiner has finished with the invoices they are returned to the chief clerk of the Appraiser's Stores, who again must see that all invoices of food products have either been stamped or tagged, as indicated above, and that all tagged invoices are properly endorsed by the examiner.

With the exception of those packages sent to the Appraiser's Stores for appraisement with respect to duty (usually one-tenth of the shipment), and from which the samples used by the food laboratory are taken, shipments of imported foods are delivered immediately to the importers. This has been found necessary as because of the magnitude of the shipments it is impracticable to store them until the completion of the analysis. Moreover, the expense of cartage and storage would be considerable and is to be avoided on that account. Again, importers desire to be permitted to use their goods at as early a date as possible, and it is a great convenience to them to have them in their possession in order that they may forward them to their customers, when desired, as soon as the inspection is completed. When a sample is taken for analysis the appraiser of merchandise notifies the importer of that fact and requests him to keep the goods intact until further notice. On the completion of the examination the chief of the food laboratory communicates with the importer and notifies him of the result of the inspection. If the goods have been found to comply with the law, the importer is then at liberty to dispose of them. If the examination indicates that the shipment does not comply with the law, the collector of customs is notified and immediately takes measures to secure the return of the shipment to the customs warehouse, in order that it may be shipped beyond the jurisdiction of the United States. In case of an apparent violation of the law, the importer is also notified of that fact, and that the case will be decided on a certain date at the port laboratory, when he may be present and present evidence. This date may be deferred at the request of the importer to enable him to secure such evidence. Information relative to any particular shipment is given only to the importer. No information is accessible to any importer regarding the inspection of goods consigned to a competitor. If it should happen, therefore, that a mistake were made and goods wrongly considered to be in violation of the law, the matter can be rectified by reviewing the case, in response to a protest from the importer, and no injury is done to the reputation of the importer or of the goods in question.

Immediately on complaint of an analysis from which it appears that a shipment of food is in violation of the law, a brief report of the cause on a blank form prepared for that purpose is made to the Bureau of Chemistry in Washington, accompanied by a sample of the shipment. The work is immediately repeated in Washington, the result reported to the port laboratory and the action recommended by that laboratory confirmed or reversed before the time set for final hearing of the case. In this way all possible safeguards are thrown about the work connected with the enforcement of the law and the interests of importers are protected in every practicable way.

When we consider the large number of shipments inspected in all of the ports, it is apparent that all phases of the work should be systematized as completely as possible. When a sample is received at the laboratory it is entered in an index book kept for that purpose and also on one of the several forms of report cards. The information descriptive of the sample is entered on duplicate cards, one of which is made as a carbon copy. On the completion of the analysis the analytical results are entered on the cards and one copy is filed on the port laboratory and the other forwarded to Washington with a recommendation as to the final disposition of the goods and the reason therefor endorsed on the back. The analytical results obtained in the Washington laboratory are entered on the same card in parallel column and the card filed according to the manufacturer. It is thus a simple matter to determine at any time whether a given manufacturer is shipping the same quality of goods to the different ports of entry, or whether the results obtained in the various laboratories of the Bureau upon the goods of a given manufacturer are consistent with each other. A copy of the results of the Washington laboratory is also forwarded to the port laboratory upon a separate card, with an endorsement confirming or reversing the action recommended by the port laboratory. I give below copies of the report cards employed in this work. They are so arranged as to permit the recording of a quite complete analysis of the different classes of products for which they are adapted, although in ordinary inspections a small number of determinations is made. It is only under exceptional conditions, when some circumstance casts suspicion on a shipment, that a more complete analysis is made:

Port No......... Label..
F. I. No...
Consignee ....Marks
Manufacturer......................Entry..........C. H. Inv..........
Steamer ....Received..........Action.............

	Port Lab.	Washington Lab.		Port Lab.	Washingt'n Lab.
Sp. gr. (15.50°C)			Free acids		
Index Refrac. (15.5°C)			Halphen test		
Sp. Temp. No.			Villavecchia test		
Iodin No.			Renard test		
Sap. Val.			M. Pt. Arach. acid		
M. Pt. Fat. acids			Adulterant		

Port No......... Label..

F. I. No..

Consignee ....Marks

Manufacturer.....................Entry..........C. H. Inv...........

Steamer ....Received.........Action.............

	Port Results	Washington Results		Port Results	Washingt'n Res'ts
Solids (per cent)			Reducing sugars before inv. (per cent.)		
Ash (per cent)			Reducing sugars after inv. (per cent.)		
Acidity (cc N. alk. per 100 grs.)			Coloring matter		
Polarization direct at °C degrees)			Salicylic acid		
Polarization invert at °C (degrees)			Benzoic acid		
Polarization at 86°C (degrees)			Fluorids		
Sucrose (per cent)			Sulphurous acid (per cent total)		
Glucose (per cent)					

The enforcement of this law has had a decided influence upon the character of the foods imported into the United States. Many shippers are sending a higher grade of produce than before. Others have altered their labels or added to them a declaration of some foreign substance.

When the law went into effect much of the oil imported into the country under olive oil labels was found to be mixed with peanut, sesame, or cotton seed oil. During the first few months of the enforcement of the law, numerous importations

case. In this way all possible safeguards are thrown about the work connected with the enforcement of the law and the interests of importers are protected in every practicable way.

When we consider the large number of shipments inspected in all of the ports, it is apparent that all phases of the work should be systematized as completely as possible. When a sample is received at the laboratory it is entered in an index book kept for that purpose and also on one of the several forms of report cards. The information descriptive of the sample is entered on duplicate cards one of which is made as a carbon copy. On the completion of the analysis the analytical results are entered on the cards and one copy is filed on the port laboratory and the other forwarded to Washington with a recommendation as to the final disposition of the goods and the reason therefor endorsed on the back. The analytical results obtained in the Washington laboratory are entered on the same card in parallel column and the card filed according to the manufacturer. It is thus a simple matter to determine at any time whether a given manufacturer is shipping the same quality of goods to the different ports of entry, or whether the results obtained in the various laboratories of the Bureau upon the goods of a given manufacturer are consistent with each other. A copy of the results of the Washington laboratory is also forwarded to the port laboratory upon a separate card, with an endorsement confirming or reversing the action recommended by the port laboratory. I give below copies of the report cards employed in this work. They are so arranged as to permit the recording of a quite complete analysis of the different classes of products for which they are adapted, although in ordinary inspections a small number of determinations is made. It is only under exceptional conditions, when some circumstance casts suspicion on a shipment, that a more complete analysis is made:

Port No......... Label.................. ....
F. I. No.............................. ....
Consignee ....Marks
Manufacturer.................. ....Entry..........C. H. Inv........
Steamer .... .Received..........Action.............

	Port Lab.	Washington Lab.		Port Lab.	Washingt'n Lab.
Sp. gr. (15.50°C)			Free acids		
Index Refrac. (15.5°C)			Halphen test		
Sp. Temp. No.			Villavecchia test		
Iodin No.			Renard test		
Sap. Val.			M. Pt. Arach. acid		
M. Pt. Fat. acids			Adulterant		

Port No......... Label..

F. I. No...

Consignee ....Marks

Manufacturer........................E..y..........C. H. Inv..........

Steamer ....Rece..d..........Action..............

	Port Results	Washington Results		Port Results	Washingt'n Res'ts
Solids (per cent)			educing sugars before inv. (per cent.)		
Ash (per cent)			educing sugars after inv. (per cent.)		
Acidity (cc N. alk. per 100 grs.)			oloring matter		
Polarization direct at °C degrees)			alicylic acid		
Polarization invert at °C (degrees)			enzoic acid		
Polarization at 86°C (degrees)			luorids		
Sucrose (per cent)			ulphurous acid (per cent total)		
Glucose (per cent)					

The enforcement of this law has had a decided influence upon the character of the foods imported into the United States. Many shippers are sending a higher grade of produce than before. Others have altered their labels or added to them a declaration of some foreign substanc........ ......oils

When the law went into effect ugh of the oil im— is also the country under olive oil labels was found articles peanut, sesame, or cotton seed oil ed for im— months of the enforcement of th our notice

15

of oil so adulterated were received. It was frequently necessary to notify importers of the unsatisfactory character of the goods they were importing, and in a number of cases shipments of adulterated olive oils were re-exported. At the present time such practices do not obtain. Although the inspection of imported olive oil still continues, only one recent importation was found to be adulterated with other oils. The same statement is true in a general way of many other classes of foods. It has been held that canned goods consisting of fragments of articles ordinarily imported whole or in symmetrical parts must not be labeled in such a manner as to deceive the consumer as to their nature. For instance, canned mushrooms are supposed to consist of the whole fungus, including top and stem. It has been the custom to bring into the country a grade of fragments and culls of mushrooms. These were labeled "Hotels," and sold at a low price. The Department has decided that this article should not be labeled simply "Mushrooms," but the label must also bear the words "fragments and scraps," or "pieces and stems."

The Department has decided that the label must declare the presence in the foods of all added substances having in themselves no food or condimental value. For instance, it has been decided that canned peas and other legumes greened with copper shall have the words "Colored with sulphate of copper," or "Prepared with sulphate of copper," printed on the label in prescribed type. It has been held by manufacturers that the practice of using copper for this purpose was familiar to all consumers, and that such a declaration was superfluous. In this connection, however, it is of interest to note that vegetables green with copper are not commonly used in the countries of their manufacture, but that they are chiefly prepared for the English and American trade. At the present time all importations of such products are labeled with the declaration mentioned above.

When this law went into effect it was freely predicted that its enforcement was impracticable. It was said that the enforcement of the law would antagonize manufacturers, importers, and dealers, and that numerous interests would speedily combine for its repeal. It was not to be expected that the law would command the aproval of all foreign manufacturers and

American importers. It is a great satisfaction to note, however, that the majority of importers endorse the law and approve of its enforcement, and that manufacturers are complying with its provisions without protest, only requesting an opportunity to dispose of goods that have already been packed. It was said that manufacturers would never be willing to declare the presence of preservatives or coloring matter in food. At the present time, however, the great majority of shipments coming into this country have the required declaration of these substances plainly stated on the labels, and some manufacturers have discontinued the use of preservatives and coloring matter rather than to make such a declaration, and find it entirely practicable to market their wares without such addition. The following illustrates the attitude of the majority of American importers:

A shipment of lemon oil was inspected at the port of New York and found to be adulterated with turpentine and mineral oil. The importer was notified that the shipment did not comply with the imported food law and that it should be re-shipped beyond the jurisdiction of the United States. He expressed himself as highly pleased with that disposition of the goods. He stated that he bought the oil as a result of bids upon specifications for pure lemon oil of high grade. He had supposed it to be as represented and had been surprised to have a shipment rejected by one of his customers a few days before because it was not suitable for use in the preparation of flavoring extract. The importer expressed himself as being anxious to handle only the highest grade of products and as being pleased with the protection afforded by the operation of the food law.

The enforcement of this law has called attention anew to the desirability of a Federal law relating to food products. In 1903, a shipment of alleged olive oil was refused admission into the United States because of its content of 20 per cent. of peanut oil. The next shipment from the same manufacturer consisted of four barrels of olive oil and one barrel of peanut oil. It could only be assumed that it was intended that the two oils should be mixed in this country after importation. It is also possible for importers to bottle or re-pack imported articles and label them in a manner that would not be permitted for importation. One illustration of this that came to our notice some months ago was the following: An importation of French olive oil was labeled as a California product. It was not allowed to enter. It was found, however, that the same grade of oil was imported unlabeled and the label added in this country. Under the present law these practices cannot be stopped, as the Department has no jurisdiction after the importation of a shipment. Wherever the investigations of the Department suggest the possibility of such a practice, the authorities charged with the enforcement of the food laws of the State into which the importations are made are notified, in order that they may investigate and take such action as the case will warrant.

The analytical work connected with the enforcement of this law is not by any means confined to the analysis of samples from shipments inspected. It has been found necessary to analyze large numbers of samples of known purity and of commercial products for the sake of obtaining data for standards regarding classes of foods for which no standards exist. In such cases no practical results are possible except in the nature of the standards that have been adopted or are under consideration. At the same time, the data secured will enable the Bureau to enlarge its work in the near future and include a much wider range of foods than the articles now inspected. Owing to a recent extension of the work by the establishment of the new port laboratories, the Bureau is now inspecting a much larger number of shipments than has been possible heretofore. With our present organization a much larger volume of work will be disposed of during the next year than has been possible in the past.

The shipments inspected during the first two years of the enforcement of the law, from July 1, 1903, to July 1, 1905, are given in the following tabular statement:

Statement of Imported Food Samples Received by the Bureau of Chemistry and Results of Inspection Reported for the Period Beginning July 1, 1903, and Ending June 30, 1905.

Found Contrary to Law	Wine	Meat	Olive Oil	Miscellaneous	Total
Admitted with a caution of being first offense	50	9	11	38	108
Released without prejudice to future decisions in similar cases	78	5	2	186	271
Admitted after the labels were changed to harmonize with the law	10	11	10	190	221
Required to be reshipped beyond the jurisdiction of the U. S. or destroyed .	46	5	23	15	89
Condemned but not disposed of	5	11	2	5	23
Total violation of the law .	189	41	48	434	712
Found to comply with law .	1097	196	622	949	2864
Total number of samples examined from invoices detained	1286	237	670	1383	3576
Samples taken from invoices not detained . . .	700	2	4	392	1098

ELECTROLYSIS OF GAS PIPES.

In discussing a paper on "High-pressure Gas Distribution of To-day," read by Mr. H. L. Rice before the meeting of the American Gas Light Association, at Milwaukee, Wis., on October 19, Mr. Frank S. Richardson related some experiences with electrolysis of gas pipes due to the current which escaped from a neighboring trolley railway. The power station for the railway is located at about the middle of the six-mile pipe line, and the pipe is connected to the trolley rail at this point in five places. About 600 feet of pipe has recently been removed opposite the power house on account of the damage which had been done before the pipe was connected to the rail. One length of pipe had 27 pits in it, and three other lengths were pitted so badly that they leaked. Formerly there was an insulated joint in the pipe line at intervals of 500 feet, but the pipe has recently been bonded with No. 4-0 wire, making the line a continuous conductor. Experience indicates that it is best to have the line free from insulated joints and to bond the rails wherever possible. Tests have shown that, due to the unavoidable potential drop in the rail portion of the electric railway circuit, difference in potential exists between the opposite ends of each insulating joint and current will pass from one length of pipe to the next by way of the soil. Observations have furnished conclusive evidence that insulated joints will not keep the pipe from being damaged, and that the best method is to bond the pipe joints.

Mr. Theodore Bunker stated that a difference of potential as high as 25 volts had been found between the pipe line from Camden to Trenton, N. J., which is laid parallel to a poorly bonded electric railway. A heavy copper wire was run from the pipe to the power house and after three years of operation, when the pipe was uncovered, it was found that no damage had been caused by the electrolysis.

THE PRESERVATION OF NIAGARA FALLS.

It is very evident that public interest is being aroused for the presercation of the Falls of Niagara. The extent of the power development on the Canadian side, coupled with the eager efforts made to obtain extravagant franchises from the Legislature of New York, has directed attention to the conditions surrounding Niagara. President Roosevelt gives national importance to the subject by according it place in his annual message to Congress, and at a meeting of the Canadian Section of the International Waterways Commision, recently held in Toronto, the time was passed in discussing the policy which the Canadians will endeavor to follow in dealing with the important question of the Niagara power development and the preservation of the scenic beauties of the great cataract. Beyond question much will depend upon the powers of the commission to regulate the power companies already established.

Up to this time there has been no international regulation of the rights of power developments at Niagara, for it is only recently that the subject has been considered from this standpoint, it being clear that, separately, the State of New York and the Province of Ontario can hardly be relied upon to control the situation. Franchises on the Canadian side of the river are granted by the provincial government, while those of the New York side are granted by the State Legislature. A natural rivalry exists for industrial greatness, the tendency being to overlook the sentimental thought for the preservation of the scenic spectacle. New York State gets no revenue from the power development, while the commissioners of Victoria Park have said they expected the revenue from the Candian franchises to amount to $200,000 annually. It is time that a body, vested with powers from the federal government of both countries, should step in and regulate power matters at Niagara in the interests of the people on both sides of the border.—*Iron Age.*

Last year 33 steel steamers were launched at the various shipbuilding plants along the Great Lakes. Ten of these were from the yards of the Great Lakes Engineering Works of Detroit, 22 from the yards of the American Shipbuilding Company, and one from the yard of the Toledo Shipbuilding Company, Toledo, Ohio. Detroit had the greatest number of launchings, eight from the Ecorse yards of the Great Lakes Engineering Works and four from the Wyandotte plant of the American Shipbuilding Company.

(Stated Meeting, held Thursday, February 1, 1906.)

The Analysis of Dyestuffs.

By J. Merritt Matthews, Ph. D.

Head of Chemical and Dyeing Department, Philadelphia Textile School; Lecturer on the Chemistry of Textiles and Pottery, Franklin Institute.

The analysis of dyestuffs may be viewed from several points of departure. In the first place, a dyestuff may be considered as any other organic body and its analysis may have reference merely to the determination of the amounts of the chemical elements present in it, which, of course, are determined in the usual manner of organic analysis, which presents no difference in form from that ordinarily pursued. This character of dyestuff analysis has no importance to the dyer or textile chemist in the manufacture and synthesis of dyestuffs. After these few words concerning this form of analysis we will dismiss it without further consideration.

Another form of dyestuff analysis is one which has for its purpose the determination of the money value or tinctorial power of the coloring-matter; in other words, the determination of the actual amount of coloring-matter present in the sample. This character of analysis is of special importance to both the manufacturer of the dye for the purpose of obtaining products of uniform and standard strength, and to the user of the coloring-matter who is desirous, of course, of obtaining the full value of his money. There are two general methods by which the tinctorial power of a dyestuff may be determined. The first one, and the one which is of chief importance, is that of an actual dyeing test. This method of analysis is one necessitating a comparison with an accepted standard sample, or a comparison between several samples of the same dyestuff. Such a test is carried out in the following manner: Suppose, for instance, that four samples of a dyestuff are submitted by different sellers to the chemist with certain quotations of price per pound. The chemist is to determine which of these samples is the cheapest one to purchase with reference to the amount of coloring-matter they may contain. It is of course presumed that the dyes are similar in character and are to be used on the same kind of fibre and by the same dyeing process, otherwise a comparison cannot be obtained. For purposes of the test, 0.5 gram of each of the samples is dissolved in about 200 cc. of boiling distilled water, and when solution is complete it is diluted to one litre with cold water. This solution will consequently contain 0.0005 gram of dyestuff in each cc. For acid dyes to be employed in the dyeing of wool, well scoured skeins of woolen yarn of 5 grams weight are to be used, and the test baths are prepared with 100 cc. of the dyestuff solutions, 20 per cent. of glaubers salt, and 4 per cent. of sulphuric acid, together with sufficient water to bring the volume of the bath to 300 cc. The percentages of chemicals are in terms of the weight of the wool dyed. The skeins are entered into these test baths at a temperature of 120° F., and the heating is so regulated as to bring the baths to the boiling point in a half hour; the dyeing is then continued for a further half hour with the baths at a simmer, and during the dyeing the skeins should be turned systematically from time to time. In order to obtain accurate results, it is important in these tests that the several skeins of woolen yarn employed are of the same kind of wool and that they weigh the same amount; also that the conditions of dyeing in the several baths are as nearly alike as possible; that is to say, the heating arrangements must be such as to permit of the several baths to be heated equally and uniformly, that each bath comes to the boil in the same time, and is maintained at that temperature for the same period of time.

After the dyeing has been carried out as indicated, the skeins are removed from their respective baths, the excess of liquor in the skein being squeezed back into the dyebath, and the dyed samples are well washed in cold water. Portions of the skeins are then dried and compared for color. The one showing the deepest color is taken as the standard by which to judge the others and is set aside for comparison. The other skeins are placed back in their respective dyebaths, further additions of the dye solutions are added, and the dyeings continued until the color on each skein matches in depth with that on the standard sample, record being kept in each case of the amounts of dye solution added. The values of the samples are then inversely proportional to the amounts of coloring-matter required to yield the standard color. For illustration, let us suppose that sample No. 1 required 65 cc., sample No. 2 required 78 cc., and sample No. 4 required 105 cc. of their respective solutions to match the color of No. 3, obtained with 50 cc. of its solution. Then calling the value of sample No. 3 as 100, the relative value of No. 1 would be 50/65x100=77; that of No. 2 would be 50/78x100=64; and that of No. 4 would be 50/105x100=47. These figures give the relative coloring powers and the relative money values of the different samples, from which their true values may be easily calculated with reference to the prices per pound at which they are offered.

In the carrying out of this analysis a rather nice adjustment of color matching is necessary, and this requires the chemist to have an eye trained in this line of observation. As it often happens, the different samples of dyes do not possess the same exact tone of color; that is to say, one red dye may show a more bluish tone than another, or one yellow may be of a more greenish tone than another, etc. In such cases it is sometimes rather difficult even for the practised eye to determine when the color on one sample dyeing is of just the same depth as on the other. Another point to be observed is that when the dyeings on the skeins deficient in color are continued after the first matching, care should be taken to keep the volumes of the dyebaths as near as possible equal to 300 cc., by supplying the necessary water to make up for that lost in boiling, otherwise the concentrations of the baths become greater and greater, a condition which will lead to erroneous results. No further additions of either Glauber's salt or acid should be made to the baths, as these would also lead to a change in the conditions of the different tests.

In the case of substantive or direct dyes on cotton, the tests should be carried out in the same general manner, only skeins of cotton weighing 10 grams should be employed and the baths should be made up with the dyestuff solutions, 20 per cent. of common salt, and 1 per cent. of soda ash. In the case of basic dyes on cotton, the skeins must previously be mordanted in the following manner: The necessary number of skeins are treated together in a bath containing 4 per cent. of tannic acid dissolved in sufficient water to conveniently work the material. The skeins of cotton are worked in this bath for half an hour at a temperature of 190°F., then immersed beneath the liquid and allowed to stand without further heating for two hours. The skeins are then removed, rinsed slightly, and worked for fifteen minutes in a solution containing 2 per cent. of tartar emetic at the ordinary temperature, after which they are well washed in several changes of water, when they are ready for dyeing. In preparing the dyebaths for basic colors, besides the dyestuff solution there is also added 4 per cent. of alum, and the dyeings should be started at a temperature of 100°F., and raised to 190°F. in half an hour, and maintained at that point for half an hour. When mordant dyes (alizarin colors and allied products) are to be dyed on wool, the test skeins should previously be mordanted together in a bath containing 3 per cent. of potassium bichromate and 4 per cent. of cream of tartar, treating the skeins at a boiling temperature for one hour, and then washing well in several changes of water, after which the skeins are ready for dyeing. It may be remarked that when mordating operations are required previous to the dyeing test, all of the skeins to be employed in the tests should be mordated simultaneously and together, for if the skeins are treated in separate baths with the mordants, it is likely that each skein will not have just the same amount of mordant as every other skein, a condition which would vitiate the dye tests.

When properly carried out this practical dye-test method of analysis is capable of yielding results accurate to within about three per cent., though the character of color being tested has much to do with the accuracy of the results; because the eye is sensitive to smaller differences with some colors than with others; for instance, it is very difficult to detect small differences in yellows, or in colors where yellow is the predominating factor, while it is comparatively easy to see such differences in intensity where the predominating color is blue or red.

Though the dye-test method of analysis appears to be the one best suited to the practical needs of the dyer and chemist, several chemical methods of analysis have been proposed from time to time. In the analysis of acid dyes, it has been suggested to precipitate their solutions with a standardized solution of a basic dye, such as Night Blue. Such a method would also have to be a comparitive one, for each acid dye would possess a different titre in terms of the Night Blue; that is to say, it would be impossible to give the standard solution any absolute factor of colorimetric units, and the analysis would have to be conducted by determining the relative values of different samples of the same dye in terms of the standard solution of Night Blue. In fact, this method would not have the same scope of operation as would the previously-mentioned dye-test method; for in the latter method, samples of different red dyes, for instance, of the same color and dye class could be tested, whether these dyes were of the same exact chemical constitution or not; whereas with the chemical method, a true comparison could only be obtained between different samples of identical dyes. Another disadvantage to this form of chemical analysis is the fact that the end point in the titration is very difficult to obtain with any degree of accuracy or nicety, owing to the strong color both of the solution to be tested and the solution employed for titrating; in fact, it is always necessary to filter the liquid at each test for the end reaction, and it is usually very difficult to remove all of the precipitated color, even by careful filtration.

Another method of chemical analysis which has come into notice rather recently is the use of certain strong reducing agents for the purpose of titrating solutions of dyes, the end reaction depending on the decolorization of the dye solution. Sodium hydrosulphite has been employed for this purpose, but titrations with this reagent are difficult to carry out owing to the extreme ease with which it is oxidized by the air. To obtain results of any degree of accuracy the titration must be conducted in an atmosphere of same inert gas, such as carbon dioxide or coal gas. Furthermore this method also has the disadvantage of being limited to a comparison of different samples of identical dyestuffs. For the analysis of indigo samples, however, this method is a very good one, the solution of sodium hydrosulphite being standardized on chemically pure indigotine, and a form of burette being used, which allows of ready titration without having the reagent at any time in contact with the air.

Another chemical method, based somewhat on the same idea as the preceding one, is the use of a solution of titanous chloride as the reducing reagent. This is a strong reducing agent, causing the complete decoloration of a large number of dye solutions, and in many cases giving a very satisfactory end reaction. The same precautions must be employed in the titration as mentioned in the case of sodium hydrosulphite. This method also has the same limitations as the preceding one in that it is only serviceable for the comparison of different samples of the same dye.

PRODUCTION OF STONE IN 1904.

A report on the stone industry in 1904 is among the recent publications of the United States Geological Survey. It is published as an extract from the volume of "Mineral Resources of the United States, 1904," and is intended for general distribution.

The total value of stone reported in 1904 was $74,200,360, which is a gain of $1,254,453 over the value of stone in 1903, when it amounted to $72,945,908. The corresponding gain in 1903 over 1902, when the figures were $69,830,351, was $3,115,557. In 1902 the gain over 1901, when the total value was $60,275,762, was $9,554,589; and in 1901 the gain was $12,267,023 over 1900.

In 1904, granite, marble, and limestone increased in value, while slate and sandstone decreased.

Granite showed the largest increase. In 1904 its total value, including that of trap rock, was $10,992,983; in 1903 it was $18,436,087, a gain of $7,443,104 for 1904. The granite production increased from $15,603,793 in 1903 to $17,169,437 in 1904, a gain of $1,565,644; and the trap rock from $2,732,294 in 1903 to $2,823,546 in 1904, a gain of $91,252.

Sandstone, including bluestone, decreased in value from $11,262,259 in 1903 to $10,295,933 in 1904, a loss of $966,326. The value of bluestone included in the sandstone was $1,779,457 in 1903, and $1,791,729 in 1904, an increase of $12,272. The sandstone figures decreased from $9,482,802 in 1903 to $8,504,204 in 1904, a loss of $978,598.

The value of marble increased from $5,362,686 in 1903 to $6,297,835 in 1904, a ain of $935,149.

This late output was valued at $6,256,885 in 1903, and at $5,617,195 in 1904, a loss of $639,690.

The limestone output remained nearly the same, being valued at $31,627,991 in 1903, and $31,996,415 in 1904, a gain of $368,424 in 1904.

CLAY-WORKING INDUSTRIES IN 1905.

In reviewing the progress made by the clay-working industries in 1905, Mr. Jefferson Middleton, of the United States Geological Survey, notes that the year was unusually prosperous. There has never been greater activity in building.

Bricks.—In nearly every large city in the United States the demand for structural material, especially brick, was greater than the supply. In Greater New York the enormous consumption of brick drove the price up to $10 a thousand for common brick during the height of the season. What this means will be better understood when it is stated that only a few years ago brick of the same grade sold for less than half this price. Other great cities also consumed unusually large amounts of building brick, though strikes in some of them, notably the teamsters' strike in Chicago, interfered to some extent with the building trades. As a result of the immense consumption of brick along the Hudson River, in both New York and New Jersey, was the largest ever recorded. The same stimulus was felt at points in Connecticut and Massachusetts, as well as at points in New Jersey some distance from the Hudson River.

During the year the paving brick industry was in a flourishing condition, and returns will show a notable increase over the business of 1904. Brick pavement seems to be growing more popular in the smaller cities, and as the proper method of laying it is becoming better understood there is no reason why it should not prove entirely satisfactory.

The use of face or front brick seems to be increasing. This kind of brick is used now instead of stone on many of the largest buildings.

The output of fire brick naturally reflects the condition of the iron and steel industries, and as these were phenomenally prosperous in 1905, it may be expected that the production of fire brick in that year will show a large increase over that of 1904.

Pottery.—The pottery industry thrived during 1905. The price agreement was broken early in the year and caused some disturbance in the industry, but by the end of the year matters had so adjusted themselves that normal conditions prevailed. It is the general impression among producers that the products for 1905 will be greater that for 1904, but that the value of the output will be about the same. The importers of pottery during 1905 will probably show a falling off, possibly explained by the fact that the quality of the ware made by American potters is steadily improving. The prospects for 1906 in all branches of the clay-working industries seem to be very bright.

Clay.—In the clay-mining industry the conditions in 1905 were similar to those in 1904. The fire clay mined will naturally increase with the increase in the fire brick output. The potters seem to be unable to procure satisfactory suppies of American kaolin, although a considerable quantity of this material is used in paper making.

THE PRODUCTION OF PLATINUM IN 1904.

The war between Russia and Japan is probably responsible for the fact that the output of platinum in the United States increased from 110 ounces in 1903, valued at $2080, to 200 ounces, valued at $4160 in 1904. Owing to anxiety in regard to the fate of the platinum industry in Russia, the price of platinum rose about 10 per cent. during 1904. "It should not be understood," says Dr. David T. Day, of the United States Geological Survey, in a recent report on the production of platinum and allied metals in 1904, "that the slight rise of 10 per cent. in the price of platinum would serve as any great stimulus to the placer gold miners of the West who furnish the platinum products of the United States, for these miners are comparatively indifferent to a slight change in price. The scarcity of platinum and the consequent rise in price, however, led to much energy on the part of eastern smelters of platinum in urging upon the placer miners of the West the advisability of saving platinum in cleaning up the hydraulic mines. The increase thus effected is interesting as showing what is possible in the United States in the future."

In the opinion of Dr. Day, the outlook for increased production for the year 1905 is good, not only on account of the continued high price of platinum, but because of the investigation undertaken by the Geological Survey of the black sands of the Pacific Slope and of the increased knowledge thus furnished to the miners in regard to the value of the platinum and to simple means of saving it.

The world's total supply of platinum for the year amounted to about 300 kilograms, or 9625 troy ounces from South America, and 6000 kilograms, or 192,500 troy ounces from Russia. No production of platinum from Australia was reported. A slight product of both platinum and palladium from the Sunbury copper mines continues to come on the market, but it is not profitable to extract all of the platinum and palladium which these ores could furnish. Increased interest in the occurrence of platinum in hydraulic mines and dredges of the Fraser River is due principally to the fact that the natural alloy of iron and nickel previously found in Josephine County, Oregon, and in Del Norte County, California, has also been found in commercial quantities in the Fraser River at Lillooet. An interesting and new occurrence of platinum in place in Sumatra has been noted by Prof. L. S. Hundeshagen. All the American platinum came from California and Oregon, inasmuch as operations have been suspended in the Rambler copper mines, Wyoming, which furnished some platinum the year before.

The imports of platinum into the United States during 1905 showed a decline of more than 8000 ounces due to European control of the supply, which also, of course, aided the rise in price. The present prices are the highest that platinum has commanded in recent years.

This brief paper of Dr. Day's is published as an extract from the Survey's annual volume, "Mineral Resources of the United States, 1904." Copies may be obtained, free of charge, on application to the Director of the Geological Survey, Washington, D. C.

Book Notices.

Electricity in Every-Day Life. In three volumes. By Edwin J. Houston, Ph. D. (Illustrated.) New York: P. F. Collier & Son.

The three volumes, constituting this latest contribution to electrical literature, made by the author, may be unreservedly recommended to the non-professional reader, as a reliable source of information and instruction upon the almost numberless and various applications of electricity in daily life.

The author has spent many years of his life as a teacher, and this fact doubtless explains his mastry of the art of elucidating his facts and data so plainly as to bring even the most abstruse topics within the comprehension of the non-expert reader.

The present work is very comprehensive in its scopes, and it is no exaggeration to say that, as a popular exposition of the historical evolution and present state of the electrical arts, there is nothing in our literature to compare with it. W.

Sections.

SECTION OF PHYSICS AND CHEMISTRY.—*Stated Meeting,* held Thursday, February 1, 8 P.M. Dr. R. H. Bradbury in the chair.

Present, eighteen members and visitors.

The following officers were elected for the current year, viz.:—

President—Dr. R. H. Bradbury.
Vice-Presidents—Dr. H. F. Keller, Prof. G. A. Hoadley.
Secretary—Dr. E. A. Partridge.
Conservator—Dr. Wm. H. Wahl.

The first communication of the evening was read by Dr. J. Merritt Matthews, on "The Analysis of Dyestuffs."

Mr. H. Clyde Snook presented a communication on "The Use of the Induction Coil with the Walter Schaltung," and gave an exhibition and demonstration with an apparatus of this description.

Both communications were freely discussed. The meeting passed a vote of thanks to the speakers of the evening, and adjourned.

E. A. PARTRIDGE, *Secretary.*

Special Meeting, held Thursday, February 15, 8 A.M. Dr. Edward Goldsmith in the chair.

Present, thirty-four members and visitors.

Prof. A. J. Henry, U. S. Weather Bureau, read an illustrated paper, on Weather Forecasting from Synoptic Charts."

The thanks of the meeting were voted to the speaker. Adjourned.

E. A. PARTRIDGE, *Secretary.*

MINING AND METALLURGICAL SECTION.—*Stated Meeting*, held Thursday, January 25, 8 P. M. Mr. G. H. Clamer in the chair.

Present, twenty-six members and visitors.

The following officers were elected to serve for the current year:
President—James Christie.
Vice-Presidents—G. H. Clamer, A. E. Outerbridge, Jr.
Secretary—S. S. Sadtler.
Conservator—Wm. H. Wahl.

Dr. Wm. Campbell, of Columbia University, New York, read the paper of the evening entitled, "Some Notes on the Structure of Iron and Steel," profusely illustrated with the aid of photo-micrographs. The thanks of the section were extended to the speaker. Adjourned.

WM. H. WAHL, *Sec'y pro tem.*

ELECTRICAL SECTION. *Stated Meeting*, held Thursday, February 8th, 8 P.M. Mr. Thomas Spencer in the chair.

Present, twenty-six members and visitors.

Mr. E. D. Hays, representing the Cooper-Hewitt Electrical Co., of Philadelphia, gave an exhibition and demonstration of the Cooper-Hewitt Mercury Vapor Lamp.

This communication was followed by an exhibition and description of a series of electric heating and cooking appliances by Messrs. Walker & Keppler, of Philadelphia.

The meeting thereupon adjourned.

RICHARD L. BINDER, *Secretary.*

Franklin Institute.

(Proceedings of the stated meeting held Wednesday, February 21, 1906.)

HALL OF THE FRANKLIN INSTITUTE,
PHILADELPHIA, February 21, 1906.

MR. JAMES CHRISTIE in the chair.

Present, forty-eight members and visitors.

Additions to membership since last report, fourteen.

In commemoration of the 200th anniversary of the birth of Franklin, the evening was devoted to an address by Dr. Edwin J. Houston, honorary member of the Institute and Professor Emeritus of Physics, on "Franklin as a Man of Science and Inventor."

The address will appear in due course in the *Journal.*

The speaker was accorded a unanimous vote of thanks, and the meeting was adjourned.

WM. H. WAHL, *Secretary.*

Committee on Science and the Arts.

(Abstract of Proceedings of the stated meeting held Monday, February 7, 1906.)

Dr. Edward Goldsmith in the chair.

The following reports were adopted:

(No. 2365.) *Speed-Jack*, C. J. Reed, Philadelphia.

Abstract: The invention is covered with five letters patents of the United States issued in 1904 and 1905, to the applicant.

A full description of the device with illustrations will be found in the *Journal* for November, 1904, to which reference is made for details of mechanical construction and operation.

Referring to the operative features of the apparatus the committee states in its report * * * "While the uses for the Speed-Jack under consideration are not exactly the same as those to which a variable speed countershaft can be placed * * * it is designed to accomplish the same result. * * * The jack tested by your committee gave excellent results under very trying conditions and appeals to us as being a thoroughly practical equipment, well suited to meet many of the conditions requiring mechanical speed control." The award of the John Scott Legacy Premium and Medal is recommended to C. J. Reed, the inventor. (*Sub-Committee*, Charles Day, Chairman; Kern Dodge, Chas. E. Ronaldson, Hugo Bilgram.)

(No. 2371.) *Friction Indicator*, Carl B. Weidlog, Sag Harbor, New York.

Abstract: This invention is covered by letters patent of the United States, No. 741,087, Oct. 13, 1903, granted to applicant.

It relates to an improved indicator whereby the depth to which a hole is being drilled can be accurately determined, and whereby a group or number of holes can be drilled to the same or any specified depth.

The device consists of a bracket adapted to be secured to the bearing of a drill-bar, supporting, on ball-bearings, an arbor on which is mounted a circular disc of hard fiber having a circumference of precisely four inches. The ball-bearing itself is supported on a swivel, so that by means of a spring the hard fiber disc can be pressed against the spindle bar of the drill-press, to be set in motion by the downward motion of the bar.

By means of an index finger secured to the arbor and registering with a stationary dial, the distance of the downward motion of the drill can be read off. If a hole of a given depth is to be drilled, the index is set to zero when the drill is in contact with the work. As the drilling proceeds, the increasing depth of the hole is then registered on the dial. * * *

The device was thoroughly tested in the workshop of Mr. Hugo Bilgram, and his report of its performance which proved very satisfactory is attached as an appendix to the committee's report.

The report concludes with the statement that "the device is very useful for the purpose designed; it is well made, accurate in its registration, and a time saver. The inventor is awarded the Edward Longstreth Medal of Merit. (*Sub-Committee*, Chas. E. Ronaldson, Chairman; Hugo Bilgram).

(No. 2375.) *Quartz-Glass Mercury Lamp.* W. C. Heraeus, Hanau, Germany.

In the opinion of the sub-committee investigating this invention its main value consists in the introduction of Quartz-Glass as a material for the bulbs of mercury-vapor lamps and in the truly remarkable skill displayed in the production of these quartz-glass vessels.

As these features, in the committee's opinion, constitute a very considerable step in advance in the construction of mercury lamps, particularly for experimental purposes, the award of the John Scott Legacy Premium and Medal is recommended to be made to W. C. Heraeus, of Hanau, Germany, its inventor. (*Sub-Committee*, Harry F. Keller, Chairman; Thomas Spencer, W. J. Williams.)

(No. 2376.) *Rapid-Fire Gun.* Victor P. De Knight, Washington, D. C. (An advisory report.)

(No. 2381.) *Concrete Pile.* Alexander C. Chenowith, Brooklyn, N. Y.

ABSTRACT: This invention is covered with two United States patents. No. 791,076, May 30th, 1905, and No. 797,556, Aug. 22d, 1905, the former covering the product and the latter the process of manufacture.

The details of this invention will be unintelligible without the aid of illustrations. The publication, therefore, will be made in full in the *Journal* in due course.

The conclusions of the sub-committee are as follows: In view of the apparent novelty of the method, and of the strength and durability of the product the award of the John Scott Legacy Medal and Premium is recommended to be awarded to the inventor. (*Sub-Committee*, Lewis M. Haupt. Chairman; Louis E. Levy). W.

BENJAMIN FRANKLIN—MAN OF SCIENCE AND INVENTOR

JOURNAL

OF THE

FRANKLIN INSTITUTE

OF THE STATE OF PENNSYLVANIA

FOR THE PROMOTION OF THE MECHANIC ARTS

| VOL. CLXI, No. 4 | 81ST YEAR | APRIL, 1906 |

The Franklin Institute is not responsible for the statements and opinions advanced by contributors to the *Journal*.

THE FRANKLIN INSTITUTE.

Franklin as a Man of Science and an Inventor.*

[An address delivered by Dr. Edwin J. Houston, Emeritus Professor of Physics, Franklin Institute, on February 21, 1906, on the occasion of the 200th anniversary of the birth of Benjamin Franklin.]

[In this review of Benjamin Franklin's activity in the domain of the Physical Sciences, the author presents a compendium of authoritative data on all the many aspects of his subject and elucidates the data by a comprehensive and critical analysis of its various features. Dr. Houston's work deals with the whole range of Franklin's experiments and inventions and especially with those which so signally opened the way for the development of modern electrical technology.—THE EDITOR.]

It gives me sincere pleasure, ladies and gentlemen, on returning, after an absence of so many years, to the lecture room of the Franklin Institute, where I spent so many hours of my earlier life, to be assigned the agreeable task of addressing you on Benjamin Franklin as a Man of Science and an Inventor.

---

*Copyrighted by Edwin J. Houston, 1906.

One can scarcely appreciate the magnitude of the task assigned me by the title of this address.

As you all know, there stands in front of the U. S. Post Office Building, Ninth and Chestnut Streets, in this City, a statue of Benjamin Franklin. The pedestal contains the following inscription:

<blockquote>
BENJAMIN FRANKLIN
1706—1790
Venerated
For Benevolence
Admired for Talents
Esteemed for Patriotism
Beloved for
Philanthropy
</blockquote>

It is to Benjamin Franklin, who may properly be regarded as the most distinguished man of science that this country has ever produced, that I desire to call your attention.

Franklin was a many-sided man. I know of no living man I can at all compare with Benjamin Franklin, unless it is that 20th Century wonder, Thomas Edison. The two men are alike in many respects. Benjamin Franklin, a utilitarian, an apostle of thrift, a self-educated man. Thomas Edison, eminently utilitarian, as well as self-educated. Perhaps it may suffice if I show a single respect in which Franklin and Edison agree. If you will take the trouble to look even carelessly over the ten volumes on "The Works of Franklin," by Jared Sparks, or over the still more recent volumes of Prof. Smyth, of the High School, on "Benjamin Franklin," you will be able to gain some idea of the many-sidedness of Franklin, as well as the wide intellectual field he occupied. Perhaps this alliterative statement may aid you, since it briefly describes the varied powers of the man: Benjamin Franklin: Printer; Philosopher; Patriot; Philanthropist; Plenipotentiary; Postmaster; Politician in the purest and best sense of the word. or this simple statement: Benjamin Franklin: Writer; Physicist; Inventor; Statesman; and Man of Affairs.

As we contemplate the varied work of the man, we are apt to inquire: What single quality permitted this wonderful and varied activity? It is not difficult to answer this question. Benjamin Franklin was a man who thoroughly appreciated the value of time, and who always made the most of the time he had at his disposal.

It is unnecessary for me to attempt to point out the many-sided character of Edison, since, as he lives in your time, you can hardly fail to be generally conversant with his work.

While it is far from my intention to weary you to-night with any extended dissertation respecting Franklin's economic philosophy as set forth in "Poor Richard's Almanac," yet I believe we can obtain no little insight into the cause for his success if you permit me to read a few quotations from "Poor Richard" respecting the husbanding of time:

"But dost thou love Life? Then do not squander time; for that is the stuff life is made of. How much more than is necessary do we spend in sleep!"

"The sleeping fox catches no poultry."

"There will be sleeping enough in the grave."

"If time be of all things the most precious, wasting time must be the greater prodigality."

"Lost time is never found again, and what we call time enough, always proves too little."

"He that riseth late, must trot all day, and shall scarce overtake his business at night."

"One to-day is worth two to-morrows."

"Never leave that until to-morrow which you can do to-day."

Methinks I hear some one of you say: "Must a man afford himself no leisure?" I will tell you, my friend, what "Poor Richard" says: "Employ thy time well if thou meanest to gain leisure; and since thou art not sure of a minute, throw not away an hour."

Thomas Edison holds to similar views. When once asked as to the principal inspiration that led to the wonderful results that he has achieved in applied physical science, he remarked: "It is not inspiration, but perspiration." Edison, too, holds similar views as to the loss of time that results from too much sleeping. It is not unusual, during the progress of an exacting investigation, that several consecutive days pass without his going to bed.

I warn you beforehand that, owing to the great scope of Franklin's work, covering, as it does, the fields of electricity, geographical physics, applied physics, as well as in other directions, it is only possible for me to very briefly indicate but a small portion of this work.

Franklin's first interest in the subject of electricity appears to have been excited in 1746, during a lecture on Electricity, delivered in Boston by a Dr. Spence. Franklin was, therefore, about forty years old when he first began the study of electricity.

As some of you may know, Franklin had formed a literary club, called the Junto, consisting of a few kindred spirits. This club had gathered together the few books they owned, and placed them in the meeting room at their club-house, where they used them in common. But in those days books were expensive, and were, moreover, ponderous, and not well suited for circulation; so that their library was not only small, but it was unsuited for circulation. In order to overcome these difficulties, Franklin conceived the idea of establishing a circulating library, and organized a regular association or company for this purpose. This company afterward became the great Library Company of Philadelphia.

At the Junto, essays, prepared by the club members, were read, and discussions held on the papers. On some occasions, simple experiments were performed for the purpose of illustrating the subject under discussion. Consequently when Franklin received from his friend, Peter Collinson, of London, a simple device in the form of an electrical tube, for the production of electricity by friction, not only were many experiments inaugurated before the Junto, but an extended series of investigations were made by Franklin, that resulted in obtaining for him a world-wide reputation as an experimental philosopher.

The apparatus employed by Franklin was not only limited in extent, but was of an exceedingly crude type. It is not, however, the apparatus that is of the greatest significance in the case of an experimental philosopher, but the type of brain power that directs its use, and then sits in deliberate judgment as to the most probable causes of the phenomena that have been obtained by its use.

It will be impossible to attempt any detailed description of the numerous experimental investigations made by this profound philosopher. We must content ourselves with a brief discussion of only some of the more important of them.

Practically, Franklin began his study of electricity by studying the power of points. He communicated his observations on this matter in a letter to Collinson, dated July 11, 1747. This was the second letter sent to Collinson after the receipt of the electric tube. In a former letter, dated March 28, 1747, in which he acknowledged the receipt of the tube, Franklin refers to the great interest that he took in electric experiments. In his letter he says:

"For my own part, I never was before engaged in any study that so totally engrossed my attention and my time as this has lately done, for, what with making experiments when I can be alone, and repeating them to my Friends and Acquaintances, who, from the novelty of the thing, come continually in crowds to see them, I have, during some months past, had little leisure for anything else."

The second letter of Franklin to Collinson, above referred to, contains so much interesting matter that I feel warranted in having parts of it printed in full:

July 11, 1747.

"SIR:

"In my last, I informed you that, in pursuing our electrical enquiries, we had observed some particular Phaenomena, which we looked upon to be new, and of which I promised to give you some account, though I apprehended they might possibly not be new to you, as so many hands are daily employed in electrical experiments on your side of the water, some or other of which would probably hit on the same observations.

"The first is the wonderful effect of pointed bodies, both in *drawing off and throwing off* the electrical fire. For example,

(a) "Place an iron shot of three or four inches diameter on the mouth of a clean dry glass bottle. By a fine, silken thread from the ceiling, right over the mouth of the bottle, suspend a small cork-ball, about the bigness of a marble; the thread of such a length, as that the cork-ball may rest against the side of the shot. Electrify the shot, and the ball will be repelled to the distance of four or five inches, more or less, according to the quantity of electricity. When in this state, if you present to the shot the point of a long slender sharp bodkin, at six or eight inches distance, the repellency is instantly destroyed, and the cork flies to the shot. A blunt body must be brought within an inch, and draw a spark, to produce the same effect. (b) To prove that the electrical fire is *drawn off* by the point, if you take the blade of the bodkin out of the wooden handle, and fix it in a stick of sealing-wax, and then present it at the distance aforesaid, or if you bring it very near, no such effect follows; but sliding one finger along the wax till you touch the blade, and the ball flies to the shot immediately. —— (c) If you present the point in the dark, you will see, sometimes at a foot distance, and more, a light gather upon it, like that of a fire-fly, or glow-worm; the less sharp the point, the nearer you must bring it to observe the light; and at whatever distance you see the light, you may draw off the electrical fire, and destroy the repellency. —— If a cork-ball so suspended be repelled by the tube, and a point be presented quick to it, tho' at a considerable distance, 'tis surprising to see how suddenly it flies back to the tube. Points of wood will do near as well as those of iron, provided the wood is not dry; for perfectly dry wood will no more conduct electricity than sealing-wax.

(d) "To show that points will *throw off* as well as *draw off* the electrical fire, lay a long, sharp needle upon the shot, and you cannot electrise the shot, so as to make it repel the cork-ball. —— Or fix a needle to the end of a suspended gun-barrel, or iron rod, so as to point beyond it like a little bayonet; and while it remains there, the gun-barrel, or rod, cannot by applying the tube to the other end be electrised so as to give a spark, the fire continually running out silently at the point. In the dark you may see it make the same appearace as it does in the case before mentioned.

(e) "The repellency between the cork ball and the shot is likewise destroyed. 1. By sifting fine sand on it; this does it gradually. 2. By breathing on it. 3. By making a smoke about it from burning wood. 4. By candle light, even though the candle is at a foot distance: these do it suddenly. —— The light of a bright coal from a wood fire; and the light from a red-hot iron do it likewise; but not at so great a distance. Smoke from dry rosin dropped on hot iron, does not destroy the repellency; but is attracted by both shot and cork-ball, forming proportionable atmospheres around them, making them look beautifully, somewhat like some of the figures in *Burnet's* or *Whiston's* theory of the earth.

"N. B.—This experiment should be made in a closet, where the air is very still, or it will be apt to fail.

"The light of the sun thrown strongly on both cork and shot by a looking glass for a long time together, does not impair the repellency in the least. This difference between fire-light and sun-light is another thing that seems new and extraordianry to us."

(a) Note here the simplicity of the apparatus employed. An iron shot, three or four inches in diameter, is insulated by being placed on a clean dry glass bottle; a small cork ball about the size of a boy's marble, suspended by a silk thread from the ceiling of the room, so that it rests against the side of the iron shot, is employed in order to indicate whether the shot is electrified or not, since the apparatus is capable of acting as a simple form of electroscope. Under these conditions, on the electrification of the shot, the cork is instantly repelled a distance of three or four inches. In order to see the effects produced by the approaching of metallic bodies, *i. e.*, if bodies that possess the power of rapidly conducting electricity, Franklin brings the point of a long, slender, sharp bodkin to a distance of from six to eight inches from the shot. The effect is marked. The shot almost instantly loses its charge, as is indicated from the fact that it no longer repels the cork, which now falls until it touches the shot.

Does this discharge of the electrified shot result only from the fact that it is formed of conducting material, or is it also, and possibly mainly, determined by its shape. In order to settle this question, the same experiment is repeated. Now, however, a blunt metallic rod, instead of a sharp piece of pointed metal, is approached to the shot. No result is produced until the distance is sufficiently decreased to permit a spark to pass between the electrified shot and the blunt body. Clearly, then, the shape of the approaching body, *i. e.*, its pointed condition, is also an important determining factor.

(b) The same experiment is repeated, except, however, that a blunt instead of a sharp metallic body is approached to the cork. No effect is produced, until the distance is so short that a spark passes between the shot and the blunt body. Clearly, then, the shape of the approached body, *i. e.*, its pointed condition, is an important determining factor. On the approach of the blunt body, the electrified shot loses its electricity by an apparently single discharge, that is attended by a spark and a crackling sound. What is the difference in the character of this discharge and that which occurs on the approach of the pointed conductor? Does not the electricity continually pass off from the point in small quantities, and thus reach the ground?

In order to determine this question, Franklin contrives a charming experiment which convinces him that the electricity is drawn off by the point, or, as he calls it, in the form of electrical fire. He removes the blade of the bodkin from the wooden handle, which is but a poor insulator, and replaces it with a stick of sealing wax. Then, holding the bodkin by the sealing wax, he brings it as before within six or eight inches from the shot, or even nearer, without any effect following. Here the conditions are the same as before, except that the sealing wax prevents the establishment of a conducting path to the ground. In order to convince himself that this result was due to the inability of the electric charge to pass along the bodkin to the wax support, and thus through his body to the ground, he slides one finger along the wax until the blade is touched, when instantly, a conducting path thus being established to the earth, a discharge of the shot and repulsion of the ball instantly ceases, the ball falling to the shot.

(c) Franklin repeats these experiments in the dark, and notes that a faint light can be observed as gathering on the point employed, somewhat resembling the light emitted by a fire-fly or glow-worm. He calls attention to the fact that the less sharp the point, the nearer it must be brought to the ball to be able to observe the light.

(d) But what effect would be produced if the charged body is itself pointed? To determine this, a long, sharp needle is laid on the shot. Under these conditions it is impossible to electrify the shot so as to repel the cork ball; for it parts with its charge as soon as it is received. Franklin gives to his friend, Mr. Thomas Hopkinson, the credit for first making this observation. Particular attention is called to Franklin's ingenious experiment of discharging the electrified shot by means of the smoke obtained from burning wood, or from the emanations given out by a lighted candle.

As intimated in Franklin's first letter to Collinson, these experiments, besides being tried before the Junto, were repeated to select audiences of friends and acquaintances. It is quite possible, by a careful reading of the latter part of Franklin's second letter to Collinson, to review some of the curious experiments that this ingenious man tried before these audiences. I will enumerate a few of these experiments, without any particular reference to the order in which Franklin described them:

(1) A lighted candle, when just blown out, is relighted by causing a spark to pass through the smoke between the wire and the snuffers. Here was first produced, although in a somewhat different manner, the well-known modern method of the electric ignition of gas lights by causing the spark of a sparkcoil to pass through the jet of gas as it is issuing from a gas burner.

(2) Artificial lighting (and bear in mind this was in the year 1747, before it was actually known that the lightning flash is identical with an electric discharge) was produced by a wire carrying the current over a china plate decorated with gilt flowers; or similarly, by applying the wire to the gilt frames of looking-glasses. "We represent lightning," says Franklin, "by passing the wire in the dark over a china plate that has gilt flowers, etc."

(3) Place a person on a cake of wax (Franklin's simple form of insulating stool). Cause him to hold the electrified bottle in his hand, touching only the outer coating of the bottle. Then touch the wire with your finger, and bring it near his hand or face, when sparks are produced at each approach. Franklin explains this result in accordance with the single-fluid theory, as follows: By taking the spark from the wire, the electricity inside the bottle is decreased. The outside of the bottle then draws electricity from the person holding it, leaving him in the negative state; thus, when his hand or face is touched, an equal quantity is restored to him by the person touching.

(4) A small piece of burnt cork, so as to resemble a spider, is provided with legs of linen thread, and weighted with a grain or two of lead placed in the body. The counterfeit spider is hung over a table, on which an upright wire is stuck, as high as the Leyden phial and wire, two or three inches from the suspended mock spider. The spider is then animated by setting a charged Leyden jar at the same distance on the other side. The spider immediately jumps to the wire of the Leyden jar, bends his legs in touching it; then springs off and flies to the wire on the table; then, again, to the wire of the Leyden jar, playing with his legs against both in a very entertaining manner, and appearing perfectly alive. A single charge of the Leyden jar, in dry weather, will enable him to continue this motion for an hour or longer.

(5) A book, provided with a double line of gold around the covers, is electrified on a plate of glass in the dark. On the application of a knuckle to the gilding, the fire appears everywhere on the gold, like a flash of lightning, but not on the leather, and not even if the leather is touched instead of the gold.

(6) The electrified bumper. This consists of a small, thin,

glass tumbler, nearly filled with wine, and electrified in the same manner as the Leyden jar. When brought near to the lips, a sharp electric shock is given, provided the person is closely shaved, and does not breathe on the liquor.

(7) The magic picture, an ingenious device whereby a picture, say of the King of England, constitutes in reality a Leyden jar, the outer and inner coatings consisting of concealed gold leaf placed on parts of the opposite sides of the glass plate employed for covering the picture. This was presented to a person to hold in such a manner that his hand would come in contact with one coating, while the crown of the King was placed in contact with the opposite coating. On the attempt of the person holding the picture to remove the crown from the King's head, he would receive a shock as punishment for his attempted treason. Franklin, on the other hand, holding the picture by some part not in contact with one of the coatings, could safely touch the crown.

(8) The electric jack, an ingeniously constructed electrostatic motor that was capable of rotating, even when loaded with a turkey properly prepared for cooking and placed before a fire. It was this apparatus that Franklin refers to, in a letter dated April 29, 1749, as follows: "A turkey is to be killed for our dinner by the electrical shock, roasted by the electrical jack before a fire kindled by the electrical bottle."

(9) The electrical ignition of gunpowder. The following is Franklin's description of this experiment: "A small cartridge is filled with dry powder, hard rammed, so as to bruise some of the grains; two pointed wires are then thrust in, one at each end, the points approaching each other in the middle of the cartridge till within the distance of half an inch; then, the cartridge being placed in the circle, when the four jars are discharged, the electric flame leaping from the point of one wire to the point of the other, within the cartridge amongst the powder, *fires it*, and the explosion of the powder is at the same instant with the crack of the discharge."

Franklin says, concerning this turkey, that he conceits himself that the bird killed by the shock of the Leyden jar battery was uncommonly tender. In a subsequent publication he suggests that this increase in tenderness might have been due to the rending and tearing of the flesh during the discharge, in the manner well known to occur in the case of the passage of a discharge through such solids as wood, etc. He also attributes the tendency of an animal killed by an electric shock to undergo rapid decomposition to a similar cause.

As already remarked, these experiments were made by Franklin, for the greater part, during the first half of the year 1747. It will be remembered it was towards the close of 1745 that a discovery was made in electricity that greatly excited all electricians. This was the discovery of the Leyden jar, according to some by Von Kleist, Bishop of Pomerania, and according to others by Cuneus, of Leyden, a pupil of Muschenbroeck.

Quite naturally, we find that Franklin made a somewhat extended investigation of this electric device. Franklin's theory of electricity, as we shall shortly explain, known generally as the single-fluid theory, recognized the presence of but a single electric fluid, the differences of positive and negative excitement being, in his opinion, due to a greater and a smaller amount respectively of a hypothetical electric fluid. Consequently, as he soon discovered that when the inner coating of a Leyden jar was positively electrified, the outer coating was negatively electrified, he naturally came to the conclusion that in charging the jar, the electricity passed out from the negative coating, and entered the positive coating. In another letter to Collinson, dated September 1st, 1747, we find that Franklin soon reached the conclusion that the electric charge in the conducting coating of a Leyden jar differed from the charge in a conductor quite separated and distinct from the jar, in that in the jar the electric charge is accumulated on the surfaces of the non-electric, or, quoting the language of this letter:

(a) "The non-electric contain'd in the bottle differs when electrised from a non-electric electrised out of the bottle, in this: that the electrical fire of the latter is accumulated *on its surface*, and forms an electrical atmosphere round it of considerable extent; but the electrical fire is crowded *into the substance* of the former, the glass confining it.

(b) At the same time that the wire and top of the bottle, &c., is electrised *positively* or *plus*, the bottom of the bottle is electrised *negatively* or *minus*, in exact proportion: *i. e.*, whatever quantity of electrical fire is thrown in at the top, an equal quantity goes out of the bottom. To understand this, suppose the common quantity of electricity in each part of the bottle, before the operation begins, is equal to 20; and at every stroke of the tube, suppose a quantity equal to 1 is thrown in; then, after the first stroke, the quantity contain'd in the wire and upper part of the bottle will be 21, in the bottom 19. After the second, the upper part will have 22, the lower 18, and so on, till, after 20 strokes, the upper part will have a quantity of electrical fire equal to 40, the lower part none: and then the operation ends: for no more can be thrown into the upper part, when no more can be driven out of the lower part. If you attempt to throw more in, it is spued back through the wire, or flies out in loud cracks through the sides of the bottle."

(a) The non-electric here referred to was one of the coatings of the Leyden jar. From a reading of this paragraph, it will be observed that Franklin believed that the electricity in a charged Leyden jar resided in the surface of the coating, forming an electrified atmosphere around it, and being kept in the coating in a crowded state where it was confined by the glass; i. e., that "The electrical fire is crowded into the substance of the former (the conducting coating), the glass confining it." As we shall see, Franklin subsequently corrects this error by stating that the power of the Leyden jar to give shocks is due to the electricity accumulating on the opposite surfaces of the glass itself, the conducting coatings in contact with the two surfaces only serving to collect the charges from the innumerable points on the glass.

(b) It will be noticed here that Franklin is applying his single-fluid theory of electricity to this phenomena, since, of course, if the inner coating of the jar is electrified positively or plus, the outer coating or bottom must be electrified negatively or minus; for, from his method of looking at the phenomena, whatever amount of electricity is thrown in at the top would necessarily require an equal amount to go out at the bottom, and we find him endeavoring to figure up the condition of affairs that would result when the jar was charged to as full an extent as his electric source would permit.

In another portion of this letter to Collinson, Franklin describes the following experiments he made on the Leyden jar:

"EXPERIMENT I.

(a) "Place an electrised phial on wax: a small cork-ball suspended by a dry silk thread held in your hand, and brought near to the wire, will first be attracted, and then repelled; when in this state of repellency, sink your hand, that the ball may be brought towards the bottom of the bottle: it will be there instantly and strongly attracted, 'till it has parted with its fire.

"If the bottle had a *positive* electrical atmosphere, as well as the wire, an electrified cork would be repelled from one as well as from the other.

"EXPERIMENT II.

(b) "Fig. 1 (a) From a bent wire sticking in the table, let a small linen thread (b) hang down within half an inch of the electrised phial (c). Touch the wire of the phial repeatedly with your finger, and at every touch you will see the thread instantly attracted by the bottle. (This is best done by a vinegar cruet, or some such belly'd bottle.) As soon as you draw any fire out from the upper part, by touching the wire, the lower part of the bottle draws an equal quantity in by the thread.

"EXPERIMENT III.

(c) "Fig. 2. Fix a wire in the lead, with which the bottom of the bottle is armed (d) so as that bending upwards, its ring-end may be level with the top or ring-end of the wire in the cork (e), and at three or four inches distance. Then electrise the bottle, and place it on wax. If a cork sus-

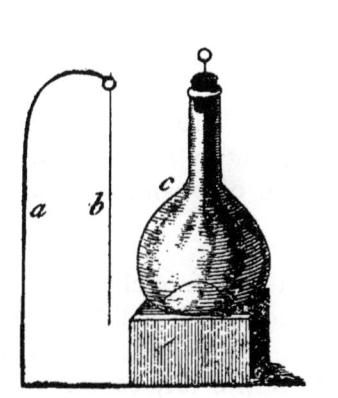

Fig. 1. Apparatus referred to in Franklin's Experiment.

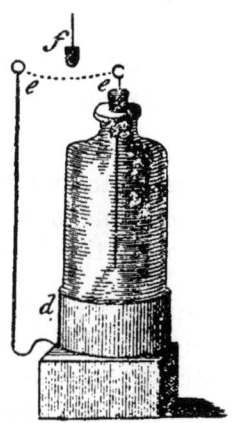

Fig. 2. Apparatus referred to in Franklin's Experiment.

pended by a silk thread (f) hang between these two wires, it will play incessantly from one to the other, 'till the bottle is no longer electrised; that is, it fetches and carries fire from the top to the bottom of the bottle, 'till the equilibrium is restored."

(a) In this manner, Franklin proves that the inside and outside of the jar, *i. e.*, its inner and outer coatings, are oppositely electrified, the inside positively and the outside negatively.

(b) Here the bent wire is electrically connected with the ground, being inserted in the top of an ordinary table. A cork ball is suspended by means of a conducting thread of linen. Every time the inner coating is touched by the hand, the thread is instantly attracted or drawn to the bottle. Franklin here considers, in accordance with his single-fluid hpyothesis, that the electric fluid is drawn out from the inner coating of the jar by touching it with the hand, an equal quantity of electricity is drawn into the lower part or outer coating of the jar when it is touched by the linen thread.

(c) In this beautiful experiment, Franklin is able to discharge a Leyden jar slowly by means of a great number of successive discharges between the inner and the outer coatings. The wire, instead of being supported on the table, as in the preceding experiment, is electrically connected with the lead that forms the outer coating of the jar. The cork is suspended by means of a silken thread midway between the two smooth metallic spheres e, e, (Fig. 2) are connected respectively with the inner and outer coatings of the jar, that it becomes midway between these two spheres. When this jar is electrified and placed on an insulating stand of wax, the cork ball continues to move between the two balls until the jar is entirely discharged.

In another letter to Collinson, dated only with the year, 1778, but apparently following the preceding letter from some reference it contains, a description is given of Franklin's further investigation of the Leyden jar.

As a result of a number of experiments made in charging and discharging the Leyden jar, which, of course, it will be impossible to discuss here for want of space, Franklin soon reaches the conclusion that the entire force of the jar, so far as its power of giving a shock is concerned, lies in the glass itself, the coating only serving the purpose of coming in contact with the sides of the glass in such a manner as to give and to receive the electricity from the several parts of the glass. This matter is referred to by Franklin as follows in this letter to Collinson, which may be spoken of as the fourth Collinson letter:

"Thus, the whole force of the bottle and power of giving a shock, is in the glass itself; the non-electrics in contact with the two surfaces, serving only to *give* and *receive* to and from the several parts of the glass; that is, to give on one side, and take away from the other.

(a) "This was discovered here in the following manner: Purposing to analyse the electrified bottle, in order to find wherein its strength lay, we placed it on glass, and drew out the cork and wire, which for that purpose had been loosely put in. Then taking the bottle in one hand, and bringing a finger of the other near its mouth, a strong spark came from the water, and the shock was as violent as if the wire had remained in it, which showed that the force did not lie in the wire. Then, to find if it resided in the water, being crowded into and condensed in it, as confin'd by the glass, which had been our former opinion, we electrified the bottle again, and placing it on glass, drew out the wire and cork as before; then taking up the bottle, we decanted all its water into an empty bottle, which likewise stood on glass; and taking up that other bottle, we expected, if the force resided in the water, to find a shock from it; but there was none. We judged then that it must either be lost in decanting, or remain in the first bottle. The latter we find to be true; for that bottle on trial gave the shock, though filled up as it stood with fresh unelectrified water from a tea-pot. —— To find, then, whether glass had this property merely as glass, or whether the form contributed anything to it; we took a pane of sash-glass, and laying it on the hand, placed a plate of lead on its upper surface; then electrified that plate, and bringing a finger to it, there was a spark and shock. We then took two plates of lead of equal dimensions, but less than the glass by two inches every way, and electrified the glass between them, by electrifying the uppermost lead; then separated the glass from the lead, in doing which, what little fire might be in the lead was taken out, and the glass being touched in the electrified parts with a finger, afforded only very small pricking sparks, but a great number of them might be taken from different places. (b) Then dextrously placing it again between the leaden plates, and compleating a circle between the two surfaces, a violent shock ensued. —— Which demonstrated the power to reside in glass as glass, and that the non-electrics in contact served only, like the armature of a loadstone, to unite the force of the several parts, and bring them at once to any point desired: it being the property of a non-electric, that the whole body instantly receives or gives what electrical fire is given to or taken from any one of its parts."

(a) The methods followed by Franklin in this investigation give a wonderful insight into the ability of the man as an experimental philosopher. He electrifies a Leyden jar, and proposes to himself to ascertain wherein its peculiar strength lay. The jar is placed on glass for the purpose of insulating it.

In order to determine whether the force lay in the wire, which, passing through the cork at the mouth of the Leyden jar, extended down into the water which formed the inner coating of the jar, Franklin removes the cork and wire, which have been loosely placed in the jar. Then, holding the bottle in one hand, thus touching its outside coating, he brings a finger of the other hand near the mouth of the jar. As soon as the hand approaches the water, a spark comes from the water, and a shock is received, as violent as if the wire had remained in the jar. Consequently, Franklin concludes that the force did not lie in the wire.

"But," reasoned this philosopher, "this occult force may be in the water, being crowded into and condensed in it."

In order to test this possibility, the bottle is again electrified and placed on the glass, the wire and cork drawn out as before. The bottle is then carefully taken in the hand, and its water poured into an empty Leyden jar, likewise placed on the glass. Franklin takes up this second jar, confidently expecting, if the force resides in the water, to get a shock from it, but there was none. If, then, the electric force had resided in the water, it must have been lost in the act of decanting. If not in the water, it must remain in the first bottle. This bottle, on trial, gives a shock, though filled up as it stood with fresh unelectrified water from a teapot. The conclusion is inevitable, that the force resides in the glass.

Our philosopher is still unsatisfied. Is the power of the glass to retain the shock dependent on its nature as a material, or is it by reason of the shape of the bottle? He is here possibly thinking of his experiments with pointed conductors. An ingenious experiment is made, whereby a piece of ordinary window glass is momentarily converted into a Leyden jar, and a shock taken from it. A more carefully constructed, modified Leyden jar is then made by placing on the opposite sides of a plate of glass two plates of lead of equal size, but so much smaller than the plates of glass as to leave a free space two inches around them. On electrifying the glass between these coatings, and separating the glass from the lead, the glass is found to be electrified. In this manner, however, the glass is only charged locally.

(b) While the glass plate remains electrified, Franklin places it between two leaden plates, and shows that the charge still remains in the glass, receiving as he did a violent shock from the combination. Consequently, it is proved that the power of the Leyden jar lies in the glass, and is independent of the shape of the glass, the non-electrics or conducting plates merely serving, like the armature of a loadstone, "To unite the force of the several parts, and bring them at once to any one point desired."

In a subsequent note, Franklin describes what he believes to be the manner in which the Leyden jar or bottle receives an electric charge, in accordance with his hypothesis or theory of electricity. This note is as follows:

"Place a thick plate or glass under the rubbing cushion, to cut off the communication of electrical fire from the floor to the cushion; then, if there be no fine points or hairy threads sticking out from the cushion, or from the parts of the machine opposite to the cushion, (of which you must be careful) you can get but a few sparks from the prime conductor, which are all the cushion will part with.

"Hang a phial then on the prime conductor (a), and it will not change though you hold it by the coating. —— But

"Form a communication by a chain from the coating to the cushion, and the phial will charge.

"For the globe then draws the electric fire out of the outside surface of the phial, and forces it through the prime conductor and wire of the phial, into the inside surface.

"Thus the bottle is charged with its own fire, no other being to be had while the glass plate is under the cushion.

"Hang two cork balls by flaxen threads to the prime conductor; then touch the coating of the bottle, and they will be electrified and recede from each other.

"For just as much fire as you give the coating, so much is discharged through the wire upon the prime conductor, whence the cork balls receive an electrical atmosphere. —— But

"Take a wire bent in the form of a C (b), with a stick of wax fixed to the outside of the curve, to hold it by; and apply one end of this wire to the coating, and the other at the same time to the prime conductor, the phial will be discharged; and if the balls are not electrified before the discharge, neither will they appear to be so after the discharge, for they will not repel each other.

"Now if the fire discharged from the inside surface of the bottle through its wire, remained on the prime conductor, the balls would be electrified, and recede from each other.

"If the phial really exploded at both ends, and discharged fire from both coating and wire, the balls would be *more* electrified, and recede *farther*; for none of the fire can escape, the wax handle preventing.

"But if the fire, with which the inside surface is surcharged, be so much precisely as is wanted by the outside surface, it will pass round through the wire fixed to the wax handle, restore the equilibrium in the glass, and make no alteration in the state of the prime conductor.

"Acordingly we find, that if the prime conductor be electrified, and the cork balls in a state of repellency before the bottle is discharged, they continue so afterwards. If not, they are not electrified by that discharge."

(a) PRIME CONDUCTOR: That conductor of the electrical

machine which collects positive electricity. This part of the frictional machine was first employed by Bose.

(b) The wire bent in the form of the letter C, represents what is well known as the discharging rod. This apparatus is represented in Fig. 3. As will be seen, the ends of the wire are provided with smooth metallic spheres soldered to them. The stick of wax employed for supporting the wires is represented at H.

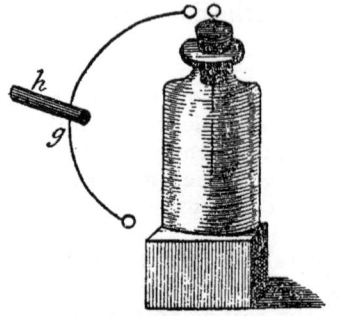

Fig. 3. Franklin's form of Discharging Rod.

At a later date, Franklin conducted some experiments with the view of ascertaining how long a Leyden bottle, properly charged and hermetically sealed, will retain its electricity:

"I formerly had an opinion that a Leyden bottle, charg'd and then seal'd hermetically, might retain its electricity forever; but having afterwards some suspicion that possibly that subtil fluid might, by slow imperceptible degrees, soak through the glass, and in time escape, I requested some of my friends, who had conveniences for doing it, to make trial, whether, after some months, the charge of a bottle so sealed would be sensibly diminished. Being at Birmingham, in September, 1760, Mr. Bolton of that place, opened a bottle that had been charged, and its long tube neck hermetically sealed in the January preceding. On breaking off the end of the neck, and introducing a wire into it, we found it possessed of a considerable quantity of electricity, which was discharged by a snap and spark. This bottle had lain near seven months on a shelf, in a closet, in contact with bodies that would undoubtedly have carried off all its electricity, if it could have come readily through the glass. Yet, as the quantity manifested by the discharge was not apparently so great as might have been expected from a bottle of that size well charged, some doubt remained whether part had escaped while the neck was sealing, or had since, by degrees, soaked through the glass. But an experiment of Mr. Canton's, in which such a bottle was kept under water a week, without having its electricity in the least impaired, seems to show, that when the glass is cold, though extremely thin, the electric fluid is well retained by it. As that ingenious and accurate experimenter made a discovery, like yours, of the effect of heat in rendering thin glass permeable by that fluid, it is but doing him justice to give you his account of it, in his own words, extracted from his letter to me in which he communicated it, dated Oct. 31, 1760, viz.

"'Having procured some thin glass balls, of about an inch and a half in diameter, with stems, or tubes, of eight or nine inches in length, I electrified them, some positively on the inside, and others negatively, after the manner of charging the Leyden bottle, and sealed them hermetically. Soon after I applied the naked balls to my electrometer, and could not discover the least sign of their being electrical; but holding them before the fire, at the distance of six or eight inches, they became strongly electrical in a very short time, and more so when they were cooling. These balls will, every time they are heated, give the electrical fluid to, or take it from other bodies, according to the plus or minus state of it within them. Heating them frequently, I find will sensibly diminish their power; but keeping one of them under water a week, did not appear in the least to impair it. That which I kept under water, was charged on the 22d of September last, was several times heated before it was kept in water, and has been heated frequently since, and yet it still retains its virtue to a very considerable degree. The breaking two of my balls accidentally, gave me an opportunity of measuring their thickness, which I found to be between seven and eight parts in a thousand of an inch.

"A down feather, in a thin glass ball, hermetically sealed, will not be affected by the application of an excited tube, or the wire of a charged vial, unless the ball be considerably heated; and if a glass pane be heated till it begins to grow soft, and in that state be held between the wire of a charged vial, and the discharging wire, the course of the electrical fluid will not be through the glass, but on the surface, round by the edge of it.'"

There were two rival hypotheses or theories of electricity that were entertained by physicists in the time of Franklin; i. e., the single-fluid hypothesis propounded by Franklin, and the double-fluid hypothesis propounded by Symmer of Dufay. Franklin's hypothesis, propounded in 1749,, asserted that all electrical phenomena were due to the presence of a single, extremely tenuous and practically weightless or imponderable electric fluid, that exists in all matter. That the particles of the electric fluid were mutually repellent, but were attracted by all kinds of matter. All kinds of matter are capable of containing a certain quantity of the electric fluid without manifesting any excitement. If, however, a body contains either a surplus or a deficit, it at once manifests electric excitement. During electrification by friction, an excess of the fluid is given to one body, thus imparting to it a positive excitement, leaving the other body with a deficit of the fluid, thereby rendering it negatively excited.

The double-fluid hypothesis assumed that all kinds of matter contained an indefinite quantity of an imponderable neutral electric fluid, formed by the combination of two separate electric fluids; i. e., the positive and the negative. That in unelectrified matter these two fluids combine and neutralize each other, the act of electrification consisting in their separation. According to this hypothesis, when a body is electrified by friction, the work done, resulting as it does in the separation of the two fluids, the rubber will retain one electric fluid and the thing rubbed the other electric fluid, so that the rubber and the thing rubbed acquire opposite electric excitements.

According to either theory, both the rubber and the thing rubbed will be electrified with opposite electrification. If, however, the thing rubbed consists of some conducting substances like metallic bodies, any electric excitement it attains will be rapidly lost by the escape of the fluid through the body of the person holding it to the ground. If, however, the piece of metal be insulated in any way, as by supporting it in a glass handle, the metal can be as readily electrified as a bit of glass or sealing wax.

Franklin formed his single-fluid hypothesis or theory of electricity very early in his experiments with the tube received from Mr. Collinson. Although this hypothesis is not credited at the present time, yet it forms a very convenient means for explaining phenomena, and we find that Franklin, in all of his subsequent work, constantly had resource to this theory in endeavoring to picture to himself and to his readers the manner in which the phenomena occurred; for example, in his second letter to Collinson, dated July 11, 1747, he speaks as follows regarding the manner in which he understands that a body is electrified by friction, and explains by this hypothesis the manner in which, as he understands it, it is possible to electrify people that are insulated from the earth; *i. e.*, while standing on cakes of wax.

(a) "We had for some time been of opinion, that the electrical fire was not created by friction, but collected, being really an element diffus'd among, and attracted by other matter, particularly by water and metals. We had even discovered and demonstrated its afflux to the electrical sphere, as well as its efflux, by means of little light windmill wheels made of stiff paper vanes, fixed obliquely and turning freely on fine wire axes. Also by little wheels of the same matter, but formed like water-wheels. Of the disposition and application of which wheels, and the various phaenomena resulting, I could, if I had time, fill you a sheet. The impossibility of electrising one's self (though standing on wax) by rubbing the tube, and drawing the fire from it; and the manner of doing it, by passing the tube near a per-

son or thing standing on the floor, etc., had also occurred to us some months before Mr. *Watson's* ingenious Sequel came to hand, and these were some of the new things I intended to have communicated to you.
—— But now I need only mention some particulars not hinted in that piece, with our reasonings thereupon; though perhaps the latter might well enough be spared.

(b) "1. A person standing on wax, and rubbing the tube, and another person on wax drawing the fire, they will both of them, (provided they do not stand so as to touch one another) appear to be electrised to a person standing on the floor; that is, he will perceive a spark on approaching each of them with his knuckle.

"2. But if the persons on the wax touch one another during the exciting of the tube, neither of them will appear to be electrised.

"3. If the touch of of one another after exciting the tube, and drawing the fire as aforesaid, there will be a stronger spark between them, than was between either of them and the person on the floor.

"4. After such strong spark, neither of them discover any electricity.

(c) "These appearances we attempt to account for thus: We suppose as aforesaid, that electrical fire is a common element, of which every one of the three persons above mentioned has his equal share, before any operation is begun with the tube. A, who stands on wax and rubs the tube, collects the electrical fire from himself into the glass: and his communication with the common stock being cut off by the wax, his body is not again immediately supply'd. B, (who stands on wax likewise) passing his knuckle along near the tube, receives the fire which was collected by the glass from A; and his communication with the common stock being likewise cut off, he retains the additional quantity received. —— To C, standing on the floor, both appear to be electrised: for he having only the middle quantity of electrical life, receives a spark upon approaching B, who has an over quantity; but gives one to A, who has an under quantity. If A and B approach to touch each other, the spark is stronger, because the difference between them is greater: After such touch there is no spark between either of them and C, because the electrical fire in all is reduced to the original equality. If they touch while electrising, the equality is never destroyed, the fire only circulating. (d) Hence, have arisen some new terms among us: we say, B, (and bodies like circumstanced) is electrised *positively*; A, *negatively*. Or rather, B is electrised *plus*; A, *minus*. And we daily in our experiments electrise bodies *plus* or *minus*, as we think proper. * * * To electrise *plus* or *minus*, no more needs to be known than this, that the parts of the tube or sphere that are rubbed, do, in the instant of the friction, attract the electrical fire, and therfore take it from the thing rubbing: the same parts immediately, as the friction upon them ceases, are disposed to give the fire they have received, to any body that has less. Thus you may circulate it, as Mr. *Watson* has shewn; as you connect that body with the rubber or with the receiver, the communication with the common stock being cut off. We think that ingenious gentleman was deceived when he imagined (in his Sequel) that the electrical fire came down the wire from the ceiling to the gun-barrel, thence to the sphere, and so electrised the machine and the man turning the wheel, etc. We suppose it was *driven off*, and not brought on through that wire; and that the machine and man, etc., were electrised *minus*; *i. e.*, had less electrical fire in them than things in common."

Here it will be noted that Franklin states he had, for a long time, believed that electricity is not produced by friction, but is collected from matter in which it has been originally diffused. In other words, the conception of the single-fluid theory of electricity evidently came with his first investigations.

He notes very clearly the fact that if two people are standing on wax at a short distance from each other, one rubbing the tube and the other taking the electric sparks that are produced by the friction, they will, provided that they do not touch each other, both appear to be electrified to a person standing on the floor; that is, either will give a spark to a person standing on the floor who approaches them with his knuckle. Neither of these people will appear electrified, although they stand on the wax, if they touch each other during the rubbing of the tube. If, however, they touch each other after the rubbing of the tube, although a spark will pass between them, yet after such spark passes, neither of them will manifest any electric excitement.

Note, now, Franklin's application of his single-fluid hypothesis to the above phenomena. Calling the persons that stand on the two cakes of wax A and B, and the person standing on the floor C, suppose A rubs the tube. According to Franklin, he simply passes the electric fluid from himself into the glass. Since his communication with the common stock of electric fluid is cut off by the insulating power of the wax, his body is not again immediately supplied with electricity. B, holding his knuckle near the tube, receives the electricity which was collected by A, on the glass, and his communication being cut off from the common supply, he retains this additional quantity. To C, standing on the floor, both appear to be electrified, for C, having only a neutral quantity of electricity, receives a spark on approaching B, who has a surplus of the fluid, and gives one to A, who has a deficit of the fluid. If A and B approach so as to touch each other, a spark passes, but after such passage, neither of them appear electrified, because they have regained the amount of electricity they originally possessed.

Franklin applies the term positively electrified to B; *i. e.*, the one who has received into his body an additional quantity of electricity, taking it, as he had, from the excited glass tube; therefore B, was also said to be electrified plus. A, from whose body the surplus of electricity received from B, was obtained, and who, therefore, had a deficit of electricity, was negatively excited, who had a minus electrification.

There can be no doubt that to Franklin is due the credit of the great discovery of the positive and negative condition of electric excitement. It is true that the honor of this discovery has been credited by the English to Dr. Watson, but Watson's paper on this subject is dated January 28th, 1748, while this letter of Franklin; *i. e.*, the second letter to Mr. Collinson, was dated July 11, 1747. Moreover, as will be noted in the beginning of this quotation, Franklin says: "We had for some time been of the opinion," etc.

At a date prior to July 11, 1747, when Franklin first published his distinction between positive and negative electric excitement, a Frenchman, Dufay, observed that two different kinds of electric excitement are produced by rubbing glass and sulphur respectively. He called these two kinds of electricity vitreous, or that produced by glass, and resinous, or that produced by resins, pointing out the fact that rock crystal, precious stones, the wool or hair of animals, together with many other bodies, likewise produce vitreous electricity, while such bodies as amber, copal, gum lac, silk, paper, thread, etc., produc resinous electricity. He, however, pointed out the fact that a body charged with vitreous electricity possessed the power of attracting all bodies charged with resinous electricity but repelled other bodies charged with vitreous electricity, and that, on the contrary, a body charged with resinous electricity attracted all other bodies charged with vitreous electricity, and repelled all bodies charged with resinous electricity.

Kindersley communicated Dufay's discovery to Franklin, a discovery which otherwise seemed to have remained unnoticed by the scientific world. Franklin studied these phenomena, and soon came to the conclusion that the vitreous and resinous electricity were, in reality, nothing other than positive and negative electricity.

At a later date, *i. e.*, 1749, Franklin published the following paper, describing more fully his single-fluid theory of electricity:

(a) Opinions and Conjectures, concerning Properties and Effects of the electrical matter, arising from Experiments and Observations, (b) made at Philadelphia, 1749.

"1. The electrical matter consists of particles extremely subtile. since it can permeate common matter, even the densest metals, with such ease and freedom as not to receive any preceptible resistance.

(c) "2 If any one should doubt whether the electrical matter passes through the substance of bodies, or only over and along their surfaces, a shock from an electrified large glass jar, taken through his own body, will probably convince him.

(d) "3. Electrical matter differs from common matter in this, that the parts of the latter mutually attract, those of the former mutually repel, each other. Hence the appearing divergency in a stream of electrified effluvia.

"4. But though the particles of electrical matter do repel each other, they are strongly attracted by all other matter.

5. From these three things, the extreme subtilty of the electrical matter, the mutual repulsion of its parts, and the strong attraction between them and other matter, arise this effect, that, when a quantity of electrical matter is applied to a mass of common matter of any bigness or length. within our observation (which hath not already got its quantity) it is immediately and equally diffused through the whole.

(e) "6. Thus common matter is a kind of a spunge to the electrical fluid. And as a spunge would receive no water if the parts of water were not smaller than the pores of the spunge; and even then but slowly, if there were not a mutual attraction between those parts and the parts of the spunge; and would still imbibe it faster, if the mutual attraction among the parts of the water did not impede, some force being required to separate them; and fastest, if, instead of attraction, there were a mutual repulsion among those parts, which would act in conjunction with the attraction of the spunge. So is the case between the electrical and common matter.

(f) "7. But in common matter there is (generally) as much of the electrical as it will contain within its substance. If more is added, it lies without upon the surface, and forms what we call an electrical atmosphere; and then the body is said to be electrified.

"8. 'Tis supposed, that all kinds of common matter do not attract and retain the electrical, with equal strength and force, for reasons to be given hereafter. And that those called electrical *per se*, as glass, &c., attract and retain it strongest, and retain the greatest quantity.

(g) "9. We know that the electrical matter is *in* common matter, because we can pump it *out* by the globe or tube. We know that common matter has near as much as it can contain, because. when we add a little more to any portion of it, the additional quantity does not enter, but forms an electrical atmosphere. And we know that common matter has not (generally) more than it can contain, otherwise all loose portions of it would repel each other, as they constantly do when they have electric atmospheres."

* * * * * * *

(h) "15. The form of the electrical atmosphere is that of the body it surrounds. This shape may be rendered visible in a still air, by raising a smoke from dry rosin, dropt into a hot tea-spoon under the electrised body, which will be attracted, and spread itself equally on all sides, covering and concealing the body. And this form it takes because it is attracted by all parts of the surface of the body, though it cannot enter the substance already replete. Without this attraction, it would not remain round the body, but dissipate in the air.

(i) "16. The atmosphere of electrical particles surrounding an electrified sphere, is not more disposed to leave it, or more easily drawn off from any one part of the sphere than from another, because it is equally attracted by every part. But that is not the case with bodies of any other figure. From a cube it is more easily drawn at the corners than at the plane sides, and so from the angles of a body of any other form, and still most easily from the angle that is most acute. Thus if a body shaped as A, B, C, D, E, in Fig. 8 (our

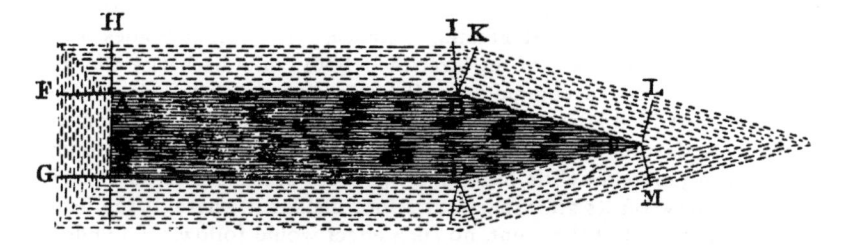

Fig. 4. Franklin's explanation of the power of points in discharging electrified body.

Fig. 4) be electrified or have an electrical atmosphere communicated to it, and we consider every side as a base on which the particles rest, and by which they are attracted, one may see, by imagining a line from A to F, and another from E to G, that the portion of the atmosphere include in F, A, E, G. has the line A E for its basis. So the portion of atmosphere include in H, A, B, I, has the line A, B, for its basis. And likewise, the portion included in K, B, C, L, has B, C, to rest on; and so on the other side of the figure. Now if you would draw off this atmosphere with any blunt, smooth body, and approach the middle of the side A, B, you must come very near, before the force of your attractor exceeds the force or power with which that side holds its atmosphere. But there is a small portion between I, B, K, that has less of the surface to rest on, and to be attracted by, than the neighbouring portions, while at the same time there is a mutual repulsion between its particles, and the particles of those portions, therefore here you can get it with more ease, or at a greater distance. Betwen F, A, H, there is a larger portion that has yet a less surface to rest on, and to attract it; here, therefore, you can get it away still more easily. But easiest of all between L, C, M, where the quantity is largest, and the surface to attract and keep it back the least. When you have drawn away one of these angular portions of the fluid, another succeeds in its place, from the nature of fluidity and the mutual repulsion beforementioned; and so the atmosphere continues flowing off at such angle, like a stream, till no more is remaining. The extremities of the portions of atmosphere over these angular parts, are likewise at a greater distance from the electrified body, as may be seen by the inspection of the above figures; the point of the atmosphere of the angle C, being much farther from C, than any other part of the atmosphere over the lines C, B, or B, A: And, besides the distance arising from the nature of the figure, where the attraction is less, the particles will naturally expand to a greater distance by their mutual repulsion. "On these accounts we suppose electrified bodies discharge their atmosphere upon unelectrified bodies more easily, and at a greater distance from their angles and points than from their smooth sides. —— Those points will also discharge into the air, when the body has too great an electrical atmosphere, without bringing any non-electric near, to receive what is thrown off:" For the air, though an electric *per se*, yet has always more or less water and other non-electric matters mixed with it: and these attract and receive what is so discharged.

(j) "17. But points have a property, by which they *draw on* as well as *throw off* the electrical fluid, at greater distances than blunt bodies can. That is, as the pointed part of an electrified body will discharge the atmosphere of that body, or communicate it farthest to another body, so the point of an unelectrified body will draw off the electrical atmosphere from an electrified body, farther than a blunter part of the same unelectrified body will do. Thus a pin held by the head, and the point presented to an electrified body, will draw off its atmosphere at a foot distance: where, if the head were presented instead of the point, no such effect would follow. To understand this, we may consider, that if a person standing on the floor would draw off the electrical atmosphere from an electrified body, an iron crow and a blunt knitting-needle held alternately in his hand, and presented for that purpose, do not draw with different forces in proportion to their different masses. For the man, and what he holds in his hand, be it large or small, are connected with the common mass of unelectrified matter; and the force with which he draws is the same in both cases, it consisting in the different proportion of electricity in the electrified body, and that common mass. But the force with which the electrified body retains its atmosphere by attracting it, is proportioned to the surface over which the particles are placed; *i. e.* four square inches of that surface retain their atmosphere with four times the force that one square inch retains its atmosphere. And as in plucking the hairs from the horse's tail, a degree of strength not sufficient to pull away a handful at once, could yet easily strip it hair by hair; so a blunt body presented cannot draw off a number of particles at once, but a pointed one, with no greater force, takes them away easily, particle by particle.

(k) "18. These explanations of the power and operation of points, when they first occurr'd to me, and while they first floated in my mind, appeared perfectly satisfactory; but now I have wrote them, and considered them more closely in black and white, I must own I have some doubts about them; yet, as I have at present nothing better to offer in their stead, I do not cross them out: for even a bad solution read, and its faults discovered, has often given rise to a good one, in the mind of an ingenious reader.

"19. Nor is it of much importance to us, to know the manner in which nature executes her laws; 'tis enough if we know the laws themselves. 'Tis of real use to know that china left in the air unsupported will fall and break; but *how* it comes to fall, and *why* it breaks, are matters of speculation. 'Tis a pleasure indeed to know then, but we can preserve our china without it."

(a) Note the modest title of this paper: "Opinions and Conjectures Concerning the Properties and Effects of Electrical Matter, Arising from Experiments and Observations, Made at Philadelphia, 1749." Franklin evidently does not desire all these statements to be regarded as worthy of the name Opinions, but states some of them as conjectures. The distinction here drawn appears to be the same as that which would be drawn to-day between theories and hypotheses.

(b) "Made at Philadelphia." Although practically the experiments made at Philadelphia were all Franklin's yet he generally uses the phrase, Experiments made at Philadelphia, rather than Experiments made by me, or by Dr. Franklin, at Philadelphia.

(c) An unquestionably convincing proof. Franklin quaintly remarks that, if any one doubts the ability of a discharge from a large glass jar to pass through one's body, let him try it, and the result "Will probably convince him."

(d) Franklin appears to have obtained a clear idea of the fact that a stream of electrified particles, called by him the electrified effluvia, passing from a charged conductor (positive), assumes the well known divergent appearance characteristic of the positive electric brush, or convective discharge, since he correctly ascribes the cause of this phenomena to the mutual repulsion of similarly charged particles. This discharge con-

Fig. 5. Appearance of positive or brush discharge (Von Marum)

sists, as now known, of streams of both electrified air particles and minute particles of the metal of the pointed electrode (Franklin's effluvia) that diverge from one another under the influence of their mutual repulsion. Fig. 5 represents the appearance of such a discharge as obtained by Von Marum, by the use of his powerful frictional electric machine.

(e) "Thus common matter is a kind of sponge to the electric fluid." Franklin clearly teaches in this hypothesis that there is a certain quantity of electric fluid which matter is capable of absorbing without manifesting its presence. He cites here the correct illustration of a sponge, which, as is well known, is capable of absorbing a certain quantity of water without parting with the same, unless it is compressed or shaken. It would clearly be impossible for the water to be absorbed by the sponge were not the particles of water smaller than the pores or openings in the sponge, and that even then, this absorption would take place but slowly did not a mutual attraction exist between the particles of the sponge and the particles of the water. Moreover, that the sponge would possess the power of absorbing the water more rapidly if the mutual attraction between the particles of the water did not prevent them from entering the sponge; that is, if what we call cohesion did not oppose itself to the particles leaving one another and entering the mass of the sponge. And finally, if, instead of the particles of the water which was being absorbed by the sponge, possessing cohesion, and to that extent, resisting entrance into the mass of the sponge, they possessed the property of mutual repulsion, this repulsion would act in the same direction as the attraction of the sponge, and would, therefore, increase the speed with which the water would enter it. As will be seen, the reasoning here is both concise and complete, characteristics common in Franklin's experimental investigation.

(f) "Forms what we call an electrical atmosphere." It will be observed that Franklin uses the term electrical atmosphere in the same sense as the electrification of a body, his conception of positive electrification being substantially as follows: That ordinary unelectrified matter contains no more of the electrical fluid than it can contain in its substance. When an additional amount is added, this additional quantity must reside on the surface. This is what is called at electrical atmosphere, and the body is, therefore, electrified (positively).

(g) Here Franklin gives his reasons for the single-fluid hypothesis. He concludes that the electric fluid resides in all common matter since it can be taken out either by means of the globe (electrical machine) or of the tube, the frictional device employed for obtaining electricity. That ordinary matter possesses about as much as it can contain of this electric fluid is shown since if a little more is added, this quantity does not enter, but forms the electric atmosphere, and that ordinary matter has not generally more than it can contain is seen from the fact that if it had, all additional or loose particles would repel one another, as all bodies do that possess electrical atmospheres.

(h) "The form of the electrical atmosphere is that of the body it surrounds." In order to render this visible, the electrified body is surrounded by a smoke, produced by the simple experiment of dropping dry rosin in a hot teaspoon held under the electrified body. This smoke, being attracted by the charged body, spreads itself around the body under the influence of the electric charge.

(i) Here again Franklin reverts to the power of points, which he had already discussed in his second letter to Collinson, dated July 11, 1747. In this paragraph he gives an excellent explanation as to the effect produced by the shape of an electrified body on the distribution of its atmosphere of electrified particles. For example, in the case of a sphere, the depth of this atmosphere should be uniform over all portions of the surface, since the mass of the sphere attracts them equally at every point. The same, however, would not be true with bodies of any other figure; for example, in case of a cube, the electrical atmosphere, or the electricity, could be more readily drawn from any of the corners than it could from the plane sides; and in general, in the case of any irregular figure, the electrical atmosphere can be drawn most easily from the sharpest or most acute angle; thus, if a body shaped as A,B,C,D,E. Fig. 4, be electrified, and a blunt body be approached to it on different sides, it will be more difficult to draw off the electrical charge when approached to a point midway between A and E, or between A and B, or E and D, since at these points the fluid is held on to the charged body by a portion that has AE, AB, or ED, for its base. If, however, a blunt smooth body be approached to the point C, there will be a greater area of such blunt body opposed to C, for at C, the surface of the body that the charge has to rest on is much less that that of the approached body. Hence, the charge will readily leave the body at this point. Moreover, as Franklin points out, at points of a charged body, as, for example, at C, in Fig. 4, the mutual repulsion of the particles of electricity constituting the electrified atmosphere may result in causing such points to discharge the electricity into the air whenever the charged body possesses too great an electrical atmosphere.

(j) "But points have a property by which they draw on, as well as throw off the electrical fluid, at greater distances than blunt bodies can." Franklin here gives a quaint explanation based on the analogy of plucking hairs from the tail of a horse, calling attention to the fact that, while a certain force exerted to pull out simultaneously the hairs from an area of the tail of the horse, equal approximately to the quarter of a square inch, might be totally inadequate to remove any of the hairs, yet that the same force would probably suffice readily to remove them one by one. In a similar manner, that if a blunt body be approached to the plane surface of an electrified body, and an endeavor is made to remove simultaneously a number of electrified particles from a certain area of such surface, say a square inch, it might not be able to overcome the attraction existing between the gross matter and the electrical atmosphere; whereas, if it attempted to remove the electrified particles a few at a time, as by approaching the point of a charged conductor, it might readily be able to overcome the attraction existing between the limited amount of matter at such point, and its electrical atmosphere.

(k) Note here the unquestioned mental integrity of the man, and his candid confession that, while the preceding explanations appeared to him, when they first suggested themselves to his mind, perfectly satisfactory, yet when he sees them written in cold black and white, he begins to doubt their probability. "I must own I have some doubts about them; yet, as I have at present nothing better to offer in their stead, I do not cross them out; for even a bad solution read, and its faults discovered, has often given rise to a good one in the mind of an ingenious reader."

Then again, note in the next paragraph his quaint reflections on the laws of nature, and the practical value that we can obtain from a knowledge of such laws. While no one can question the use of the knowledge that china, left unsupported in the air, will fall and break; "But how it comes to fall, and why it breaks, are matters of speculation;" and this statement would remain true to the present day. " 'Tis a pleasure indeed," says Franklin, "to know them, but we can preserve our china without it."

Note in a similar manner the following quotations taken from a letter written in September, 1755, to Collinson:

"These thoughts, my dear friend, are many of them crude and hasty; and if I were merely ambitious of acquiring some reputation in philosophy, I ought to keep them by me, till corrected and improved by time and farther experience. But since even short hints and imperfect experiments in any branch of science, being communicated, have oftentimes a good effect, in exciting the attention of the ingenious to the subject, and so become the occasion of more exact disquisition, and more complete discoveries. You are at liberty to communicate this paper to whom you please; it being of more importance that knowledge should increase, than that your friend should be thought an accurate philosopher."

But leaving the above important paper by Franklin on the properties and effects of electric matter, especially as regards the influence of points on the discharge of electrified bodies, careful reference should be made to some additional statements in this article. It is especially important since these statements have a bearing on a discussion that has recently existed among scientific circles as to whether Franklin should be given the credit for first actually demonstrating the identity of lightning and electricity, or whether such credit should be assigned to Dalibard, De Romas, or others. The additional matter above referred to, as contained in the article written in 1749, is as follows:

(a) "20. Thus in the present case, to know this power of points, may possibly be of some use to mankind, though we should never be able to explain it. The following experiments, as well as those in my first paper, shew this power. I have a large prime conductor, made of several thin sheets of clothier's pasteboard, form'd into a tube, near ten feet long and a foot diameter. It is cover'd with *Dutch* emboss'd paper, almost totally gilt. This large metallic surface supports a much greater electrical atmosphere than a rod of iron of 50 times the weight would do. It is suspended by silk lines, and when charged will strike at near two inches distance, a pretty hard stroke so as to make one's knuckle ache. Let a person standing on the floor present the point of a needle at 12 or more inches distance from it, and while the needle is so presented, the conductor cannot be charged, the point drawing off the fire as fast as it is thrown on by the electrical globe. Let it be charged, and then present the point at the same distance, and it will suddenly be discharged. In the dark you may see a light on the point, when the experiment is made. And if the person holding the point stands upon wax, he will be electrified by receiving the fire at that distance. Attempt to draw off the electricity with a blunt body, as a bolt of iron round at the end, and smooth (a silversmith's iron punch, inch thick), is what I use, and you must bring it within the distance of three inches before you can do it, and then it is down with a stroke and crack. As the pasteboard tube hangs loose on silk lines, when you approach it with the punch iron, it likewise will move towards the punch, being attracted while it is charged; but if, at the same instant, a point be presented as before, it retires again, for the point discharges it. Take a pair of large brass scales, of two or more feet beam, the cords of the scales being silk. Suspend the beam by a packthread from the ceiling, so that the bottom of the scales may be about a foot from the floor: The scale will move around in a circle by the untwisting of the pack-thread. Set the iron punch on the end upon the floor, in such a place as that the scales may pass over it in making their circle: Then electrify one scale, by applying the wire of a charged phial to it. As they move round, you see that scale draw nigher to the floor, and dip more when it comes over the punch; and if that be placed at a proper distance, the scale will snap and discharge its fire into it. But if a needle be stuck on the end of the punch, its point upwards, the scale, instead of drawing nigh to the punch, and snapping, discharges its fire silently through the point, and rises higher from the punch. Nay, even if the needle be placed upon the floor near the punch, its point upwards, the end of the punch, tho' so much higher than the needle, will not attract the scale and receive its fire, for the needle will get it and convey it away, before it comes nigh enough for the punch to act. And this is constantly observable in these experiments, that the greater quantity of electricity on the pasteboard tube, the farther it strikes or discharges its fire, and the point likewise will draw it off at a still greater distance.

(b) "Now, if the fire of electricity and that of lightning be the same, as I have endeavored to shew at large, in a former paper, this pasteboard tube and these scales may represent electrified clouds. If a tube of only ten feet long will strike and discharge its fire on the punch at two or three inches distance, an electrified cloud of perhaps 10,000 acres may strike and discharge on the earth at a proportionably greater distance. The horizontal motion of the scales over the floor, may represent the motion of the clouds over the earth; and the erect iron punch, a hill or high building; and then we see how electrified clouds passing over hills or high buildings at too great a height to strike, may be attracted lower till within their striking distance. And lastly, if a needle fixed on the punch with its point upright, or even on the floor below the punch, will draw the fire from the scale silently at a much greater than the striking distance, and so prevent its descending towards the punch; or if in its course it would have come nigh enough to strike, yet being first deprived of its fire it cannot, and the punch is thereby secured from the stroke. (c) I say, if these things are so, may not the knowledge of this power of points be of use to mankind, in preserving houses, churches, ships, &c., from the stroke of lightning, by directing us to fix on the highest parts of those edifices, upright rods of iron, made sharp as a needle, and gilt to prevent rusting, and from the foot of those rods a wire down the outside of the building into the ground, or down round one of the shrouds of a ship, and down her side till it reaches the water? Would not these pointed rods probably draw the electrical fire silently out of a cloud before it came nigh enough to strike, and thereby secure us from that most sudden and terrible mischief?

(d) "21· To determine the question, whether the clouds that contain lightning are electrified or not, I would propose an experiment to be try'd where it may be done conveniently. On the top of some high tower or steeple, place a kind of sentry-box (as in Fig. 9) (our Fig. 6) big enough to contain a man and an electric stand. From the middle of the stand let an iron rod rise and pass bending out of the door, and then upright 20 or 30 feet, pointed very sharp at the end. If the electrical stand be kept clean and dry, a man standing on it when such clouds are passing low, might be electrified and afford sparks, the rod drawing fire to him from a cloud. If any danger to the man should be apprehended (though I think there would be none) let him stand

Fig. 6. The forerunner of the lightning rod.

so the sparks, if the rod is electrified, will strike from the rod to the wire, on the floor of his box, and now and then bring near to the rod the loop of a wire that has one end fastened to the leads, he holding it by a wax handle; and not affect him.

(e) "22· Before I leave this subject of lightning, I may mention some other similarities between the effects of that, and those of electricity. Lightning has often been known to strike people blind. A pigeon that we struck dead to appearance by the electrical shock, recovering life, drooped about the yard several days, eat nothing, though crumbs were thrown to it, but declined and died. We did not think of it being deprived of sight; but afterwards a pullet struck dead in like manner, being recovered by repeatedly blowing into its lungs, when set down on the floor, ran headlong against the wall, and on examination appeared perfectly blind. Hence we concluded the pigeon also had been absolutely blinded by the shock. The biggest animal we have yet killed, or tried to kill, with the electrical stroke, was a well-grown pullet.

(f) "23. Reading in the ingenious Dr. Miles's account of the thunder storm at Stretham, the effect of the lightning in stripping off all the paint that had covered a gilt moulding of a panel of wainscot, without hurting the rest of the paint, I had a mind to lay a coat of paint over the filletting of gold on the cover of a book, and try the effect of a strong electrical flash sent through that gold from a charged sheet of glass. But having no paint at hand, I pasted a narrow strip of paper over it; and when dry, sent the flash through the gilding, by which the paper was torn off from end to end, with such force, that it was broken in several places, and in others brought away part of the grain of the Turkey-leather in which it was bound; and convinced me, that had it been painted, the paint would have been stript off in the same manner with that on the wainscot at Stretham.

"24. Lightning melts metals, and I hinted in my paper on that subject, that I suspected it to be a cold fusion; I do not mean a fusion by force of cold, but a fusion withou t heat. (g) We have also melted gold, silver, and copper, in small quantities, by the electrical flash. The manner is this: Take leaf gold, leaf silver, or leaf gilt copper, commonly called leaf brass, or *Dutch gold*; cut off from the leaf long narrow strips, the breadth of a straw. Place one of these strips between two strips of smooth glass that are about the width of your finger. If one strip of gold, the length of the leaf, be not long enough for the glass, add another to the end of it, so that you may have a little part hanging out loose at each end of the glass. Bind the pieces of glass together from end to end with strong silk thread; then place it so as to be part of an electrical circuit, (the ends of gold hanging out being of use to join with the other parts of the circuit) and send the flash through it, from a large electrified jar or sheet of glass. Then if your strips of glass remain whole, you will see that the gold is missing in several places, and instead of it a metallic stain on both the glasses; the stains on the upper and under glass exactly similar in the minutest stroke, as may be seen by holding them to the light; the metal appeared to have been not only melted, but even vitrified, or otherwise so driven into the pores of the glass as to be protected by it from the action of the strongest Aqua Fortis, or Aqua Regia. I send you enclosed two little pieces of glass with these metallic stains upon them, which cannot be removed without taking part of the glass with them. Sometimes the stain spreads a little wider than the breadth of the leaf, and looks brighter at the edge, as by inspecting closely you may observe in these. Sometimes the glass breaks to pieces; once the upper glass broke into a thousand pieces, looking like coarse salt. These pieces I send you were stain'd with *Dutch gold*. True gold makes a darker stain, somewhat reddish; silver, a greenish stain. We once took two pieces of thick looking-glass, as broad as a *Gunter's* scale, and six inches long; and placing leaf-gold between them, put them between two smoothly plain'd pieces of wood, and fix'd them tight in a book-binder's small press; yet though they were so closely confined, the force of the electrical shock shivered the glass into many pieces. The gold was melted, and stain'd into the glass, as usual. The circumstances of the breaking of the glass differ much in making the experiment, and sometimes it does not break at all: but this is constant, that the stains in the upper and under pieces are exact counterparts of each other. And though I have taken up the pieces of glass between my fingers immediately after this melting, I never could perceive the least warmth in them.

"25. In one of my former papers, I mentioned, that, gilding on a book, though at first it communicated the shock perfectly well, yet failed after a few experiments, which we could not account for. We have since found that one strong shock breaks the continuity of the gold in the filletting and makes it look rather like dust of gold, abundance of its parts being broken and driven off; and it will seldom conduct above one strong shock. Perhaps this may be the reason: When there is not a perfect continuity in the circuit, the fire must leap over the vacancies: There is a certain distance which it is able to leap over according to its strength; if a number of small vacancies, though each be very minute, taken together exceed that distance, it cannot leap over them, and so the shock is prevented."

(a) A careful reading of this paragraph shows that Franklin had, at this early date, very clear ideas concerning the probable identity of electric discharges and lightning flashes. Here we find him actually constructing two kinds of movable charged conductors, that he believed correctly represented movable thunder-clouds, devising means by which these artificial clouds could readily move so that they could be brought within striking distance of earth-connected objects.

One of these movable clouds consists of a tube of pasteboard covered on the outside with gilt paper. By suspending this tube from the ceiling of the room by means of silk threads, he not only insulated the cloud, but also rendered it, within certain limits, readily movable, so as to permit it to approach earth-connected objects. Although this movable pasteboard cloud was only ten feet in length and one foot in diameter, yet its electric charge was capable of giving a sharp blow to the knuckle of a hand, approached sufficiently near to discharge it. He proved that such a cloud, when charged, was capable of being silently discharged by means of a needle point, held in the hands of a person standing on the floor of a room, when brought within a distance of twelve inches or so from the tube.

But while the needle point is thus capable of silently discharging the cloud, a blunt body, such as a smooth iron bolt, is unable to do so until brought much nearer to the tube, say within three inches instead of twelve inches. As soon as a blunt object is sufficiently near, the tube is disruptively discharged with a snap or crack. The phenomena to which Franklin calls especial attention in these experiments is the ability of the cloud to be silently discharged by the needle point, but that when the cloud is approached by a blunt object before the discharge takes place, an attraction occurs, whereby the cloud is moved towards the blunt object until the distance is sufficiently small to permit of its disruptive discharge. It is in this manner, he believes, that the real clouds in the sky approach the tops of tall objects on the earth until the distance is sufficiently small to permit the bolt to strike.

Another form of movable cloud was obtained by charging one of the scale pans of an ordinary balance. This experiment is practically the same as that of the pasteboard, except that during it, he showed the possibility of discharging the scale pan by means of a needle point placed on the floor of the room immediately below the balance, with its point directed upwards.

(b) "Now if the fire of electricity and lightning be the same." Franklin here acknowledges that an objection may fairly be made that the pasteboard cloud is puny when compared with the clouds that occur in thunder storms. In this connection however, he properly calls attention to the fact that, if the pasteboard cloud, with the limited area possessed by a cylinder of but one foot in diameter and ten feet in length, is capable of discharging across a spark gap of two or three inches, might not a cloud consisting of probably some 10,000 acres of electrified particles of moisture be able to strike the earth across a much greater gap. Moreover, if a needle point, approached to within a distance of say twelve inches of the pasteboard cloud, be capable of silently discharging it, might not the same thing be done with thunder clouds, and thus prevent their disastrous strokes to the earth.

(c. "I say, if these things are so, may not the knowledge of this power of points be of use to mankind in preserving houses, churches, ships, etc., from strokes of lightning." Here is a clear statement of Franklin's great invention of the lightning rod, a matter that we shall discuss later on. The point to which I especially desire to call your attention is the fact that Franklin invented the lightning rod long before he actually drew electricity from the clouds.

(d) "To determine the question whether the clouds that contain lightning are electrified or not, I would propose an experiment to be tried, where it can be done conveniently." Remember that these words were written in 1749. To Franklin's philosophical mind, there did not appear to exist any reasonable doubt but that the clouds that contained lightning were electrified, so that his artificial cloud, consisting as it did of a movable gilded pasteboard tube, appeared to resemble an actual cloud, save only in two important respects: first, as regards its size; and second, as regards its distance from the surface of the earth, or from such a point as he could approach it sufficiently to draw from it some of its electrical charge. It was an easy matter to discharge his artificial cloud to the earth through the metallic punch, at a distance of two or three inches, but it was another thing to get sufficiently near an actual cloud as to successfully draw from it some of its charge. Franklin conceived the idea that, if he could climb to the top of a church steeple or tower, thereby coming that much nearer to the cloud, he could treat it as he did the electrified tube; but, unfortunately for Franklin, there were no tall towers or church steeples existing in Philadelphia at that time, which could serve his purpose.

Consequently, with the spirit of a true philosopher, he describes the experiment that he would like to have tried by others "Where it may be done conveniently." So he directs the construction of the sentry box represented in Fig. 6, large enough to hold an observer and an electrical stand; i. e., a cake of wax. At the middle of this cake of wax, an iron rod passes by a bend out of the door, and so thirty or forty feet above the top of the box, thus getting that much nearer to the clouds. Franklin's idea was, and that it it was correct the use of this apparatus by others clearly showed, that if a cloud should pass sufficiently near to the upper part of the rod, which was provided with a sharp point at the end, it would become electrified, so that sparks could be drawn from it, just as he did afterwards from his kite. Franklin expresses the opinion that the man would be exposed to no danger if he stood on the floor of his box, and occasionally brought near to the rod a loop of wire, the lower end of which was earthed by being fastened to the metallic leads of the building, he holding the other end by means of an insulating handle formed of wax. It was this suggestion, as we shall afterwards see, that Dalibard made use of in his drawing electricity from the sky.

(e) In the balance of this portion of the paper that we have quoted, Franklin cites some of the resemblances between lightning and electricity. The lightning has been known, he says, to strike people blind. He recalls an experiment he made with a pigeon, that was apparently killed by an electrical shock, but which was partially resuscitated, and died a few days afterwards. Also on an experiment he made with a pullet, which was afterwards resuscitated by artificial respiration, and was apparently perfectly blind. He, therefore, from this experiment, concluded that the pigeon had probably also been blinded by the shock, as it had refused food.

(f) Note here the ingenious method employed by Franklin to duplicate the effects of a lightning stroke that occurred during a thunder storm at Stretham, referred to by Mr. Miles. In this storm the lightning bolt stripped off all the paint that had covered a gilded moulding on the panel of a wainscoat, without, however, affecting the rest of the paint. Franklin conceived the idea of reproducing this effect by seeing what would happen is he sent a strong electrical discharge through some gilding on a book. Not having any paint at hand, he employs in its place a narrow strip of paper, pasted over the gilding. As soon as this had thoroughly dried, a discharge was sent through the gilding, with the result that the paper was torn off, from end to end, so violently that it was broken in several places. Indeed, in some places it took away a portion of the Turkey leather with which the book was bound.

(g) In the next paragraph, Franklin refers to some curious

results in the melting of metal by a process that he characterizes as cold fusion. Though his philosophy here is bad, his technique, that is, the method that he employs for melting metals by his so-called cold fusion, is excellent. It will be observed that in the method employed he obtained a continuous circuit of thin gold leaf between two plates of glass that were tightly bound together by silk thread, in such a manner as to prevent the breaking of the film by the accidental moving of the glass. The conditions were such as to permit the passage of a powerful Leyden jar discharge through this circuit by means of the two ends of gold leaf projected beyond the glass plate. Under these circumstances, the gold was deflagrated; i. e., not only fused, but fused and volatilized by the great heat of the discharge; so that a series of metallic stains were produced on the glass, these stains being exactly duplicated to the minutest details on both pieces of glass that were in contact with the metal. Franklin corrected his erroneous idea concerning cold fusion at a later date. In order, however, to prevent confusion at this point we defer its discussion to another part of this article.

In this connection, Franklin refers to some experiments he had made in sending a discharge through the gilding on a book. He remarks that a careful examination of the book after the passage of a disruptive discharge showed that a single strong shock resulted in breaking the continuity of the gilding, so that, instead of remaining in the shape of a metallic leaf, it was shattered to metallic dust.

As is clearly indicated by the preceding experiments, Franklin had given considerable thought respecting the probable identity of a disruptive electric discharge and a lightning flash. It appears that he was not only in the habit of making notes containing descriptions of the experiments he had tried, but also of those which he desired to try. In these latter notes he generally placed the reasons he had for trying the experiments. It was from these notes that his letters were subsequently prepared.

There is an unfortunate absence of exact data in the case of many of Franklin's experiments. This is especially true as regards his great experiment of drawing electricity from the sky. In a letter written to Dr. L, of Charlestown, South Carolina, dated May 18, 1755, he says:

"Your question, how I came first to think of proposing the experiment of drawing down the lightning, in order to ascertain its sameness with the electric fluid, I cannot answer better than by giving you an extract from the minutes I used to keep of the experiments I made, with memorandums of such as I purposed to make, the reasons for making them, and the observations that arose upon them, from which minutes my letters were afterwards drawn. By this extract you will see that the thought was not so much 'an out-of-the-way one,' but that it might have occurred to any electrician.

"'Nov. 7, 1749. Electrical fluid agrees with lightning in these 'particulars: 1. Giving light. 2. Colour of the light. 3. Crooked direc-'tion. 4. Swift motion. 5. Being conducted by metals. 6. Crack or noise 'in exploding. 7. Subsisting in water or ice. 8. Rending bodies it passes 'through. 9. Destroying animals. 10. Melting metals. 11. Firing in-'flammable substances. 12. Sulphureous smell. * * * The electric fluid 'it attracted by points. * * * We do not know whether this property is 'in lightning. * * * But since they agree in all the particulars wherein 'we can already compare them, is it not probable they agree likewise in 'this? * * * Let the experiment be made.'"

Franklin, apparently growing tired of waiting for a church steeple or tower to be erected in the City of Philadelphia, conceived the bold idea of reaching the higher regions of the atmosphere by means of a kite, and by this means actually succeeds in drawing the electricity from the clouds, and thus establishing the identity between the disruptive discharge and the lightning bolt.

In Vol. I, page 108, of the Complete Works in Philosophy, Politics and Morals of the late Benjamin Franklin, published in London in 1806, by Dr. Stuber, a Philadelphian, and intimate friend of Franklin, the following statement is made concerning this kite:

"While Franklin was waiting for the erection of a spire, it occurred to him that he might have more ready access to the region of clouds by means of a common kite. He prepared one by fastening two cross-sticks to a silk handkerchief, which would not suffer so much from the rain as paper. To the upright stick was affixed an iron point. The string was, as usual, of hemp, except the lower end, which was silk. Where the hempen string terminated, a key was fastened. With this apparatus, on the appearance of a thunder-gust approaching, he went out into the common accompanied by his son, to whom alone he communicated his intentions, well knowing the ridicule which, too generally for the interest of science, awaits unsuccessful experiments in philosophy. He placed himself under a shed to avoid the rain. His kite was raised. A thunder cloud passed over it. No sign of electricity appeared. He almost despaired of success, when suddenly he observed the loose fibres of his string move toward an erect position. He now presented his knuckle to the key and received a strong spark. Repeated sparks were drawn from the key, the phial was charged, a shock, and all the experiments made which are usually performed with electricity."

As regards the construction of this kite, Franklin himself gives the following description in a letter to his friend Collinson, dated Philadelphia, October 19, 1753:

"As frequent mention is made in public papers from *Europe* of the success of the *Philadelphia* experiment for drawing the electric fire from clouds by means of pointed rods of iron erected on high buildings, &c., it may be agreeable to the curious to be informed that the same experiment has succeeded in *Philadelphia*, though made in a different and more easy manner, which is as follows:

"Make a small cross of two light strips of cedar, the arms so long as to reach to the four corners of a large thin silk handkerchief when extended; tie the corners of the handkerchief to the extremities of the cross, so you have the body of a kite; which being properly accommodated with a tail, loop, and string, will rise in the air, like those made of paper; but this being of silk, is fitted to bear the wet and wind of a thunder-gust without tearing. To the top of the upright stick of the cross is to be fixed a very sharp pointed wire, rising a foot or more above the wood. To the end of the twine, next the hand, is to be tied a silk ribbon, and where the silk and twine join, a key may be fastened. This kite is to be raised when a thunder gust appears to be coming on, and the person who holds the string must stand within a door or window, or under some cover, so that the silk ribbon may not be wet; and care must be taken that the twine does not touch the frame of the door or window. As soon as any of the thunder clouds come over the kite, the pointed wire will draw the electric fire from them, and the kite, with all the twine, will be electrified, and the loose filaments of the twine will stand out every way, and be attracted by an approaching finger. And when the rain has wet the kite and twine, so that it can conduct the electric fire freely, you will find it stream out plentifully from the key on the approach of your knuckle. At this key the phial may be charged; and from electric fire thus obtained, spirits may be kindled, and all the other electric experiments be performed, which are usually done by the help of a rubbed glass globe or tube, and thereby the sameness of the electric matter with that of lightning completely demonstrated."

It is not certain just at what portion of the City of Philadelphia Franklin flew his kite. In the opinion of some who have studied this matter carefully, it was probably in the neighborhood of the high ground at about 18th and Spring Garden Streets, since this situation would most probably have afforded him a better exposure to the wind, and, at the same time, would have given him that seclusion which, for the time being, he desired. Franklin was evidently accustomed to walking over all the outskirts of the city which lay between the Delaware and the Schuylkill, especially in the portions near the latter river, for, he mentions in one of his letters experiments made in igniting sulphur by a wire laid across the Schuylkill River. Succeeding in raising his kite on the approach of a thunder storm at some place, near 18th and Spring Garden Streets, or somewhere in this neighborhood, he at last has the satisfaction of being able to draw from the key attached to the lower end of the kite string, sparks, which he recognizes at once as being similar to those produced by the electrical machine. In this manner was realized one of the grandest discoveries in the domain of physical science, and the actual identity of the lightning flash and electric discharges was at last demonstrated beyond any possible doubt.

It must not for a moment be supposed that, having in this manner succeeded in drawing electricity from the clouds, Franklin was willing to rest his great discovery. On the contrary, we find him making repeated experiments, obtaining electricity from the clouds under varying conditions. At a somewhat later date, he was able to draw electricity from a rod erected from the roof of a house, and with this apparatus we find him actually testing the character of the electric excitement that was to be found in charged thunder-clouds. Generally speaking, he found such clouds to possess a negative charge, but he also found that their charge was sometimes positive.

Although the following description of this original experiment, prepared shortly afterward, is not unlike the one given by Stuber, yet literature on the subject is so comparatively limited that I thought it advisable to annex it:

"Furnished with this apparatus, on the approach of a storm, he went out upon the commons near Philadelphia, accompanied by his son, to whom alone he communicated his intentions, well knowing the ridicule which would have attended the report of such an attempt should it prove to be unsuccessful. Having raised the kite, he placed himself under a shed, that the ribbon by which it was held might be kept dry, as it would become a conductor of electricity when wetted by rain, and so fail to afford that protection for which it was provided. A cloud, apparently charged with thunder, soon passed directly over the kite. He observed the hempen cord; but no bristling of its fibres was apparent, such as was wont to take place when it was electrified. He presented his knuckle to the key, but not the smallest spark was perceptible. The agony of his expectation and suspense can be adequately felt by those only who have entered into the spirit of such experimental researches. After the lapse of some time he saw that the fibres of the cord near the key bristled, and stood on end. He presented his knuckle to the key, and received a strong, bright spark. It was lightning. The discovery was complete, and Franklin felt that he was immortal."

Probably one of the most curious circumstances connected with this phase of the early history of electric science is the difficulty in fixing an actual date when this memorable experiment was made by Franklin. It seems almost incredible that no one of the accounts of so important an experiment, that was at once heralded over the entire scientific world, should have failed to give the precise date at which it was made. In Stuber's account of this experiment, there is the same uncertainty, although we are here given a little more accurate information, viz.:"It was not until the summer of 1752 that Franklin was enabled to complete his grand and unparallelled discovery by experiment."

Priestley, in his "History of Electricity," (Vol. I, London, 1775), says: "This happened in June, 1752, a month after the electricians in France had verified the same theory, but before he had heard of anything they had done."

But in the meanwhile the memorable letter sent by Franklin to Collinson, containing a full account of his (Franklin's) numerous electrical experiments made at Philadelphia in 1749, but dated Philadelphia, July 29, 1750, and among them descriptions of his artificial clouds and his proposals to protect buildings by means of pointed conductors, was brought to the attention of the scientific world, and began to have its effect. Collinson recognized that this letter contained discoveries and inventions of more than ordinary value. Therefore, he brought it at once to the attention of the Royal Society, but the paper was received with actual derision. The members of the Society were apparently unwilling to acknowledge that a provincial could make a discovery of such a nature at to be worthy their attention.

Collinson, provoked at the treatment his friend had thus received at the hands of the Royal Society, endeavored to obtain elsewhere the publication of this valuable paper. He offered it to Cave, the publisher of the Gentlemen's Magazine. While unwilling to permit Franklin's paper to appear as a part of his regular journal, yet Cave consented to print it as a separate publication. This was done in 1751. This publication, however, appears to have been received coldly in England, owing doubtless to the indifferent manner in which the paper was received by the Royal Society.

Fortunately, however, one of these publications of Franklin's paper was sent to Count de Buffon, at Paris. De Buffon, at once recognizing the very great value of the paper, took it to D'Alibard, a well-known French botanist, and requested him to translate it into French. Although this translation was far from being good, yet the paper attracted such attention throughout France that it had an enormous sale, creating, as it did, no little excitement in intellectual circles. The probability of Franklin's views as to the cause of lightning being correct, became a matter of general discussion, not only in scientific society, but throughout the Court of France. In this manner, Franklin's experiments and suggestions came to the knowledge of the King, who requested that these experiments be shown him. This was done, and greatly excited his interest.

Noting the great interest the King took in these experiments, three Frenchmen, *i. e.*, D'Alibard, M. de Lor, an instructor in Physics, and Count de Buffon, consulted together and came to the conclusion that none of the experiments made or suggested by Franklin would so interest the King as would the successful trying of Franklin's proposed experiment of drawing down lightning from the clouds by means of a pointed rod. Each of these gentlemen determined to try the experiment independently of the others, and all of them succeeded, D'Alibard on the 10th of May, De Lor on the 18th of May, and De Buffon on the 19th of May, 1752.

D'Alibard prepared for his experiment by erecting an iron rod, forty feet high, in a garden at Marly-la-Ville, eighteen miles from Paris. In accordance with Franklin's directions, this rod was insulated at its base, which rested on a table arranged within a small cabin, from the posts of which the rod was insulated by silken ropes.

When the first thunder storm to which this apparatus was subjected occurred, D'Alibard was not in the neighborhood. He had, however, placed a trustworthy soldier, named Corffier, in whom he had confidence, in charge of the apparatus, and had provided him with a brass wire, insulated by being mounted in a glass bottle for a handle, and had given him instructions to draw off any sparks from the rod should it become electrified. Some three days after the erection of the rod, while Corffier was on guard, a storm occurred, and Corffier was able to draw long and very noisy discharges from the insulated rod on the 10th of May. The dangerous appearances of these discharges so frightened the soldier that he sent for the village priest, one Raulet, under the full conviction that the marked disturbance, accompanied as it was with what he believed was a sulphurous odor, belonged to things of the lower world. The Priest, accompanied by many villagers, arrived before the storm had passed, and the Priest, not fearing the power of the Evil One, was willing to make experiments for himself by drawing sparks with a brass wire. He describes these experiments in a letter he sent to D'Alibard as follows: "I repeated this experiment at least six times in about four minutes, in the presence of many persons, and every time the experiment lasted the space of a pater and an eve." At one time he accidentally touched the rod, and got a very severe shock.

Before he left the place where the experiment was made the Priest sent the letter above referred to, to D'Alibard by Corffier. D'Alibard at once prepared a memoir on the subject, which he communicated to the French Academy of Sciences three days later; *i. e.*, May 13.

The following translation from this paper will show that D'Alibard did not hesitate to give to Franklin the credit of this experiment:

"From all the experiments and observations contained in the present paper, and more especially from the recent experiment made at Marly-la-

Ville, it is shown beyond doubt that the matter made of lightning is the same as that of electricity; it has become a reality, and I believe that the more we realize what he (Franklin) has published on electricity, the more will we acknowledge the great debt that physical science owes to him."

The relations existing between D'Alibard and Franklin were of the most cordial character. In a letter from D'Alibard to Franklin, dated Paris, March 31, 1752, D'Alibard says:

"We are all waiting with the greatest eagerness to hear from you. I beg that you will let me have letters as soon and often as possible. Your name is venerated in this country as it deserves to be. There are but few electricians, like the Abbe Nollet, whose jealousy is excited by the honor your discoveries have obtained.

"With great respect and esteem I am, &c.,

DALIBARD."

But that restless kind of philosopher, the so-called higher critic, who, in so many cases, unable to do anything original themselves, rest content with calling into question the work of better men, have sought to detract from the merit of Franklin's great discovery by the statement that even if Franklin did draw electricity from the clouds, which some of them have been foolish enough to doubt, yet in reality one De Romas, a French barrister, residing in Nerac, some seventy-five miles south of Bordeaux, antedated Franklin, as some say, in the conception of the kite; as others say, in the date at which it was employed for drawing lightning from the air; and as still others say, in both of these respects.

Without going any further into this subject, it suffices to say that Franklin's kite was successfully employed, as already mentioned, some time in June, 1752, while the kite of De Romas was not raised until the 14th of May, 1753. At this date his experiment was unsuccessful, but at a later date, June 7, 1753, having made the string of his kite a better conductor by wrapping around the outside of its entire length a fine copper wire, he was successful in his attempt.

Priestly gives the following description of his kite in Vol. I of his book on Electricity already referred to:

"The greatest quantity of electricity that was ever brought from the clouds, by any apparatus prepared for that purpose, was by Mr. De Romas, assessor to the presideal of Nerac. This gentleman was the first who made use of a wire interwoven in the hempen cord of an electrical kite, which he made seven feet and a half high, and three feet wide, so as to have eighteen square feet of surface. This cord was found to conduct the electricity of the clouds more powerfully than a hempen cord would do, even though it was wetted; and, being terminated by a cord of dry silk, it enabled the observer (by a proper management of his apparatus) to make whatever experiments he thought proper, without danger to himself.

"By the help of this kite, on the 7th of June, 1753, about one in the afternoon, when it was raised 550 feet from the ground and had taken 780 feet of string, making an angle of near forty-five degrees with the horizon; he drew sparks from his conductor three inches long and a quarter of an inch thick, the snapping of which was heard about 200 paces. Whilst he was taking these sparks, he felt, as it were, a cob-web on his face, though he was above three feet from the string of the kite; after which he did not think it safe to stand so near, and called aloud to all the company to retire, as did himself about two feet."

"Thinking himself now secure enough, and not being incommoded by any body very near him, he took notice of what passed among the clouds which were immediately over the kite; but could perceive no lightning either there or anywhere else, nor scarce the least noise of thunder, and there was no rain at all. The wind was West, and pretty strong, which raised the kite 100 feet higher, at least than in the other experiments.

"Afterwards, casting his eyes on the tin tube, which was fastened to the string of the kite, and about three feet from the ground, he saw three straws, one of which was about one foot long, a second four or five inches, and performing a circular dance, like puppets, under the tin tube, without touching one another.

"This little spectacle, which much delighted several of the company, lasted about a quarter of an hour; after which, some drops of rain falling, he again perceived the sensation of the cob-web on his face, and at the same time heard a continual rustling noise, like that of a small forge bellows. This was a farther warning of the increase of electricity; and from the first instant that Mr. De Romas perceived the dancing straws, he thought it not advisable to take any more sparks, even with all his precautions; and he again entreated the company to spread themselves to a still greater distance.

"Immediately after this came on the last act of the entertainment, which De Romas acknowledged made him tremble. The longest straw was attracted by the tin tube, upon which followed three explosions, the noise of which greatly resembled that of thunder. Some of the company compared it to the explosion of rockets, and others to the violent crashing of large earthen jars against a pavement. It is certain that it was heard into the heart of the city, notwithstanding the various noises there.

"The fire that was seen at the instant of the explosion had the shape of a spindle eight inches long and five lines in diameter. But the most astonishing and diverting circumstances was produced by the straw, which had occasioned the explosion, following the string of the kite. Some of the company saw it at forty-five or fifty fathoms distance, attracted and repelled alternately, with this remarkable circumstance, that every time it was attracted by the string flashes of the fire were seen, and cracks were heard, though not so loud as at the time of the former explosion.

"It is remarkable, that, from the time of the explosion to the end of the experiment, no lightning at all was seen, nor scarce any thunder heard. A smell of sulphur was perceived, much like that of the luminous electric effluvia issuing out of the end of an electrified bar of metal. Round the string appeared a luminous cylinder of light, three or four inches in diameter: and this being in the day-time Mr. De Romas did not question but that, if it had been in the night, that electric atmosphere would have appeared to be four or five feet in diameter. Lastly, after experiments were over, a hole was discovered in the ground, perpendicularly under the tin tube, an inch deep, and half an inch wide, which was probably made by the large flashes that accompanied the explosions.

"An end was put to these remarkable experiments by the falling of the kite, the wind being shifted into the East, and rain mixed with hail coming on in great plenty. Whilst the kite was falling, the string came foul of a penthouse; and it was no sooner disengaged, than the person who held it felt such a stroke in his hands, and such a commotion through his whole body, as obliged him instantly to let it go; and the string, falling on the feet of some other persons, gave them a shock also, though much more tolerable."

The extraordinary interest awakened by the successful drawing of electricity from the clouds caused these experiments to be repeated in different parts of the world, with apparatus of various types. To a great extent, this apparatus consisted of rods extending upwards into the air. Some of this apparatus was exceedingly simple and crude in its construction, as may be seen from the following description given by Priestley:

"The most accurate experiments made with these imperfect instruments, were those of Mr. Monnier. He was convinced that the high situation in which the bar of iron had commonly been placed was not absolutely necessary for this purpose: for he observed a common speaking-trumpet, suspended upon silk five or six feet from the ground, to exhibit very evident signs of electricity. He also found that a man placed upon cakes of rosin, and holding in his hand a wooden pole, about eighteen feet long, about which an iron wire was twisted, was so well electrified when it thundered, that very lively sparks were drawn from him; and that another man, standing upon non-electrics, in the middle of a garden, and only holding up one of his hands in the air, attracted, with the other hand, shavings of wood which were held to him."

An Englishman named Mr. Canton, succeeded in drawing an electric discharge from the clouds on the 20th of July, 1752, as will be seen from a letter written by him, dated Spital-Square, July 21, 1752:

"I had yesterday, about five in the afternoon, an opportunity of trying
"Mr. Franklin's experiment of extracting the electrical fire from the
"clouds; and succeeded, by means of a tin tube, between three and four
"feet in length, fixed to the top of a glass one, of about eighteen inches. To
"the upper end of the tin tube, which was not so high as a stack of chimnies
"on the same house, I fastened three needles with some wire; and to the
"lower end was folded a tin cover to keep the rain from the glass tube,
"which set upright in a block of wood. I attended this apparatus as soon
"after the thunder began as possible, but did not find it in the least electri-
"fied, till between the third and fourth clap; when applying my knuckle to
"the edge of the cover, I felt and heard an electrical spark; and approach-
"ing it a second time, I received the spark at the distance of about half an
"inch, and saw it distinctly. This I repeated four or five times in the space
"of a minute; but the sparks grew weaker and weaker; and in less than two
"minutes the tin tube did not appear to be electrified at all. The rain con-
"tinued during the thunder, but was considerably abated at the time of mak-
"ing the experiment."

At a somewhat later date, August 12, 1752, another Englishman, a Mr. Wilson, succeeded, by the use of an exceedingly crude apparatus, in drawing electricity from the clouds. The following description of Wilson's apparatus experiment is from a letter of Mr. Watson to the Royal Society:

"Mr. Wilson likewise of the Society, to whom we are much obliged for the trouble he has taken in these pursuits, had an opportunity of verifying Mr. Franklin's hypothesis. He informed me, by a letter from near Chelmsford in Essex, dated August 12, 1752, that, on that day about noon, he perceived several electrical snaps, during, or rather at the end of a thunder storm, from no other apparatus than an iron curtain rod, one end of which he put into the neck of a glass phial, and held this phial in his hand. To the other end of the iron he fastened three needles with some silk. This phial, supporting the rod, he held in one hand, and drew snaps from the rod with a finger of his other. This experiment was not made upon any eminence, but in the garden of a gentleman, at whose house he then was."

Shortly after his successful experiment with the electrical kite, Franklin erected a rod on his house, of which he gives the following description:

In September, 1752, I erected an iron rod to draw the lightning down into my house, in order to make some experiments on it, with two bells to give notice when the rod should be electrifi'd: A contrivance obvious to every electrician.

"I found the bells rang sometimes when there was no lightning or thunder, but only a dark cloud over the rod: that sometimes after a flash of lightning they would suddenly stop; and, at other times, when they had not rang before, they would, after a flash, suddenly begin to ring; that the electricity was sometimes very faint, so that when a small spark was obtain'd, another could not be got for sometime after; at other times the sparks would follow extremely quick, and once I had a continual stream from bell to bell, the size of a crow-quill: Even during the same gust, there were considerable variations."

In a letter dated April 18th, 1754, to Collinson, Franklin describes some of the results obtained by this apparatus:

"Since September last, having been abroad on two long journeys, and otherwise much engag'd, I have made but few observations on the *positive* and *negative* state of electricity in the clouds. But Mr. Kinnersley kept his rod and bells in good order, and has made many.

"Once this winter the bells rang a long time, during a fall of snow, tho' no thunder was heard, or lightning seen. Sometimes the flashes and cracks of the electric matter between bell and bell were so large and loud as to be heard all over the house: but by all his observations, the clouds were constantly in a negative state, till about six weeks ago, when he found them once to change in a few minutes from the negative to the positive. About a fortnight after that he made another observation of the same kind; and last Monday afternoon, the wind blowing hard at S. E. and veering round to N. E. with many thick driving clouds, there were five or six successive changes from negative to positive, and from positive to negative, the bells stopping a minute or two between every change. Besides the methods mentioned in my paper of *September* last, of discovering the electrical state of the clouds the following may be us'd: When your bells are ringing, pass a rubb'd tube by the edge of the bell, connected with your pointed rod: if the cloud is then in a negative state, it will continue, and perhaps be quicker. Or, suspend a very small cork-ball by a fine silk thread, so that it may hang close to the edge of the rod-bell: then whenever the bell is electrified, whether positively or negatively, the little ball will be repell'd, and continue at some distance from the bell. Have ready a round-headed glass stopper of a decanter, rub it on your side till it is electrified, then present it to the cork-ball. If the electricity in the ball is positive, it will be repell'd from the glass stopper as well as from the bell. If negative, it will fly to the stopper."

It seldom falls to the lot of any scientific man, no matter how great his ability, that he is able to follow a road so absolutely new that it has never been traversed before. This unquestionably was the case with Franklin in his grand discovery of drawing the lightning from the sky, thus robbing Jove of his thunder-bolts.

Very long, indeed, before Franklin's time, the general resemblance between lightning and electrical discharges had been noted. Indeed, some go so far as to say that before lightning had been known to be a form of electricity, some bold investigators had actually succeeded in purposely drawing it to the earth. The fabled Prometheus, who is credited with having stolen the sacred fire from Heaven, is claimed by some as being in reality Franklin's predecessor; that the sacred fire was lightning, purposely drawn down from the clouds by Prometheus. There is, of course, no actual evidence of this being the case. It would seem to me that the fable of Prometheus was intended to immortalize the act of that master mind who first taught the human race the art of cooking, thus increasing the number of food products that could be safely employed after the material had been subjected to the action of heat.

Another ancient who has been brought forward as a predecessor of Franklin was one of the earlier Romans, Numa Pompilius, who, it is claimed, on several occasions drew down the sacred fire with entire safety. Not so fortunate, however, was Tillius Hostilius, who, it is claimed, having read some notes left by Numa respecting the sacred art of worshipping Jupiter, rashly attempted to repeat the worship, but, departing from the rules of the sacred rite, was struck dead by a bolt from the irate Jupiter.

Ovid thus refers to one of the Kings of Alba, who was struck dead by a bolt from Heaven while performing a similar ceremony:

"Fulmineo periit imitator fulminis ictu."
("In imitating thunder, the thunderer perished.")

But the prior claimants for the honor of antedating Franklin in his great discovery are not limited to these vague beliefs. There are a number of unquestioned printed records in which, at least the close resemblance between electrical discharges and lightning flashes are referred to. Some of these are as follows:

In 1705, Hawkesbee, while experimenting on the luminous phenomena produced by permitting mercury to fall from the top of a glass tube in which a partial vacuum is maintained, noted that there were thus "Produced flashes resembling lightning;" and at a later date, the same hilosopher, noting the luminous effects produced by the friction of a woolen cloth against a glass tube, said that he "Observed light to break from the agitated glass in as strange a form as lightning." Hawkesbee did not know that the luminous effects were in reality electric discharges, so that this can scarcely be regarded as an anticipation of Franklin's work.

In the same year, however, Gray calls attention, in the Philosophical Transactions, to the resemblances that exist between the effects of electric discharges and those of lightning and thunder.

In 1708, Wall, in the Philosophical Transactions, calls attention to the resemblance between the crackling and the flash accompanying the rubbing of amber, and thunder and lightning, remarking: "This light and crackling seem in some degree to represent thunder and lightning."

At a much later date; *i. e.*, in 1748, the Abbe Nollet, in his "Lessons in Physics," says:

"If any one should undertake to prove, as a clear consequence of the phenomenon, that thunder is in the hands of nature what Electricity is in ours * * * that those wonders which we dispose at our pleasure are only imitations on a small scale of those grand effects which terrify us, and that both depend on the same mechanical agents, if it were made manifest that a cloud prepared by the effects of the wind, by heat, by a mixture of exhalations, etc., is in relation to a terrestrial object what an electrified body is in relation to a body near it not electrified, I confess that this idea, well supported, would please me much; and to support it how numerous and specious are the reasons which present themselves to a mind conversant with Electricity. The universality of the electric matter, the readiness of its actions, its instrumentality and its activity in giving fire to ohter bodies, its property of striking bodies externally and internally, even to their smallest parts (the remarkable example we have of this effect even in the Leyden-jar experiment, the idea which we might truly adopt in supposing a greater degree of electric power), all these points of analogy which I have been for some time meditating, begin to make me believe that one might, by taking electricity for the model, form to one's self in regard to thunder and lightning more perfect and more probable ideas than hitherto proposed."

As will be seen, however, this suggestion is general, and does not appear to have ever been carried out by Nollet, nor can it for a moment properly be regarded as competing with the publication of Franklin before referred to, coupled, as Franklin's paper was, with a clear description of the manner in which a mimic lightning cloud discharged a bolt to neighboring objects, together with full directions as to how such a discharge might be avoided by the use of a suitably constructed pointed rod.

The following extracts from a letter written to Mr. Collinson give an account of Franklin's hypothesis for explaining the phenomena of thunder-gusts. This letter appears to have been written some time between April 29th, 1749, and July 29th, 1750. As the paper is a long one, I will only give extracts from it. The first portion of the paper refers to the manner in which electricity is capable of being produced by the friction of air against water.

"Observations and Suppositions, towards forming a new Hypothesis, for explaining the several Phenomena of Thunder-Gusts:

"9. The ocean is a compound of water, a non-electric, and salt an electric per se.

"10. When there is a friction among the parts near its surface the electrical fire is collected from parts below. It is then plainly visible in the night; it appears at the stern and in the wake of every sailing vessel; every dash of an oar shews it, and every surf and spray: In storms the whole sea seems on fire. * * * The detach'd particles of water then repelled from the electrified surface, continually carry off the fire as it is collected; they rise and form clouds, and those clouds are highly electrified, and retain the fire till they have an opportunity of communicating it.

"11. The particles of water rising in vapours, attach themselves to particles of air."

   *   *   *   *   *   *   *

"26. Hence clouds formed by vapours raised from fresh waters within land, from growing vegetables, moist earth, &c., more speedily and easily deposite their water, having but little electrical fire to repel and keep the particles separate. So that the greatest part of the water raised from the land to the sea are dry; there being little use for rain on the sea, and to rob the land of its moisture, in order to rain on the sea, would not appear reasonable.

"27. But clouds formed by vapours raised from the sea, having both fires, and particularly a great quantity of the electrical, support their water strongly, raise it high, and being moved by winds, may bring it over the middle of the broadest continent from the middle of the widest ocean.

"28. How these ocean clouds, strongly supporting their water, are made to deposite it on the land where it is want'd is next to be considered.

"29. If they are driven by winds against mountains, those mountains being less electrified attract them, and on contact take away their electrical fire (and being cold, the common fire also;) hence the particles close towards the mountains and towards each other. If the air was not much loaded, it only falls in dews on the mountain tops and sides, forms springs, and descends to the vales in rivulets, which united, make larger streams and rivers. If much loaded, the electrical fire is at once taken from the whole cloud; and, in leaving it, flashes brightly and cracks loudly; the particles instantly coalescing for want of that fire, and falling in a heavy shower.

"30. When a ridge of mountains thus dams the clouds, and draws the electrical fire from the cloud first approaching it; that which next follows, when it comes near the first cloud, now deprived of its fire, flashes into it, and begins to deposit its own water; the first cloud again flashing into the mountains; the third approaching cloud, and all the succeeding ones, acting in the same manner as far back as they extend, which may be over many hundred miles of country.

"31. Hence the continual storms of rain, thunder, and lightning on the east side of the *Andes*, which running north and south, and being vastly high, intercept all the clouds brought against them from the *Atlantic* ocean by the trade winds, and oblige them to deposit their waters, by which the vast rivers *Amazons*, *La Plata*, and *Oroonoko* are formed, which return the water into the same sea, after having fertilized a country of very great extent.

"32. If a country be plain, having no mountains to intercept the electrified clouds, yet it is not without means to make them deposit their water. For if an electrified cloud coming from the sea, meets in the air a cloud raised from the land, and therefore not electrified; the first will flash its fire into the latter, and thereby both clouds shall be made suddenly to deposit water.

"33. The electrified particles of the first cloud close when they lose their fire, the particles of the other cloud close in receiving it: in both, they have thereby an opportunity of coalescing into drops. * * * The concussion or jerk given to the air, contributes also to shake down the water, not only from those two clouds, but from others near them. Hence the sudden fall of rain immediately after flashes of lightning.

*   *   *   *   *   *   *

"35. Thus when sea and land clouds would pass at too great a distance from the flash, they are attracted towards each other till within that distance; for the sphere of electrical attraction is far beyond the distance of flashing.

"36. When a great number of clouds from the sea meet a number of clouds raised from the land, the electrical flashes appear to strike in different parts; and as the clouds are jostled and mixed by the winds, or brought near by the electrical attraction, then continue to give and receive flash after flash, till the electrical fire is equally diffused."

*   *   *   *   *   *   *

"40. When the air, with its vapours raised from the ocean between the tropics, comes to descend in the polar regions, and to be in contact with the vapours arising there, the electrical fire they brought begins to be communicated, and is seen in clear nights, being first visible where 'tis first in motion, that is, where the contact begins, or in most northern part; from thence the streams of light seem to shoot southerly, even up the zenith of northern countries. But tho' the lights seem to shoot from the north southerly, the progress of the fire is really from the south northerly, its miction beginning in the north being the reason that 'tis there first seen.

"For the electrical fire is never visible but when in motion, and leaping from body to body, or from particle to particle thro' the air. When it passes thro' dense bodies 'tis unseen. When a wire makes part of the circle, in the explosion of the electrical phial, the fire, though in great quantity, passes in the wire invisibly: but in passing along a chain, it becomes visible as it leaps from link to link. In passing along leaf gilding 'tis visible; for the leaf-gold is full of holes; hold a leaf to the light and it appears like a net, and the fire is seen in its leaping over the vacancies. * * * And as when a long canal filled with still water is opened at one end, in order to be discharged, the motion of the water begins first near the opened end, and proceeds towards the close end, tho' the water itself moves from the close towards the open end: so the electrical fire discharged into the polar regions, perhaps from a thousand leagues length of vaporised air, appears first where 'tis first in motion, *i.e.*, in the most northern part, and the appearance proceeds southward, tho' the fire really moves northward. This is supposed to account for the *Aurora Borealis*.

"41. When there is great heat on the land, in a particular region (the sun having shone on it perhaps several days, while the surrounding countries have been screen'd by clouds) the lower air is rarified and rises, the cooler denser air above descends; the clouds in that air meet from all sides, and join over the heated place; and if some are electrified, others not, lightning and thunder succeed, and showers fall. Hence thunder-gusts after heats, and cool air after gusts; the water and the clouds that bring it, coming from a higher and therefore a cooler region.

"42. An electrical spark, drawn from an irregular body at some distance is scarce ever straight, but shows crooked and waving in the air. So do the flashes of lightning; the clouds being very irregular bodies.

"43. As electrified clouds pass over a country, high hills and high trees lofty towers, spires, masts of ships, chimneys, &c., as so many prominencies and points, draw the electrical fire, and the whole cloud discharges there."

It is interesting to note that Franklin's theory of thunder-gusts as contained in the above quotation, was published before he had actually drawn electricity from the clouds. This theory was, consequently, crude in many respects. Nevertheless, it is surprising how great an insight into the true theory of the thunder storm Franklin seemed to have obtained. This will be seen from the following general statements contained in this remarkable paper:

(1) He traces thunder-gusts to their true source; *i. e.*, the formation of clouds by vapors rising from the ocean.

(2) He points out that the most marked effects of lightning are produced when charged clouds are attracted to mountains or other objects on the earth; when, losing their electricity, the particles of water that were prevented from condensing by reason of the repulsion produced by that charge, fall to the ground. Here he appears to have lost sight of the fact that the loss of the charge follows the condensation rather than precedes it. This general truth he appears to have realized at a later period.

(3) That oppositely-charged clouds are capable of discharging into each other when they approach sufficiently near.

(4) That the frequent thunder storms which occur on the eastern plateau of the Andes are due to the discharge of the clouds on their coming in contact with the slopes of the mountains.

But Franklin goes further than this, and ascribes the cause of the aurora borealis to the effects of electrical discharges in regions where the discharges are no longer disruptive, as in the case of lightning, but take the form of quiet discharges.

It will be seen from an examination of paragraphs 9, 10 and 11, as given in the first part of the quotation, that while Franklin points out at least one of the ways in which the vapor obtains its electrical charge; *i. e.*, by the friction between the air and the surface of the ocean, he is in error in believing that the glow of light which is frequently distinctly visible during dark nights at the bow and in the wake of a sailing vessel, is an evidence of the presence of such an electrical charge, since, as it is now known, this glow is due to phosphorescence produced in an entirely different manner. At a later date; *i. e.*, January, 1752, in a letter to James Bowdoin, dated Philadelphia, January 24, 1752, and read before the Royal Society, May 27, 1756, he refers to this matter as follows:

"My supposition, that the sea might possibly be the grand source of lightning, arose from the common observation of its luminous appearance in the night, on the least motion; an appearance never observed in fresh water. Then I knew that the electric fluid may be pumped up out of the earth, by the friction of a glass globe, on a non-electric cushion; and that. notwithstanding the surprising activity and swiftness of that fluid, and the non-electric communication between all parts of the cushion and the earth, yet quantities would be snatched up by the revolving surface of the globe, thrown on the prime conductor, and dissipated in air. How this was done, and why that subtile, active spirit did not immediately return again from the globe into some part or other of the cushion, and so into the earth, was difficult to conceive; but, whether from its being opposed by a current setting upwards to the cushion, or from whatever other cause, that it did not so return was an evident fact. Then I considered the separate particles of water as so many hard spherules, capable of touching the salt only in points and imagined a particle of salt could therefore no more be wet by a particle of water, than a globe by a cushion; that there might therefore be such a friction between these original constituent particles of salt water, as in at sea of globes and cushions; that each particle of water on the surface might obtain from the common mass, some particles of the universal diffused, much finer, and more subtile electric fluid, and, forming to itself an atmosphere of those particles, be repelled from the then generally electrified surface of the sea, and fly away with them into the air. I thought, too, that possibly the great mixture of particles electric *per se*, in the ocean water might, in some degree, impede the swift motion and dissipation of the electric fluid through it to the shores, &c. But, having since found, that salt in the water of an electric phial does not lessen the shock; and having endeavored in vain to produce that luminous appearance from mixture of salt and water agitated; and observed, that even the sea-water will not produce it after some hours standing in a bottle; I suspect it to proceed from some principle yet unknown to us (which I would gladly make some experiments to discover, if I lived near the sea), and I grow more doubtful of my supposition, and more ready to allow weight to that objection (drawn from the activity of the electrical fluid, and the readiness of water to conduct), which you have indeed stated with great strength and clearness.

"In the meantime, before we part with this hypothesis, let us think what to substitute in its place. I have sometimes queried, whether the friction of the air, an electric *per se*, in violent winds, among trees, and against the surface of the earth, might not pump up, as so many glass globes, quantities of the electrical fluid, which the rising vapors might receive from the air, and retain in the clouds they form; on which I should be glad to have your sentiments. An ingenious friend of mine supposes the land clouds more likely to be electrified than the sea clouds. I send his letter for your perusal, which please return to me."

But paragraph 41, of the quotation, entitled Observations and Suppositions towards forming a new Hypothesis for Explaining the several Phenomena of Thunder-Gusts, is most worthy of notice in pointing out the cause of thunder-gusts. Here Franklin refers to any highly-heated area on the surface of the earth as the source from which the thunder-gust starts or has its origin. "When there is great heat on the land in a particular region (the sun having shone on it perhaps several days, while the surrounding countries have been screen'd by clouds), the lower air is rarefied and rises, the cooler, denser air above descends; the clouds in that air meet from all sides and join over the heated place; and if some are electrified, others not, lightning and thunder descends, and showers fall; hence thunder-gusts after heats, and cool air after gusts, the water, and the clouds that bring it, coming from the higher, and therefore a cooler, region."

Franklin's idea that the electricity of the thunder cloud comes from the vapors that rise from the waters of the ocean was in accord with views that have been set forth after his time. Volta, for example, at a later date, was, perhaps, the first to assert that the free electricity of the air was due to the evaporation of water. Volta's idea was that evaporation alone, unattended by any wind or friction, could produce an electrical charge. These two statements therefore differed from each other in the fact that Franklin apparently regarded the cause of the charge to be due to the friction of the air against the particles of water, while Volta recognized that the mere molecular friction, such as attended the tearing apart of the molecules of the liquid water so as to permit them to vaporize, was also a cause of the charge.

Pouillet made the assertion that the evaporation of ocean water, in which, as is well known, a large percentage of common salt exists in solution, produced the most marked electrification; that under such circumstances the vapor was positively charged, while the vessel containing the water was negatively charged.

It is interesting to note in this connection a modern theory of atmospheric electricity proposed by Prof. Oliver Lodge, who has given considerable attention to these phenomena. Lodge questions whether it is probable that the electricity of the atmosphere is obtained by the friction of air against water, although he thinks it possible that a spray of mist driven by a storm wind against the earth's surface, may be able to produce electric charges. He suggests that since, as is well known. all thunder storms are attended by severe atmospheric disturbances, in which a rotary or whirling motion of the wind exists, that if it be acknowledged that in thunder-gusts the axis of this rotary motion is horizontal, there would be produced in this manner a species of huge natural cylindrical frictional machine, with the earth as the rubber, and the upper conducting regions of the atmosphere as the prime conductor; that the air electrified by friction against the earth carries the charges so produced into the upper regions of the air, into which it discharges them.

Before passing from theory to facts, it will be interesting to review some of the attendant phenomena of thunder storms.

As is well known, thunder storms are generally divided into three classes; i. e.,—

(1.) Cyclonis Thunder Storms.
(2.) Heat Thunder Storms.
(3.) Winter Thunder Storms.

The cyclonic thunder storms attend areas of low barometer, and have a progressive motion, similar to that of cyclones. Under certain circumstances these storms produce tornadoes.

Heat thunder storms are those that are produced by the local heating of the lower air. A quiet day, when the air is filled with moisture, is favorable for the formation of storms of this character. They require for their production a sufficient heating of the air to carry it by means of strong ascending currents into the higher regions. It was mainly storms of this character that were studied by Franklin.

Winter thunder storms occur more frequently at night and in the higher latitudes. These storms are more frequent near the coasts of the continents than in the interior.

As is well known a thunder storm is characterized by a rotary motion of the atmosphere, with a progressive motion in which the area around which the air rotates moves bodily over the surface of the earth. Thunder storms only occur when there is an unusual quantity of moisture in the air. Consequently, they are generally accompanied by rainfall. The following description, condensed from Davis, gives an excellent idea of the different phenomena of the storm. In the temperate latitudes of the earth the beginning of the storm is heralded by hot day, of a fore-running layer of fleecy, feathery clouds, known as cirro-status clouds, that are arranged in horizontal bands or layers in the higher regions of the atmosphere. The forward edge of this layer is thin, fibrous and hazy in appearance. As the cloud advances it grows thicker at its opposite end, from which ragged festoons of clouds slowly descend and dissolve from its lower surface as the great rain-bearing cloud mass approaches. The fore-running cloud may at times be from ten to fifty miles in advance of the approaching rain cloud. The air, which is oppressively hot before the storm, grows slightly cooler as the fore-running cloud hides the sun. Thunder heads, or lurid rounded dark clouds, can generally be seen rising in the west an hour or so before the fore-running sheet of clouds. Distant thunder is heard as these clouds approach, when below their level base, b, Fig. 7, a gray rain curtain, r, is to be seen, trailing over the ground and hiding all objects behind it. Small detached clouds, d, frequently form in front of the main cloud mass, rapidly increasing in size, and finally merging with the

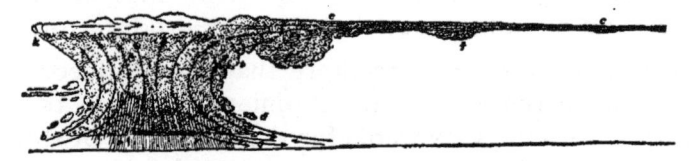

Fig. 7. Phenomena of approaching thunder gust. (Davis)

storm cloud, which, moving more rapidly towards the east, at length overtakes them. A ragged squall cloud, s, light gray in color, rolls beneath the dark mass of clouds, somewhat back of its forward edge. The mass of storm cloud advances broadside with a velocity of twenty to fifty miles an hour. The short-lived outrushing wind squall, g, carrying with it clouds of dust, moves below the clouds and in front of the rain. A marked decrease in temperature occurs during the storm, this decrease frequently being from ten to twenty degrees in less than half an hour. The first falling drops of rain are large, rapidly changing, however, to a heavy downpour, that is sometimes accompanied by hail. The moisture in the air increases, vivid lightning flashes occur, with loud thunder, and, as the storm center comes more nearly overhead, the lightning flashes and thunder succeed each other more rapidly. The dark sheet in front of the storm grows less marked, and, as the center passes over, the lightning flashes become less frequent, until the storm passes, the rain ceases, the clouds break in the west, blue sky appears, the air grows cooler, dryer and clearer.

In continuation of Franklin's theory of thunder storms and lightning generally, the following letter, written to Collinson from Philadelphia, September, 1753, and consequently after he had drawn the electricity from the clouds, gives an account of some experiments he made in order to determine whether clouds are electrified positively or negatively:

"In the winter following I conceived an experiment, to try whether the clouds were electrifi'd positively or negatively; but my pointed rod, with its apparatus, becoming out of order, I did not refit it till towards the spring when I expected the warm weather would bring on more frequent thunder-clouds.

"The experiment was this: To take two phials; charge one of them with lightning from the iron rod, and give the other an equal charge by the electric glass globe, thro' the prime conductor: When charg'd to place them on a table within three or four inches of each other, a small cork ball being suspended by a fine silk thread from the ceiling, so as it might play between the wires. If both bottles then were electrified positively, the ball being attracted and repelled by one, must be also repell'd by the other. If the one positively, and the other negatively; then the ball would be attracted and repelled—alternately by each, and continue to play between them as long as any considerable charge remained.

"Being very intent on making this experiment, it was no small mortification to me, that I happened to be abroad during two of the greatest thunder-storms we had early in the Spring, and tho' I had given orders in my family, that if the bells rang when I was from home, they should catch some of the lightning for me in electrical phials, and they did so, yet it was mostly dissipated before my return, and in some of the other gusts, the quantity of lightning I was able to obtain was so small, and the charge so weak, that I could not satisfy myself: Yet I sometimes saw what heighten'd my suspicions, and inflamed my curiosity.

"At last, on the 12th of April, 1753, there being a smart gust of some continuance, I charged the phial pretty well with lightning, and the other equally, as near as I could judge, with electricity from my glass globe; and, having placed them properly, I beheld, with great surprise and pleasure, the cork ball played briskly between them; and was convinced that one bottle was electrised negatively.

"I repeated this experiment several times during the gust, and in eight successive gusts, always with the same success: and being of opinion (for reasons I formerly gave in my letter to Mr. Kinnersly, since printed in London) that the glass globe electrises positively, I concluded that the clouds are always electrised negatively, or have always in them less than their natural quantity of the electric fluid.

"Yet notwithstanding so many experiments, it seems I concluded too soon; for at last, June 6th, in a gust which continued from five o'clock, P. M to seven, I met with one cloud that was electrised positively, tho' several that pass'd over my rod before, during the same gust, were in the negative state. This was thus discovered.

"I had another concurring experiment, which I often repeated, to prove the negative state of the clouds, viz. While the bells were ringing, I took the phial charged from the glass globe, and applied its wire to the erected rod, considering, that if the clouds were electrised positively, the rod, which received its electricity from them, must be so too; and then the additional positive electricity of the phial would make the bells ring faster. * * * * But if the clouds were in a negative state, they must exhaust the electric fluid from the rod, and bring that into the same negative state with themselves, and then the wire of a positively charg'd phial, supplying the rod with what it wanted, (which it was obliged otherwise to draw from the earth by means of the pendulous brass ball playing between the two bells) the ringing would cease until the bottle was discharg'd.

"In this manner I quite discharged into the rod several phials that were charged from the glass globe, the electric fluid streaming from the wire to the rod, 'till the wire would receive no spark from the finger; and during this supply to the rod from the phial the bells stopt ringing; but by continuing the application of the phial wire to the rod, I exhausted the natural quantity from the inside surface of the same phials, or, as I call it, charged them negatively.

"At length, while I was charging a phial by my glass globe, to repeat this experiment, my bells, of themselves, stopt ringing, and, after some pause, began to ring again. * * * But now, when I approached the wire of the charg'd phial to the rod, instead of the usual stream that I expected from the wire to the rod, there was no spark; not even when I brought the wire and the rod to touch; yet the bells continued ringing vigorously, which proved to me, that the rod was then positively electrify'd, as well as the wire of the phial, and equally so; and, consequently, that the particular cloud then over the rod, was in the same positive state. This was near the end of the gust.

"But this was a single experiment, which, however, destroys my first too general conclusion, and reduces me to this: *That the clouds of a thundergust are most commonly in a negative state of electricity, but sometimes in a positive state.*

"The latter I believe is rare; for tho' I soon after the last experiment, set out on a journey to Boston, and was from home most part of the summer, which prevented my making farther trials and observations; yet Mr. Kinnersley returning from the islands just as I left home, pursued the experiments during my absence, and informs me that he always found the clouds in the negative state.

"So that, for the most part, in thunder-strokes, *'tis the earth that strikes into the clouds and not the clouds that strike into the earth.*"

The above quotation is of such a nature as to need no comment. Quite early in his electrical investigations, Franklin proposed a theory as to the cause of the aurora borealis. The aurora, the "morning hour," takes its name, as is well known, from the fact that the light, seen near the horizon at the beginning of the phenomena, presents an appearance not unlike that of the dawn of day. During the prevalence of the aurora, an arch or corona, as it is called, is to be seen in the northern sky, with its highest part situated immediately under that part which is occupied by the north magnetic pole of the earth. The height of this arch varies in different latitudes, being higher in the higher latitudes than it is in the regions near the Equator.

As the aurora progresses, the arch or corona rises higher in the sky and streams of light of varying colors, white, red and purplish, and sometimes, though more rarely, of the

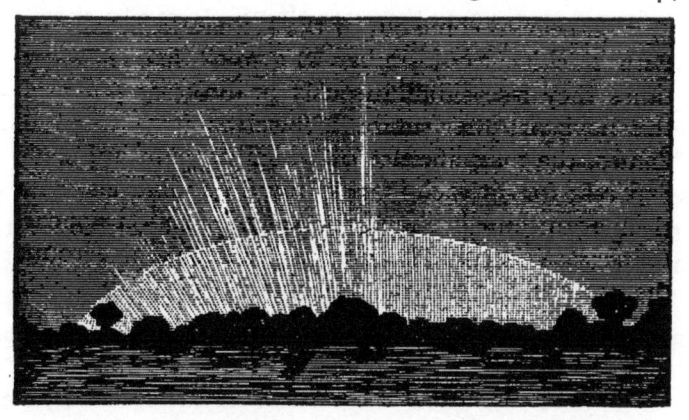

Fig. 8. Aurora Arch (Electricity in Every Day Life—Houston)

other colors of the spectrum, are suddenly to be seen darting in a weird manner upwards from the arch: Sometimes a single streamer will start up, increasing rapidly in size and brilliancy, moving over the sky, and rapidly fading away. At other times, a series of streamers follow one another in rapid succession. The general appearance of the auroral arch is seen in Fig. 8.

As we have already seen, Franklin suggested the cause of the aurora borealis in his history of thunder-gusts. In this paper he calls attention to the fact that the electric fire is never visible unless it is in motion, leaping from body to body, or from particle to particle through the air; that when it is passing in dense bodies it is invisible.

Again, in a letter to Cadwalader Colden, written at Philadelphia, April 23, 1752, and read before the Royal Society, November 11th, 1756, he says:

"Your conception of the electrical fluid, that is incomparably more subtile than air, is undoubetdly just. It pervades dense matter with the greatest ease; but it does not seem to mix or incorporate willingly with mere air, as it does with other matter. It will not quit common matter to join with air. Air obstructs, in some degree, its motion. An electric atmosphere cannot be communicated at so great a distance, through intervening air, by far, as through a vacuum. Who knows, then, but there may be, as the ancients thought, a region of this fire above our atmosphere, prevented by our air, and is own too great distance for attraction, from joining our earth? Perhaps where the atmosphere is rarest, this fluid may be densest, and nearer the earth, where the atmosphere grows denser, this fluid may be rarer; yet some of it be low enough to attach itself to our highest clouds, and thence they becoming electrified, may be attracted by, and descend towards the earth, and discharge their watery contents, together with that ethereal fire. Perhaps the aurorae boreales are currents of this fluid in its own region, above our atmosphere, becoming from their motion visible. There is no end of conjectures. As yet we are but novices in this branch of natural knowledge."

Franklin's most complete paper on the aurora borealis was read at the Royal Academy of Sciences, at Paris, at some time immediately before Easter, 1779. It was as follows:

"AURORA BOREALIS.
"Suppositions and Conjectures towards forming an Hypothesis for its Explanation.

"1. Air heated by any means becomes rarefied and specifically lighter than other air in the same situation not heated.

"2. Air being thus made lighter rises, and the neighbouring cooler, heavier air takes its place.

"3. If in the middle of a room you heat the air by a stove, or pot of burning coals near the floor, the heated air will rise to the ceiling, spread there over the cooler air till it comes to the cold walls; there being condensed and made heavier, it descends to supply the place of that cool air which had moved towards the stove or fire, in order to supply the place of the heated air which had ascended from the space around the stove or fire.

"4. Thus there will be a continual circulation of air in the room, which may be rendered visible by making a little smoke; for that smoke will rise and circulate with the air.

"5. A similar operation is performed by nature on the air of the globe. Our atmosphere is of a certain height, perhaps at a medium   miles. Above that height it is so rare as to be almost a vacuum. The air heated between the tropics is continually rising, and its place is supplied by northerly and southerly winds which come from those cool regions.

"6. The light, heated air, floating above the cooler and denser, must spread northward and southward, and descend near the two poles, to supply the place of the cooler air which had moved towards the equator.

"7. Thus a circulation of air is kept up in our atmosphere as in the room above mentioned.

"8. That heavier and lighter air may move in currents of different and even opposite directions, appears sometimes by the clouds that happen to be in these currents, as plainly as by the smoke in the experiment above mentioned. Also in opening a door between two chambers, one of which has been warmed, by holding a candle near the top, near the bottom, and near the middle, you will find a strong current of warm air passing out of the warmed room above, and another of cool air entering it below, while in the middle there is little or no motion.

"9. The great quantity of vapor rising between the tropics forms clouds, which contain much electricity.

"Some of them fall in rain, before they come to the polar regions.

"10. If the rain be received in an isolated vessel, the vessel will be electrified; for every drop brings down some electricity with it.

"11. The same is done by snow and hail.

"12. The electricity so descending in temperate climates, is received and imbibed by the earth.

"13. If the clouds are not sufficiently discharged by this means, they sometimes discharge themselves suddenly by striking into the earth, where the earth is fit to receive their electricity.

"14. The earth in temperate and warm climates is generally fit to receive it, being a good conductor.

"15. A certain quantity of heat will make some bodies good conductors, that will not otherwise conduct.

"16. Thus wax rendered fluid, and glass softened by heat, will both of them conduct.

"17. And water, though naturally a good conductor, will not conduct well when frozen into ice by a common degree of cold; not at all where the cold is extreme.

"18. Snow falling upon frozen ground has been found to retain its electricity; and to communicate it to an isolated body, when after falling, it has been driven about by the wind.

"19. The humidity, contained in all the equatorial clouds that reach the polar regions, must there be condensed and fall in snow.

"20. The great cake of ice that continually covers those regions may be too hard frozen to permit the electricity, descending with that snow, to enter the earth.

"21. It will therefore be *accumulated upon that ice.*

"22. The atmosphere being heavier in the polar regions, than in the equatorial, will there be lower; as well from that cause, as from the smaller effect of the centrifugal force; consequently the distance to the vacuum above the atmosphere will be less at the poles than elsewhere; and probably much less than the distance (upon the surface of the globe) extending from the pole to those latitudes in which the earth is so thawed as to receive and imbibe electricity; the frost continuing to latitude 80, which is 10 degrees or 600 miles from the pole, while the height of the atmosphere there, of such density as to obstruct the motion of the electric fluid, can scarce be estimated above   miles.

"23. The vacuum above is a good conductor.

"24. May not then the great quantity of electricity brought into the polar regions by the clouds, which are condensed there, and fall in snow,

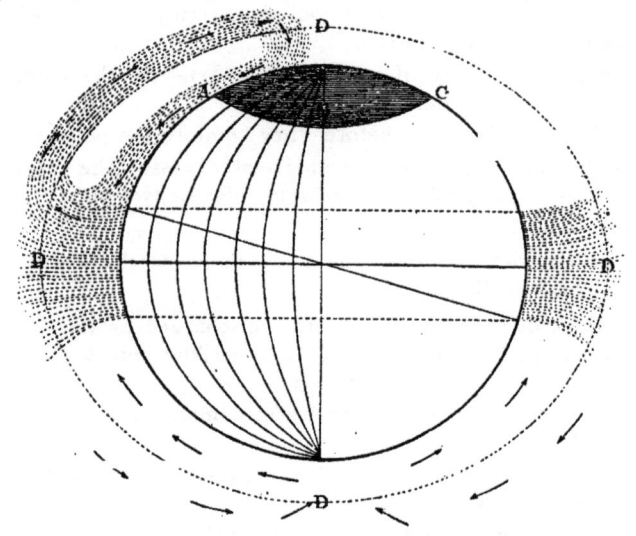

The Arrows represent the general Currents of the Air.
A.B.C. the great Cake of Ice & Snow in the Polar Regions.
D.D.D.D. the Medium Height of the Atmosphere.
The Representation is made only for one Quarter and one Meridian of the Globe: but is to be understood the same for all the rest.

Fig. 9. Franklin on the aurora borealis.

which electricity would enter the earth, but cannot penetrate the ice; may it not, I say *(as a bottle overcharged)* break through that low atmosphere and run along in the vacuum over the air towards the equator, diverging as the degrees of longitude enlarge, strongly visible where densest, and becoming less visible as it more diverges; till it finds a passage to the earth in more temperate climates, or is mingled with their upper air?

"25. If such an operation of nature were really performed, would it not give all the appearances of an aurora borealis? See (Plate XI. Fig. 1) Fig C. (Our Figure 9.)

"26. And would not the aurorae become more frequent *after the approach of winter*: not only because more visible in longer nights; but also because in summer the long presence of the sun may soften the surface of the great ice cake, and render it a conductor, by which the accumlation of electricity in the polar regions will be prevented?

"27. The *atmosphere of the polar regions* being made more dense by the extreme cold, and all the moisture in that air being frozen, may not any great light arising therein, and passing through it, render its density in some degree visible during the night-time, to those who live in the rarer air of more southern latitudes? And would it not, in that case, although in itself a complete and full circle, extending perhaps ten degrees from the pole, appear to spectators so placed (who could see only a part of it) *in the form of a segment*, its chord resting on the horizon, and its arch elevated more or less above it, as seen from latitudes more or less distant, *darkish in color*, yet sufficiently *transparent* to permit some stars to be seen through it?

"28. The rays of electric matter issuing out of a body, diverge by mutually repelling each other, unless there be some conducting body near to receive them; and if that conducting body be at a greater distance, they will *first diverge*, and then *converge*, in order to enter it. May not this account for some of the varieties of figures seen at times in the motions of the luminous matter of the aurorae; since it is possible, that, in passing over the atmosphere from the north, in all directions or meridians, towards the equator, the rays of that matter may find, in many places portions of cloudy region, or moist atmosphere under them, which (being in the natural or negative state) may be fit to receive them, and towards which they may therefore converge; and when one of those receiving bodies is more than saturated, they may *again* diverge from it, towards other surrounding masses of such humid atmosphere, and thus form the *crowns*, as they are called, and other figures, mentioned in the histories of this meteor?

"29. If it be true, that the clouds which go to the polar regions carry thither the vapors of the equatorial and temperate regions, which vapors are condensed by the extreme cold of the polar regions, and fall in snow or hail; the winds which come from those regions ought to be generally dry, unless they gain some humidity by sweeping the ocean in their way; and, if I mistake not, the winds between the northwest and northeast are for the most part dry, when they have continued some time."

Considering the early date at which this paper must have been written, it shows the wonderful insight that Franklin had into the causes of the aurora borealis. Note the following points to which he calls attention:

(1) That at a certain height above the level of the sea, the earth's atmosphere is so rare as to be practically a vacuum, and, therefore, a conductor of electricity.

2. That a circulatory movement takes place in the earth's atmosphere between the equator and the polar regions, the lighter air rising and spreading over the colder and heavier air.

(3) That the vapors rising in the tropical regions contain a great quantity of electricity.

(4) That the electricity brought into the polar regions by the vapors as they are condensed is unable readily to reach the earth, by reason of the non-conducting coating of ice and snow that exists in those regions, and therefore "Break through that low atmosphere and run along in the vacuum over the air towards the Equator, diverging as the degrees of longitude enlarge, strongly visible where densest, and becoming less visible as it more diverges, till it finds a passage to the earth in more temperate climate, or is mingled with the upper air."

"If such an operation of nature were really performed, would it not give all appearance of an aurora borealis?"

(5) That the greater frequency of auroras at the approach of winter is due both to the greater length of the nights, in which such phenomena are visible, as well as to the greater extent and thickness of the ice sheet.

(6) That the movements of the luminous matter of the aurora are probably due to the mutual repulsions of the streams of electrified particles.

The general principles contained in Franklin's paper as regards the cause of aurora borealis agree fairly well with somewhat later and more modern views. Consider, for example, the views of the French philosopher De la Rive, as to the cause of the aurora borealis, that he expresses in a paper published in the Philosophical Magazine:

"We have seen that the atmosphere is constantly charged with positive electricity. * * * electricity furnished from the vapors that rise from the sea, essentially in tropical regions, and that, on the other hand, the earth is negatively electrized; the recomposition or neutralization of the two contrary electricities of the atmosphere and of the terrestrial globe is brought about by means of the greater or less moisture with which the lower strata of the air are impregnated. But it is especially in the polar regions, where the eternal ices, that reign there constantly, condense the aqueous vapors under the form of haze, that this recomposition must be brought about; the more so as the positive vapors are carried thither and accumulated by the tropical current, which, setting out from the equatorial regions, where it occupies the most elevated regions of the atmosphere, descends in proportion as it advances toward the higher latitudes, until in the neighborhood of the poles, where it comes into contact with the earth. It is there then that the discharge between the positive electricity of the vapors and the negative electricity of the earth must essentially take place, with accompaniment of light, when it is sufficiently intense; if, as is almost always the case near the poles, and sometimes in the higher parts of the atmosphere, it meets on its route with extremely small icy particles, which constitute the hazes and the very elevated clouds."

* * * * * * *

"Now, if we examine what ought to take place in the portion of the luminous haze, which is nearest to the terrestial globe, and consequently to the polar regions, we shall find that the magnetic pole should exercise over this electrized matter, which is a veritable movable conductor, traversed by a succession of discharges, an action analogous to that which is exercised in the experiment that we have described when engaged with the luminous effects of electricity, by the pole of an electro-magnet over the jets of electric light that are made to converge in extremely rarefied air. We have seen that, as soon as the soft iron cylinder, which serves as an electro-magnet, is magnetized, the electric light, instead of coming out indifferently from the divers points of the upper surface, that serve as a pole, as had taken place before the magnetization, comes out only from all the points of the circumference of this surface, so as to form around it as it were a continuous luminous ring. This ring possesses a movement of rotation around the magnetized cylinder, sometimes in one direction, sometimes in another, according to the direction of the discharge and the direction of the magnetization. Finally some more brilliant jets seem to come out from this luminous circumference, without being confounded with the rest of the group."

Another of Franklin's contributions to geographical physics, partly electrical in character, was his theory or hypothesis concerning the origin of waterspouts. Waterspouts are phenomena that occur during the progress of a tornado, or, whirling progressive motion of the wind over the ocean. During this movement a funnel-shaped mass of cloud appears over the surface of the ocean. The upper part of the cloud consists of an inverted cone, the base of which rests in the cloud. As the phenomenon progresses, a mass of clouds is projected downward from the lower part of the funnel-shaped mass, and at the same time a column of spray is frequently formed from the ocean, the two uniting in the form of a double inverted cone. The waterspout is due to the rarefaction of the air that is produced by the centrifugal force which drives some of the air from the center of the rotating column, the motion of which is spirally inward and upward. As soon as the moisture laden air enters this central region, where the air pressure is reduced, the temperature of the air is lowered below its dew point. and the vapor is condensed into a cloud. The distance to which water is sucked up from the surface of the ocean is not very high, most of the spout being due to the moisture condensed from the cooler air.

In a paper read June 3, 1756, before the Royal Society, Franklin thus refers to the waterspouts:

(a) "When the air descends with violence in some places, it may rise with equal violence in others, and form both kinds of whirlwinds.

"When air in its whirling motion receding every way from the center or axis of the trumpet, leaves there a vacuum; which cannot be filled through the sides. the whirling air, as an arch, preventing: it must then press in at the open ends.

"The greatest pressure inwards must be at the lower end, the greatest weight of the surrounding atmosphere being there. The air entering, rises within, and carries up dust, leaves, and even heavier bodies that happen in its way, as the eddy, or whirl, passes over land.

"If it passes over water, the weight of the surrounding atmosphere forces up the water into the vacuity, part of which, by degrees, joins with the whirling air, and adding weight. and receiving accelerated motion, recedes still farther from the center or axis of the trump, as the pressure lessens; and at least, as the trump widens, is broken into small particles, and so united with air as to be supported by it, and become black clouds at the top of the trump.

"Thus these eddies may be whirlwinds at land. waterspouts at sea. A body of water so raised may be suddenly let fall. when the motion, &c.. has not strength to support it, or the whirling arch is broken so as to let in the air; falling in the sea, it is harmless, unless ships happen under it. But if in the progressive motion of the whirl. it has moved from the sea, over the land. and there breaks, sudden, violent, and mischievous torrents are the consequences."

(a) An excellent statement of the effect produced by a rapid whirling motion of the atmosphere; *i. e.*, the formation of a vacuum at the center of the axis of the whirl, the air only being able to enter at the open ends. Owing to the greater pressure at the bottom than at the top, the maximum pressure inwards is at the bottom, so that the entering air rises, carrying up such readily movable objects as dust, leaves, etc.; and, when passing over the land, and the velocity of the whirl is great, even heavier bodies, thus producing the phenomena of the whirlwind. As soon, however, as the rotating wind passes over a body of water, the pressure forces the water to enter the empty space at the bottom, most of this water acquiring a rotating motion, the velocity of the whirl increases, thus widening the diameter of the axis. This water, becoming broken into small particles, joins the black clouds at the upper portion of the spout.

In a letter to Dr. ———, of Boston, dated Philadelphia, February 4th, 1753, he gives the following description as to the cause of waterspouts, which is remarkably complete considering the date:

"I ought to have written to you, long since in answer to yours of October 16, concerning the water-spout; but business partly, and partly a desire of procuring further information, by enquiry among my sea-fearing acquaintance, induced me to postpone writing, from time to time, till I am now almost ashamed to resume the subject, not knowing but you may have forgot what has been said upon it.

(a) "Nothing, certainly, can be more improving to a searcher into nature, than objections judiciously made to his opinion, taken up, perhaps, too hastily.. For such objections oblige him to re-study the point, consider every circumstance carefully, compare facts, make experiments, weigh arguments, and be slow in drawing conclusions. And hence a sure advantage results; for he either confirms a truth, before too slightly supported; or discovers an error, and receives instruction from the objector.

(b) "In this view I consider the objections and remarks you sent me, and thank you for them sincerely: But, how much soever my inclinations lead me to Philosophical inquiries, I am so engaged in business, public and private, that those more pleasing pursuits are frequently interrupted, and the chain of thought, necessary to be closely continued in such disquisitions, so broken and disjointed, that it is with difficulty I satisfy myself in any of them: And I am now not much nearer a conclusion, in this matter of the spout, than when I first read your letter.

"Yet, hoping we may, in time, sift out the truth between us, I will send you may present thoughts, with some observations on your reasons, on the accounts in the *Transactions*, and on other relations I have met with. Perhaps, while I am writing, some new light may strike me, for I shall now be obliged to consider the subject with a little more attention.

(c) "I agree with you, that, by means of a vacuum in a whirlwind, water cannot be supposed to rise in large masses to the region of the clouds; for the pressure of the surrounding atmosphere could not force it up in a continued body, or column, to a much greater height than thirty feet. But, if there really is a vacuum in the center, or near the axis of whirlwinds, then, I think, water may rise in such vacuum to that height, or to less height, as the vacuum may be less perfect.

"I had not read *Stuart's* account, in the *Transactions*, for many years, before the receipt of your letter, and had quite forgot it; but now, on viewing his draughts, and considering his descriptions, I think they seem to favour *my hypothesis*; for he describes and draws columns of water, of various heights, terminating abruptly at the top, exactly as water would do, when forced up by the pressure of the atmosphere into an exhausted tube.

(d) "I must, however, no longer call it *my hypothesis*, Since I find *Stuart* had the same thought, though somewhat obscurely expressed, where he says 'he imagines this phaenomenon may be solv'd by suction (improperly 'so called) or rather pulsion, as in the application of a cupping glass to the 'flesh, the air first voided by the kindled flax.'

(e) "In my paper, I supposed a whirlwind and a spout to be the same thing, and to proceed from the same cause; the only difference between them being, that the one passes over land, the other over water. I find, also, in the *Transactions*, that *M. de la Pryme* was of the same opinion; for he there describes two spouts, as he calls them, which were seen at different times, at *Hatfield* in *Yorkshire*, whose appearances in the air were the same with those of the spouts at sea, and effects the same with those of real whirlwinds.

'Whirlwinds have, generally, a progressive, as well as a circular motion; so had what is called the spout, at *Topsham* * * * *(See the account of it in the Transactions)*, which also appears, by its effects described, to have been a real whirlwind. Water-spouts have, also, a progressive motion; this is sometimes greater, and sometimes less; in some violent, in others barely perceivable. The whirlwind at *Warrington* continued long in *Acrement-Close*.

"Whirlwinds generally arise after calms and great heats: The same is observed of water-spouts, which are, therefore, most frequent in the warm latitudes. The spout that happened in cold weather, in the *Downs*, described by *Mr. Gordon* in the *Transactions*, was, for that reason, thought extraordinary; but he remarks withal, that the weather, though cold when the spout appeared, was soon after much colder; as we find it, commonly, less warm after a whirlwind.

"You agree, that the wind blows every way towards a whirlwind, from a large space round. An intelligent whaleman of Nantucket, informed me, that three of their vessels, which were out in search of whales, happening to be becalmed, lay in sight of each other, at about a league distance, if I remember right, nearly forming a triangle: After some time, a water-spout appeared near the middle of the triangle, when a brisk breeze of wind sprung up, and every vessel made sail; and then it appeared to them all, by the setting of the sails, and the course each vessel stood, that the spout was to the leeward of every one of them; and they all declared it to have been so, when they happened afterwards in company, and came to confer about it. So that in this particular likewise, whirlwinds and water spouts agree.

"But, if that which appears a water-spout at sea, does sometimes, in its progressive motion, meet with and pass over land, and there produce all the phaenomena and effects of the whirlwind, it should thence seem still more evident that the whirlwind and a spout are the same. I send you, herewith, a letter from an ingenious physician of my acquaintance, which gives one instance of this, that fell within his observation.

"A fluid, moving from all points horizontally, towards a center, must, at that center, either ascend or descend. Water being in a tub, if a hole be opened in the middle of the bottom, will flow from all sides to the center, and there descend in a whirl. But, air flowing on and near the surface of land or water, from all sides, towards a center, must, at that center, ascend; the land or water hindering its descent.

"If these concentring currents of air be in the upper region, they may, indeed, descend in the spout or whirlwind; but then, when the united current reached the earth or water, it would spread, and, probably, blow every way from the center. There may be whirlwinds of both kinds, but, from the commonly observed effects, I suspect the rising one to be the most common: When the upper air descends, it is, perhaps, in a greater body, extending wider, as in our thunder-gusts, and without much whirling; and, when air descends in a spout, or whirlwind, I should rather expect it would press the roof of a house *inwards*, or force *in* the tiles, shingles, or thatch; force a boat down into the water, or a piece of timber into the earth, than that it would lift them up, and carry them away.

"It has so happened, that I have not met with any accounts of spouts, that certainly descended; I suspect they are not frequent. Please to communicate those you mention. The apparent dropping of a pipe from the clouds towards the earth or sea, I will endeavor to explain hereafter.

"The augmentation of the cloud, which as I am informed, is generally, if not always the case, during a spout, seems to shew an ascent, rather than a descent of the matter of which such cloud is composed; for a descending spout, one would expect, should diminish a cloud. I own, however, that cold air descending, may, by condensing the vapours in a lower region, form and increase clouds; which I think, is generally the case in our common thunder-gusts, and, therefore, do not lay great stress on this argument.

"Whirlwinds, and spouts, are not always, though most commonly, in the day time. The terrible whirlwind which damaged a great part of Rome, June 11, 1749, happened in the night of that day. The same was supposed to have been first a spout, for it is said to be beyond doubt, that it gathered in the neighbouring sea, as it could be tracked from *Ostia* to *Rome*. I find this in *Pere Boschovich's* account of it, as abridg'd in the *Monthly Review* for December, 1750.

"In that account, the whirlwind is said to have appeared as a very black, long, and lofty cloud, discoverable, notwithstanding the darkness of the night, by its continually lightning or emitting flashes on all sides, pushing along with a surprising swiftness, and within three or four feet of the ground. Its general effects on houses, were, stripping off the roofs, blowing away chimneys, breaking doors and windows, *forcing up the floors and unpaving the rooms*, (some of these effects seem to agree well with a supposed vacuum in the center of the whirlwind) and the very rafters of the houses were broke and dispersed, and even hurled against houses at a considerable distance, &c.

"It seems, by an expression of *Pere Boschovich's*, as if the wind blew from all sides towards the whirlwind; for, having carefully observed its effects, he concludes of all whirlwinds, 'that their motion is circular, and their action attractive.'

"He observes, on a number of histories of whirlwinds, &c., 'that a com-'mon effect of them is, to carry up into the air tiles, stones, and animals 'themselves, which happen to be in their course, and all kinds of bodies un-'exceptionably, throwing them to a considerable distance, with great im-'petuosity.' Such effects seem to shew a rising current of air.

(f) "I will endeavor to explain my conception of this matter by figures, representing a plan and an elevation of a spout or whirlwind.

"I would only first beg to be allowed two or three positions, mentioned in my former paper.

"1. That the lower region of air is often more heated, and so more rarified, than the upper; consequently, specifically lighter. The coldness of the upper region is manifested by the hail which sometimes falls from it on a hot day.

"2. That heated air may be very moist, and yet the moisture so equally diffus'd and rarified, as not to be visible, till colder air mixes with it, when it condenses, and becomes visible. Thus our breath, invisible in summer, becomes visible in winter.

"Now, let us suppose a tract of land, or sea, of perhaps sixty miles square, unscreened by clouds, and unfanned by winds, during great part of a summer's day, or, it may be, for several days successively, till it is violently heated, together with the lower region of air in contact with it, so that the said lower air becomes specifically lighter than the superincumbent higher region of the atmosphere, in which the clouds commonly float: Let us suppose, also, that the air surrounding this tract has not been so much heated during those days, and, therefore, remains heavier. The conseqence of this should be, as I conceive, that the heated lighter air, being pressed on all sides, must ascend, and the heavier descend; and, as this rising cannot be in all parts, or the whole area of the tract at once, for that would leave too extensive a vacuum, the rising will begin precisely in that column that happened to be the lightest, or most rarified; and the warm air will flow horizontally from all points to this column, where the several currents meeting, and joining to rise, a whirl is naturally formed, in the same manner as a whirl is formed in the tub of water, by the descending fluid flowing from all sides of the tub, to the hole in the center.

"And, as the several currents arrive at this central rising column, with a considerable degree of horizontal motion, they cannot suddenly change it to a vertical motion; therefore, as they gradually, in approaching the whirl, decline from right to curve or circular lines, so, having joined the whirl, they *ascend* by a spiral motion; in the same manner as the water *descends* spirally through the hole in the tub before-mentioned.

April, 1906.] *Franklin as a Man of Science and an Inventor.* 315

"Lastly, as the lower air, and nearest the surface, is most rarified by the heat of the sun, that air is most acted on by the pressure of the surrounding cold and heavy air, which is to take its place; consequently, its motion towards the whirl is swiftest, and so the force of the lower part of the whirl, or trump, strongest, and the centrifugal force of its particles greatest; and hence the vacuum round the axis of the whirl should be greatest

Fig. 10. Plan of Water Spout. (Franklin)

near the earth or sea, and be gradually diminished as it approaches the region of the clouds, till it ends in a point, as at A, in Fig. II, forming a long and sharp cone.

(g) "In Fig. I (D) which is a plan or ground-plat of a whirlwind, the circle V represents the central vacuum. (Our figure 10.)

"Between $aaaa$ and $bbbb$ I suppose a body of air condensed strongly by the pressure of the currents moving towards it, from all sides without, and by its centrifugal force from within; moving round with prodigious swiftness, (having, as it were, the momenta of all the current united in itself) and with a power equal to its swiftness and density.

"It is this whirling body of air between $a\ a\ a\ a$ and $b\ b\ b\ b$ that rises spirally; by its force it tears buildings to pieces, twists up great trees by the roots, &c., and, by its spiral motion, raises the fragments so high, till the pressure of the surrounding and approaching currents diminishing, can no longer confine them to the circle; or their own centrifugal force increasing, grows too strong for such pressure, when they fly off in tangent lines, as stones out of a sling, and fall on all sides, and at great distances.

"If it happens at sea, the water under and beneath $a\ a\ a\ a$ and $b\ b\ b\ b$ will be violently agitated and driven about, and parts of it raised with the spiral current, and thrown about, so as to form a bushlike appearance.

"This circle is of various diameters, sometimes very large.

"If the vacuum passes over water, the water may rise in it in a body, or column, to near the height of thirty-two feet.

"If it passes over houses, it may burst their windows or walls outwards, pluck off the roofs, and pluck up the floors, by the sudden rarefaction of the air contained within such buildings; the outward pressure of the atmosphere being suddenly taken off: So the stopp'd bottle of air bursts under the exhausted receiver of the air-pump.

(To be continued.)

LEAD IN BRONZE BEARING METALS.

Bronzes containing lead are much used for bearings. The lead does not alloy directly with the bronze, but remains in separate particles, which are seen under the microscope. The lead increases the plasticity of the alloy to a marked degree. There is a tendency to the separation of the lead by liquidation, as it has a much lower solidifying point than the bronze. Even if the metal is not poured until it has acquired a pasty consistency, there is a tendency for the heavy particles to sink in the mass. It is difficult to be assured of an equal distribution of the metals in such alloys, and the exact composition at the rubbing surface is unknown.—ENGINEERING AND MINING JOURNAL.

CONCRETE BLOCKS IN BUILDING.

The prizes offered by *Engineering News* and the *Cement Age* of New York for the best papers on "The Manufacture of Concrete Blocks and Their Use in Building Construction" have just been awarded by the jury, which was composed of Robert W. Lesley, past president of the American Cement Manufacturers' Association; Richard L. Humphrey, president of the Cement Users' Association and Prof. Edgar Marburg, secretary of the American Society for Testing Materials. The first prize of $250 was won by H. H. Rice, Denver, Col., secretary of the American Hydraulic Stone Company. The second prize of $100 was given to Wm. M. Torrance, New York City, assistant engineer in charge of concrete steel design for the Hudson Tunnel companies.

TALBOT CONTINUOUS OPEN-HEARTH PROCESS INSTALLATIONS.

It is now estimated that the steel producing capacity employing the Talbot process amounts to nearly 1,000,000 tons a year in the United States and Great Britain. The installations now in operation or under construction are as follows:

One 75-ton furnace, built in 1899 by the Pensoyd Iron Company, Pencoyd, Pa. Now in operation.

One 200-ton furnace, built in 1902 by Jones & Laughlin Steel Company, Pittsburgh, Pa. Now in operation.

One 200-ton furnace, built in 1904 by Jones & Laughlin Steel Company, Pittsburgh, Pa. Now in operation.

Three 200-ton furnaces, built in 1905 by Jones & Laughlin Steel Company, Pittsburgh, Pa. Now in operation.

One 100-ton furnace, built in 1902 by Frodingham Iron Company, Frod-
One 130-ton furnace, built in 1905 by Frodingham Iron Company, Frodingham, England. Now in operation.
ingham, England. Now in operation.

One 175-ton furnace, built in 1903 by Guest, Keen & Company, Cardiff, South Wales. Now in operation.

Two 175-ton furnaces, built in 1905 by the Cargo Fleet Iron Company, Limited, Middlesboro, England. Now in operation.

One 75-ton furnace under construction by the Cargo Fleet Iron Company, Limited, Middlesboro, England.

Two 175-ton furnaces under construction by the South Durham Steel & Iron Company, Limited, Middlesboro, England.

Two 175-ton furnaces under construction at the Malleable Works, Stockton-on-Tees, England.

Two 175-ton furnaces under construction by the Palmer Shipbuilding Company, Jarrow-on-Tyne, England.

Two 175-ton furnaces under construction at Longwy, France.

It is expected that the producing capacity under the Talbot process will be increased materially in 1906.—*Iron Age.*

EXPLOSION OF RADIUM TUBES.

In the issue of *Physikalische Zeitschrift* for January 15, 1906, Mr. J. Precht describes some experiences with radium which tend to show the enormous force of the ommitted gaseous products when they are confined to a limited space. He reports that a small glass tube with walls 0.5 mm. thick, containing 25 milligrammes of purest radium bromide, exploded after 11 months' use with a loud report. Three minutes before the explosion it had been removed from a bath of liquid air and placed on a wooden table. It had been used several times before. The force of the explosion was such that the glass was shivered into almost microscopic particles which were strewn all over the room, the radium being seen in the dark like a starry sky. The pressue in the tube must have been at least 20 atmospheres, and was no doubt due to the evolution of gaseous decomposition proucts of radium.

ALUMINOSILICIDES.

According to E. Vigouroux *(Comptesrend.,* 1905, CXLI, 951-953), pure silicon and aluminum, whether fused together, or whether the silicon be formed from silica by the thermite process in presence of excess of aluminum, refuse to combine, but when they are in presence of a third metal, double silicides of aluminum and the metal, or aluminosilicides of the metal, are formed. These are definite, crystaline substances, with metallic luster, dense, hard and brittle; some are attacked by dilute acids, but most of them resist all acids, even when concentrated, save hydrofluoric acid, and none of them is affected by solutions of alkali. They are formed by heating the three elements together in an atmosphere of hydrogen, or by the thermite method from mixtures of silica and a metallic oxide, using excess of aluminum, or by acting with aluminum on a mixture of the metal or its oxide or sulphide with potassium silicofluoride. In view of these facts, clay vessels should be avoided in the preparation of metals which can form aluminosilicides.

THE BATTLESHIP DREADNOUGHT, which will cost completed $7,500,000, was launched by King Edward at Portsmouth, Eng., February 10. On October 2, 1905, work was begun on the vessel, and it is expected that the Dreadnought will be commissioned at the end of another year, making a record six months less than the shortest time in which a battleship has been built. The tonnage of the Dreadnought will be 18,000, the length exceeding 500 feet and the beam 80 feet. Turbine engines are provided, with 23,000 horse-power, and the guaranteed speed is 21 knots. Armor plates are 12 inches thick, extending almost the entire length and 7 feet below the water line. The armament will consist of ten 12-inch guns in five turrets. No previous warship has carried more than four 12-inch guns.

PRODUCTION OF ALUMINUM IN EUROPE.

The output of aluminum by the Pittsburg Reducing Company in Canada and the United States is placed at 4,200 tons. The high grade of copper has recently developed new interest in the production of the metal, whose price has risen in spite of the growing quantity available, and which has come into large demand in the electric field. United States Consul Guenther, of Frankfort, sends the following German report on the European situation:

"The companies at present producing aluminum are: 1. Aluminium Industrie A. G., in Neuhausen, Switzerland, which also has works at Lend-Gastein, Austria, and at Rheinfelden. Baden. The total annual production of the three works of this company was estimated at 3,675 tons (at 2,204 metric pounds), employing 88 bureau officials and 661 workmen. It paid last year on its stock an 18 per cent. dividend, and recently raised its share capital from 16 to 26 million francs. The new shares were taken by a bankers' syndicate at 250 per cent."

AMERICAN LOCOMOTIVES EXCEL.

The superiority of American passenger locomotives over those of the best foreign build as regards speed, economy and power is shown in results of exhaustive tests made under the supervision of some of the leading American and European engineers, just made public.

The work was begun by the Pennsylvania Railroad Company at the St. Louis Exposition and covered a period of more than six months. The tests were most elaborate and involved an expenditure of $250,000.

The tests were made under the direction of an Advisory Committee, consisting of three members each of the American Society of Mechanical Engineers and of the American Railway Master Mechanics' Association. These were Dr. W. F. M. Goss, dean of the Schools of Engineering. Purdue University; Edwin M. Herr, vice-president of the Westinghouse Electric and Manufacturing Company: E. Sague, first vice-president of the American Locomotive Company; F. H. Clark, general superintendent of motive power of the Chicago, Burlington and Quincy Railroad; C. H. Quereau, superintendent of electrical equipment of the New York Central and H. H. Vaughan, superintendent of the motive power of the Canadian Pacific Railway. The affiliated members of the committee included John A. F. Aspinall, general manager of the Lancaster and Yorkshire Railway of England, and Karl Steinbiss, director of the Royal Prussian Railway, Germany.

The American passenger engines tested were one of the New York Central high speed type used for hauling the Empire State Express and one of the Atchinson, Topeka & Santa Fe Road's balanced compound. Several freight engines were also submitted to test.

One of the foreign locomotives was the much-heralded de Glehn balanced compound, built by the Société Alsacienne de Constructions Mecaniques at Belfort, France. This type of locomotive is used on the Chemins de Fer de Nord, between Calais and Paris, and makes one of the fastest runs in the world. It is also in use on the Great Western Railway of England. This engine is regarded by European experts as the acme of high speed passenger service perfection on the other side. The other European locomotive tested came from the Hannoversche Machinenbau-Actien-Gesellschaft, Linden Vor Hannaver, Germany. This engine was built for the directors of the Royal Prussian Railway Administration.—IRON AGE.

COMPARATIVE ECONOMY OF STEAM AND TURBINE ENGINES.

A saving in coal of about 9 er cent., at a speed of between 19 and 20 knots, was computed for the turbine steamers belonging to the Midland Railway (England), as compared with similar steamers of the same company propelled by reciprocating engines. The principal inferiority of the turbine vessels seems to have consisted in a difficulty in maneuvring in narrow channels, which difficulty it is believed may be readily overcome by the expedient of putting more power into the turbines used for backing. The difference in weight of machinery and in initial cost was found to favor the turbine driven ships by margins of 6 and 1½ per cent., respectively.

BARYTES MINING IN VIRGINIA.

In connection with some of the barytes deposits of Virginia, the mining problem is very simple. In fact the problem is not one of mining, but rather of handling the material by industrial railways and other similar appliances. However, the deposits are so scattered and so small that it makes the problem rather difficult, inasmuch as the proper appliance to handle the dirt and ore to the best advantage would clean out some deposits in a few weeks.

The best and most available barytes occurs in a sort of loamy dirt. An ideal way to handle this would be to put in a steam shovel and log washer, somewhat as brown ore is handled in the South, but as a thousand tons of barytes from one lease is an exceptionally large amount, it would be impractical to instal such a plant. The usual method is to make an open cut, follow the barytes where it leads and reject the dirt. Below the dirt the barytes runs into the limestone, and if that is in such a place that cheap transportation is available it pays to work.

In some places, far from the railroad, the cost of teaming is greater than the cost of mining. The limestone deposits might be worked more easily by the use of jigs. The uncertainty of the business, however, has prevented such innovations. In the Virginia district it is rather hard to accumulate sufficient barytes to run a good sized mill. The soft nature of the barytes is rather against mechanical handling, and even if there were a sufficient amount to warrant putting in a steam shovel and washer, the waste might be too great. The best and most abundant deposits seem to be the farthest from the railroad. If they were all contracted at one point, barytes mines would be a good proposition.—ENGINEERING AND MINING JOURNAL.

HARDENING STEEL.

R. A. Hadfield, Sheffield, Eng., has been granted a patent in the United States on a process of hardening steel. The first claim is on a method of hardening steel which consists in raising the temperature of the steel to about 975° centigrade, permitting it to cool slowly, reheating the cooled steel to a temperature of about 500° centigrade and permitting the steel to cool slowly after such reheating. The remaining claims introduce variations in the above temperatures, the heat employed varying according to the amount of carbon in the steel.

TESTS HAVE BEEN MADE recently on the corrosion of boiler-tubes by forcing air through tubes made wet with distilled water. In 16 weeks the loss in weight was .315 gram per sq. in. When the water was made alkaline the loss in the same time was reduced to only .0997 gram. It apears that if the water in the boiler is made slightly alkaline the corrosion of tubes may be materially reduced.—(London) NATURE.

STATUE OF FRANKLIN

in front of Post Office, Philadelphia. Designed by John H. Boyle, and presented to the City by Justice C. Strawbridge.

JOURNAL

OF THE

FRANKLIN INSTITUTE

OF THE STATE OF PENNSYLVANIA

FOR THE PROMOTION OF THE MECHANIC ARTS

| VOL. CLXI, No. 5 | 81ST YEAR | MAY, 1906 |

The Franklin Institute is not responsible for the statements and opinions advanced by contributors to the *Journal.*

THE FRANKLIN INSTITUTE.

Franklin as a Man of Science and an Inventor.*

[An address delivered by Dr. Edwin J. Houston, Emeritus Professor of Physics, Franklin Institute, on February 21, 1906, on the occasion of the 200th anniversary of the birth of Benjamin Franklin.]

(Concluded from vol. clxi, p. 316)

"Fig. 11 is to represent the elevation of a water-spout. wherein, I suppose P P P to be the cone, at first a vacuum, till W W, the rising column of water, has filled so much of it. S S S S, the spiral whirl of air surrounding the vacuum, and continued higher in a close column after the vacuum ends in the point P, till it reaches the cool region of the air. B B, the bush described by Stuart, surrounding the foot of the column of water.

(h) "Now, I suppose this whirl of air will, at first, be as invisible as the air itself, though reaching, in reality, from the water, to the region of cold air, in which our low summer thunder-clouds commonly float; but presently, it will become visible at its extremities. *At its lower end,* by the agitation of the water under the whirling part of the circle, between P and S, forming Stuart's bush, and by the swelling and rising of the water, in the beginning vacu-

---

*Copyrighted by Edwin J. Houston, 1906.

Fig. 11. Elevation of Water Spout. (Franklin)

um, which is, at first, a small, low, broad cone, whose top gradually rises and sharpens, as the force of the whirl increases. *At its upper end* it becomes visible, by the warm air brought up to the cooler region, where its moisture begins to be condensed into thick vapour, by the cold, and is seen first at A, the highest part, which being now cooled, condenses what rises next at B, which condenses that at C, and that condenses what is rising at D, the cold operating by the contact of the vapors faster in a right line downwards, than the vapours themselves can climb in a spiral line upwards; they climb, however, and as by continual addition they grow denser, and, concentrating currents that compose the whirl, they fly off, spread, and form a cloud.

"It seems easy to conceive, how, by this successive condensation from above, the spout appears to drop or descend from the cloud, though the materials of which it is composed are all the while ascending.

"The condensation of the moisture contained in so great a quantity of warm air as may be supposed to rise in a short time in this prodigiously rapid whirl, is, perhaps, sufficient to form a great extent of cloud, though the spout should be over land, as those at *Hatfield*; and if the land happens not to be very dusty, perhaps the lower part of the spout will scarce become visible at all; though the upper, or what is commonly called, the descending part, be very distinctly seen.

"The same may happen at sea, in case the whirl is not violent enough to make a high vacuum, and raise the column, &c. In such case, the upper part A B C D only will be visible, and the bush, perhaps, below.

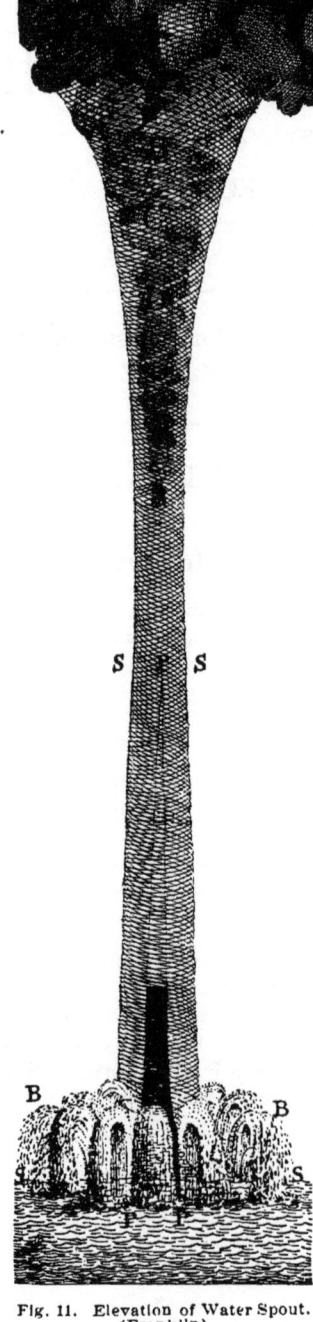

"But if the whirl be strong, and there be much dust on the land, and the column W W be raised from the water, then the lower part becomes visible, and sometimes even united to the upper part. For the dust may be carried up in the spiral whirl, till it reach the region where the vapour is condensed, and rise with that even to the clouds: And the friction of the whirling air, on the sides of the column W W, may detach great quantities of its water, break it into drops, and carry them up in the spirial whirl mixed with the air; the heavier drops may, indeed, fly off, and fall, in a shower, round the spout; but much of it will be broken into vapour, yet visible; and thus, in both cases, by dust at land, and, by water at sea, the whole tube may be darkened and rendered visible.

"As the whirl weakens, the tube may (in appearance) separate in the middle; the column of water subsiding, and the superior condensed part drawing up the cloud. Yet still the tube, or whirl of air, may remain entire, the middle only becoming invisible, as not containing visible matter."

(a) Here we have the ring of the true philosopher: "Nothing, certainly, can be more improving to a searcher into nature than objections, judiciously made, to his opinion, taken up, perhaps, too hastily." Franklin welcomes the fair criticisms of the Boston Doctor.

(b) "I am so engaged in business, public and private." To all who are acquainted with the immense amount of work performed by Franklin, it is not surprising that he finds but comparatively little time for the more pleasing investigations of science. But it was then, as it is now, that it is the busiest man who finds time for all necessary work, and Franklin certainly found time in this case to prepare the most excellent scientific paper he sent to the Boston Doctor.

(c) Even if the vacuum were complete, the height of the column that is pressed inwards by the weight of the atmosphere could not greatly exceed thirty feet.

(d) "I must, however, no longer call it my hypothesis." A generous statement, since Stuart's explanation was certainly very obscure.

(e) Franklin is quite correct in this supposition. The whirlwind and the waterspout are the same phenomena, being due to the same causes, with, however, the difference that the whirlwind is produced by a whirling column of air passing over the land, while the waterspout is produced by this column passing over the water. Note here the clear and logical statement as to their resemblances.

(1) Both waterspouts and whirlwinds possess a progressive as well as a rotary or circular motion.

(2) Both waterspouts and whirlwinds occur after periods of great atmospheric heat, when the air has been free from winds.

(3) The wind blows in all directions from the extended space surrounding both water-spouts and whirlwinds directly towards the water-spout or the whirlwind.

(4) When waterspouts, by reason of their progressive motion, leave the sea and move over the land, they produce all the characteristic effects of whirlwinds, thus showing them to be the same.

(5) Both waterspouts and whirlwinds occur most commonly during the day time.

(f) Franklin now proceeds to apply his theory as regards the formation of waterspouts, producing for this purpose a plan and an elevation of the spout as he conceives it to be produced. He bases his theory on two assumptions, that all should be willing to admit; *i. e.*,—

(1) A higher temperature in the lower regions of the atmosphere than in the upper regions, and, consequently, a more rarefied condition near the surface of the earth than in the upper regions. Such a condition would, of course, necessitate an absence of wind.

(2) An exceedingly moist condition of the atmosphere.

Franklin then draws a picture of an extended area of land or sea, of, as he says, perhaps sixty miles square. Under conditions of prolonged calm, and with no clouds in the sky to prevent the sun's heat from freely reaching the earth's surface, these conditions perhaps continuing for several days, the lower strata of air become intensely heated. He then pictures the surrounding air as being relatively much colder, and as, therefore, remaining much heavier than that over the heated area. Under these circumstances, there would necessarily be produced a rising or ascending current of the lower air, accompanied by the descent of the colder and heavier air. But this rising does not immediately take place alike over all portions of the heated area. It begins over that portion of the area which is the most highly heated, the remaining portion of the warm air flowing horizontally over all portions of the heated area towards this column. In this way, the whirl is formed pretty much as Franklin remarks, like the funnel-shaped depression produced in the surface of the water in a tub, that is discharging its water through a hole or opening in the bottom of the tub.

Franklin then points out the fact that since inflowing horizontal currents possess considerable motion, when they reach the central rising column, they are unable to suddenly change their direction to that of the vertical motion, so that they join the ascending column by means of a spiral motion.

Franklin then points out the fact that the velocity of the inflowing current will necessarily be greatest in those portions where the temperature of the air is greatest; *i. e.*, near the highly-heated surface. Consequently, it is here that the whirling motion is the swiftest, and, therefore, the vacuum must be greatest near the earth's surface, decreasing as the column rises.

(g) Fig. 10 is clear, and needs no explanation. The rising of the water from the surface of the sea is a natural result of the passage of the vacuous area.

(h) The whirling motion of the air is at first invisible, but forms a mass of clouds as soon as the moist air is condensed by the cold. It is for this reason that the spout may seem to drop or descend from the clouds, although the moisture of which it is formed has been continually ascending.

In addition to the valuable papers prepared by Franklin on geographical physics, to which we have already referred; *i. e.*, the paper as to the cause of thunder-storms, the aurora borealis and waterspouts, there yet remains to be discussed an observation of very great value respecting the cause of the great northeast storms of the United States. Since, as is now well known, the greater part of the work of the United States Weather Bureau as regards the preparation of forecasts of coming changes in the weather, is based on the peculiarities of the movements of our great northeast storms, it will readily be seen that this discovery of Franklin's should be ranked among the most important of his researches in geographical physics. While from a popular standpoint, this matter is not so attractive as his demonstration of the identity of disruptive electric discharges and lightning, yet from a scientific standpoint, it should contribute to his merited reputation as a philosopher, as much, if not more, that his famous kite experiment. Then, too, from a practical standpoint, while it might seem that the invention of the lightning rod was of more direct benefit to mankind, yet the saving of life and property that would frequently result from the timely warning of the approach of a dangerous northeast storm would probably be immensely greater than would ever be possible by protection afforded by lightning rods.

Franklin informs us just how he came to think of the causes of the great northeast storms in this country as starting in an area of low barometer somewhere in the west and progressing generally towards the northeast. It appears that an eclipse of the moon was to be visible at Philadelphia on a certain Friday at 9 P. M. Franklin made preparation for the proper observing of this eclipse, but, unfortunately, that night a storm visited Philadelphia, approaching the city from the northeast, and continuing violently all that night and the next day, prevented any observations of the eclipse from being made. To Franklin's great astonishment, the newspapers contained an account of the fact that this eclipse had been observed in the city of Boston. Since this storm apparently approached the city of Philadelphia from the northeast, it would seem that it should have reached Boston before it reached Philadelphia, Boston being, as is well known, northeast of the city of Philadelphia. Writing to his brother, who lived in the city of Boston, he ascertained the fact that the eclipse was over at least one hour before the storm commenced. This caused Franklin to make further inquiries, when he found that, as a rule, the great northeast storms of this country begin to the leeward; *i. e.*, start somewhere to the southwest, then moving in a general northeast path across the country.

In a letter to the Rev. Jared Eliot, dated Philadelphia, July 16, 1747, Franklin says:

"We have freqently, along this North American coast, storms from the northeast, which blow violently sometimes three or four days. Of these I have had a very singular opinion some years, viz., that, though the course of the wind is from northeast to southwest, yet the course of the storm is from southwest to northeast; that is, the air is in violent motion in Virginia before it moves in Connecticut, and in Connecticut before it moves at Cape Sable, &c. My reasons for this opinion, (if the like have not occurred to you,) I will give in my next."

In another letter to the same gentleman, dated Philadelphia, February 13, 1749-50 (year uncertain), he fulfills the promise referred to in the preceding letter, and at fairly full length regarding his views of the motion of the northeast storms:

"You desire to know my thoughts about the northeast storms beginning to leeward. Some years since, there was an eclipse of the moon at nine o'clock in the evening, which I intended to observe: but before night a storm blew up at the northeast, and continued violent all night and all next day; the sky thick-clouded, dark and rainy, so that neither moon nor stars could be seen. The storm did a great deal of damage all along the coast, for we had accounts of it in the newspapers from Boston, Newport, New York, Maryland, and Virginia; but what surprised me was, to find in the Boston newspapers an account of an observation of that eclipse made there; for I thought, as the storm came from the northeast, it must have begun sooner at Boston than with us, and consequently have prevented such observation. I wrote to my brother about it, and he informed me, that the eclipse was over there an hour before the storm began. Since which I have made inquiries from time to time of travellers, and of my correspondents northeastward and southwestward, and observed the accounts in the newspapers from New England, New York, Maryland, Virginia, and South Carolina; and I find it to be a constant fact, that northeast storms begin to leeward; and are often more violent there than farther to windward. Thus the last October storm, which with you was on the 8th, began on the 7th in Virginia and North Carolina, and was most violent there.

"As to the reason of this, I can only give you my conjectures. Suppose a great tract of country, land and sea, to wit, Florida and the Bay of Mexico, to have clear weather for several days, and to be heated by the sun, and its air thereby exceedingly rarefied. Suppose the country northeastward, as Pennsylvania, New England, Nova Scotia, and Newfoundland, to be at the same time covered with clouds, and its air chilled and condensed. The rarefied air being lighter must rise, and the denser air next to it will press into its place: that will be followed by the next denser air, that by the next, and so on. Thus, when I have a fire in my chimney, there is a current of air constantly flowing from the door to the chimney: but the beginning of the motion was at the chimney, where the air being rarefied by the fire rising, its place was supplied by the cooler air that was next to it, and the place of that by the next, and so on to the door. So the water in a long sluice or mill-race, being stopped by a gate at one end, to let it out, the water next the gate begins first to move, that which is next to it follows; and so, though the water proceeds forward to the gate, the motion which began there runs backwards, if one may so speak, to the upper end of the race, where the water is last in motion. We have on this continent a long ridge of mountains running from northeast to southwest; and the coast runs the same course. These may contribute towards the direction of the winds, or at least influence them in some degree. If these conjectures do not satisfy you, I wish to have yours on the subject."

At a later date; *i. e.*, May 12, 1760, in a letter to Alexander Small, of London, Franklin writes as follows:

"Agreeable to your request, I send you my reasons for thinking that our North-East storms in North America begin first, in point of time, in the South-West parts: That is to say, the air in Georgia, the farthest of our colonies to the South-West, begins to move South-Westerly before the air of Carolina, which is the next colony North-Eastward; the air of Carolina has the same motion before the air of Virginia, which lies still more North-Eastward; and so on North-Easterly through Pennsylvania, New York, New England, &c., quite to Newfoundland.

"These North-East storms are generally very violent, continue sometimes two or three days, and often do considerable damage in the harbours along the coast. They are attended with thick clouds and rain.

"What first gave me this idea, was the following circumstance. About twenty years ago, a few more or less, I cannot from my memory be certain, we were to have an eclipse of the moon at Philadelphia, on a Friday evening, about nine o'clock. I intended to observe it, but was prevented by a North-East storm, which came on about seven with thick clouds as usual, that quite obscured the whole hemisphere. Yet when the post brought us the Boston newspaper, giving an account of the effects of the storm in those parts, I found the beginning of the eclipse had been well observed there, though Boston lies N. E. of Philadelphia about 400 miles. This puzzled me, because the storm began with us so soon as to prevent any observation, and being a N. E. storm, I imagined it must have began rather sooner in places farther to the North Eastward, than it did at Philadelphia. I therefore mentioned it in a letter to my broher, who lived in Boson; and he informed me that the storm did begin with them till near eleven o'clock, so that they had a good observation of the eclipse: and upon comparing all the other accounts I received from the several colonies, of the time beginning of the same storm and since that of other storms of the same kind, I found the beginning to be always later the farther North-Eastward. I have not my notes with me here in England, and cannot from memory, say the proportion of time to distance, but I think it is about an hour to every hundred miles.

"From thence I formed an idea of the cause of these storms, which I would explain by a familiar instance or two. * * * Suppose a long canal of water stopp'd at the end of a gate. The water is quite at rest till the gate is open, then it begins to move out through the gate; the water next the gate is first in motion, and moves towards the gate; the water next to that first water moves next, and so on successively, till the water at the head of the canal is in motion, which is last of all. In this case all the water moves indeed towards the gate, but the successive times of beginning motion are the contrary way, viz. from the gate backwards to the head of the canal. Again, suppose the air in a chamber at rest, no current through the room till you make a fire in the chimney. Immediately the air in the chimney being rarefied by the fire, rises; the air next the chimney flows in to supply its place, moving towards the chimney; and, in consequence, the rest of the air successively, quite back to the door. Thus to produce our North-East storms, I suppose some great heat and rarefaction of air in or about the Gulph of Mexico; the air thence rising has its place supplied by the next more northern, cooler, and therefore denser and heavier, air; that, being in motion, is followed by the next more northern air, &c., &c., in a successive current, to which current our coast and inland ridge of mountains give the direction of North-East, as they lie N. E. and S. W.

"This I offer only as an hypothesis to account for this particular fact; and, perhaps, on farther examination, a better and truer may be found. I do not suppose all storms generated in the same manner. Our North-West thunder-gust in America I know are not; but of them I have written my opinion fully in a paper which you have seen."

While unfortunately Franklin did not in this, as in many other of his papers, give the exact date of the eclipse, yet, since this date fixes the time when his attention was first directed to the fact that the northeast storms of this country start in the southwest, Prof. Bache, of the University of Pennsylvania, by means of a careful study of all the eclipses of the moon known to have occurred at about this time, has definitely fixed the date at October 21, 1743. This is a matter of no little scientific importance, since a claim has been made that the law of the movements of the northeast storms of the United States was first published by another before the time referred to in the letter to Eliot. It seems from this fact, therefore, that Franklin was the first discoverer of the important facts concerning the movements of the northeast storms.

There remains yet another important subject in geographical physics that attracted Franklin's attention at a comparatively early date. I allude to the existence and causes of that great moving mass of heated water off the eastern coast of the United States, known as the Gulf Stream. He gives an account of this stream in a letter to David Leroy, at Paris, the letter bearing date of August, 1785:

"Vessels are sometimes retarded, and sometimes forwarded in their voyages, by currents at sea, which are often not perceived. (a) About the year 1769 or 1670, there was an application made by the Board of Customs at Boston, to the Lords of the Treasury in London, complaining that the packets between Falmouth and New York were generally a fortnight longer in their passages, than merchant ships from London to Rhode Island, and proposing that for the future they should be ordered to Rhode Island instead of New York. Being then concerned in the management of the American post-office, I happened to be consulted on the occasion; and it appearing strange to me, that there should be such a difference between two places scarce a day's run asunder, especially when the merchant ships are generally deeper laden, and more weakly manned than the packets, and had from London the whole length of the river and channel to run before they left the land of England, while the packets had only to go from Falmouth, I could not but think the fact misunderstood or misrepresented. (b) There happened then to be in London a Nantucket sea captain of my acquaintance, to whom I communicated the affair. He told me he believed the fact might be true; but the difference was owing to this, that the Rhode Island captains were acquainted with the Gulf Stream, which those of the English packets were not. 'We are well acquainted with that stream,' says he, 'because in our pursuit of whales, which keep near the sides of it, but are not to be met with in it, we run down along the sides, and frequently cross it to change our sides; and in crossing it have sometimes met and spoke with those packets, who were in the middle of it, and stemming it. We have informed them that they were stemming a current that was against them to the value of three miles an hour; and advised them to cross it and get out of it; but they were too wise to be counselled by simple American fishermen. When the winds are but light,' he added, 'they are carried back by the current more than they are forwarded by the wind; and, if the wind be good, the subtraction of seventy miles a day from their course is of some importance.' (c) I then observed it was a pity no notice was taken of this current upon the charts, and requested him to mark it out for me, which he readily complied with, adding directions for avoiding it in sailing from Europe to North America. I procured it to be engraved by order from the general post-office, on the old chart of the Atlantic, at Mount and Page's, Tower Hill; and copies were sent down to Falmouth for the captains of the packets, who slighted it however: but it is since printed in France, of which edition I hereto annex a copy. (See Fig. 12.)

(d) "This stream is probably generated by the great accumulation of water on the eastern coast of America between the tropics, by the trade winds which constantly blow there. It is known, that a large piece of water ten miles broad and generally only three feet deep, has by a strong wind had its waters driven to one side and sustained so as to become six feet deep, while the windward side was laid dry. This may give some idea of the quantity heaped up on the American coast, and the reason of its running down in a strong current through the islands into the Bay of Mexico, and from thence issuing through the Gulf of Florida, and proceeding along the coast to the banks of Newfoundland, where it turns off towards and runs down through the Western Islands. Having since crossed this stream several times in passing between America and Europe, I have been attentive to sundry circumstances relating to it, by which to know when one is in it; and besides the gulf weed with which it is interspersed, I find that it is always warmer than the sea on each side of it, and that it does not sparkle in the night. (e) I annex hereto the observations made with the thermometer in two voyages, and possibly may add a third. It will appear from them, that the thermometer may be a useful instrument to a navigator, since currents coming from the northward into southern seas will probably be found colder than the water of those seas, as the currents from the southern seas into the northern are found warmer. And it is not to be wondered, that so fast a body of deep warm water, several leagues wide,

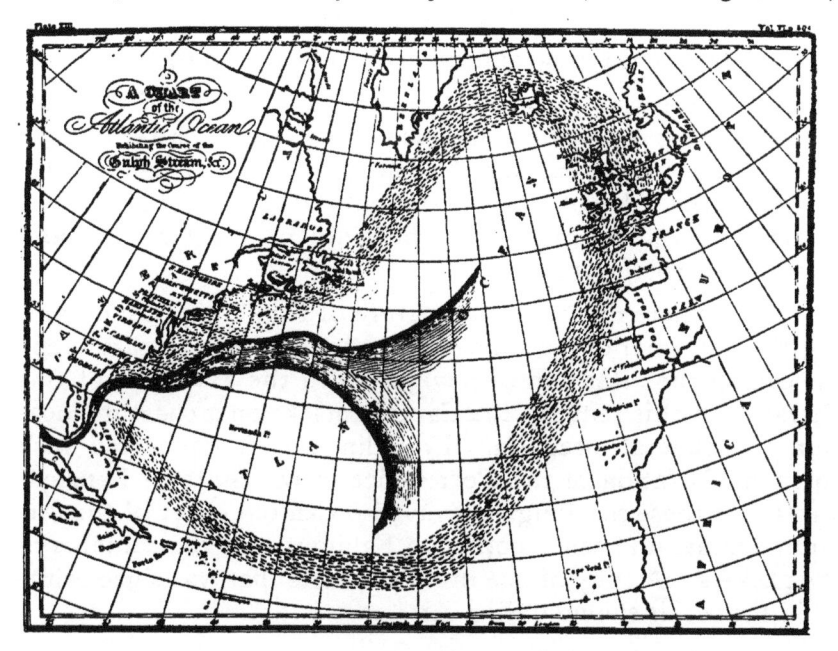

Fig. 12. Franklin's Early Chart of the Gulf Stream.

coming from between the tropics and issuing out of the gulf into the northern seas should retain its warmth longer than the twenty or thirty days required to its passing the banks of Newfoundland. The quantity is too great, and it is too deep to be suddenly cooled by passing under a cooler air. (f) The air immediately over it, however, may receive so much warmth from it as to be rarefied and rise, being rendered lighter than the air on each side of the stream; hence those airs must flow in to supply the place of the rising warm air, and, meeting with each other, form those tornadoes and waterspouts frequently met with, and seen near and over the stream; and as the vapor from a cup of tea in a warm room, and the breath of an animal in the same room, are hardly visible, but become sensible immediately when out in the cold air, so the vapor from the Gulf Stream, in warm latitudes, is scarcely visible, but when it comes into the cool air from Newfoundland, it is condensed into the fogs, for which those parts are so remarkable.

"The power of wind to raise water above its common level in the sea is known to us in America, by the high tides occasioned in all our seaports when a strong northeaster blows against the Gulf Stream.

"The conclusion from these remarks is, that a vessel from Europe to North America may shorten her passage by avoiding to stem the stream, in which the thermometer will be very useful; and a vessel from America to Europe may do the same by the same means of keeping in it. It may have often happened accidentally, that voyages have been shortened by these circumstances. It is well to have the command of them."

(a) An evident typographical error, from 1769 to 1770.

(b) A Nantucket Captain knew of the existence of the Gulf Stream. This old seaman asserted that it was a matter of general information among the American whalers that the whales kept near the sides of the Stream, but did not enter it. These whalers, therefore, ran along the sides of the Gulf Stream, sometimes crossing it.

(c) Note here the utilitarian nature of Franklin. On obtaining this information he immediately takes the necessary steps for calling the attention of navigators to the existence of this Stream. At his request, the Captain marks on a chart its location and general direction. Franklin, who is connected with the administration of the Post Office in the State of Pennsylvania, has these markings engraved on an old chart of the Atlantic Ocean, and sends copies to Falmouth for the captains of the packets. A copy of one of these charts that was since printed in France, is shown in Fig. 12.

(d) Bearing in mind the time that this paper was written, Franklin's explanation as to the causes of the Gulf Stream may be regarded as excellent.

(e) Note here Franklin's practical use of the thermometer in mapping out the position of the boundaries of the Gulf Stream, as well as his valuable suggestions as to the aid the thermometer is capable of affording navigators in such cases. Franklin published an account of measurements of the temperature of the Gulf Stream made during a subsequent voyage, while on the Pennsylvania Packet, Captain Osborne, from London to Philadelphia, in April and May, 1775. During these measurements, some interesting observations were made as to deep sea temperatures. In one instance, during perfectly calm weather an empty bottle, tightly corked, was sent down to the depth of some twenty fathoms. On drawing it up, it was still found to be empty. When again let down to a depth of thirty-five fathoms, the pressure of the water had been sufficiently great to force the cork into the bottle, so that when the bottle was drawn to the surface it was found to be filled with water at a temperature six degrees colder than that at the surface. In a somewhat similar manner, experiments made with an empty cask showed that, although some leakage occurred during the drawing up of the cask, the water it contained was at a temperature of some 12 degrees colder than the surface water.

(f) The explanation here given as to the probable cause of the fogs so common off the coast of Newfoundland is the one that is still generally held.

It was during one of the many voyages between America and Europe that Franklin invented an important device known as the swimming anchor. This device is of use in preventing a ship from driving to leeward in deep water, where there is no sounding, and where, consequently, an ordinary anchor cannot be employed. Franklin names the following characteristics which should be possessed by this type of anchor:

(1) Its surface should be of such a size that, when at the end of the hawser in the water, and placed perpendicularly, it should so hold as to bring the ship's head to the wind, in which situation the wind has least power to drive the vessel.

(2) It should be able, by its resistance, to prevent the ship's receiving way.

(3) It should be capable of being placed below the heave of the sea, but not below the undertow.

(4) It should not take up much room in the ship.

(5) It should be capable of being easily thrown into the water and assuming therein its desired position, and should afterwards be easy to take into the ship and stow away.

Franklin devised two of these anchors. One of them was made in the form of a kite, Fig. 13, while the other had the shape of an umbrella. Franklin thus describes the umbrella type of anchor:

"The other machine for the same purpose is to be made more in the form of an umbrella, as represented in figure 14. The stem of the umbrella, a square spar of proper length, with four movable arms, of which two are represented C, C, figure 14. These arms to be fixed in four joint cleats, as D, D, &c., one on each side of the spar, but so as that the four arms may open by turning on a pin in the joint. When open they form a cross on which a four-square canvass sail is to be extended, its corners fastened to the ends of the four arms. Those ends are also to be stayed by ropes fastened to the stem or spar, so as to keep them short of being at right angles with it; and to the end of one of the arms should be hung the small

Fig. 13. Franklin's Swimming Anchor.

bag of ballast, and to the end of the opposite arm the empty keg. This, on being thrown into the sea, would immediately open; and when it had performed its function, and the storm over, a small rope from its other end being pulled on, would turn it, close it, and draw it easily home to the ship. This machine seems more simple in its operation, and more easily manageable than the first, and perhaps may be as effectual."

Another important invention belonging to applied physics is to be found in what is known as Franklin's Pennsylvania Fire-Place. The following description is given by Franklin:

"An Account
of the New-Invented
Pennsylvanian Fire-Place;
wherein
Their construction and manner of operation is particularly explained; their

advantages above every other method of warming rooms demonstrated; and all objections that have been raised against the use of them answered and obviated. With directions for putting them up, and for using to the best advantage. And a copper-plate, in which the several parts of the machine are exactly laid down, from a scale of equal parts.
Philadelphia;
Printed and sold by B. Franklin, 1744."

*   *   *   *   *   *   *

"In these Northern Colonies the inhabitants keep fires to sit by, generally seven months in the year; that is, from the beginning of October, to the end of April; and, some winters, near eight months, by taking in part of September and May.

"Wood, our common fuel, which within these hundred years might be had at every man's door, must now be fetched near one hundred miles to some towns, and makes a very considerable article in the expence of families.

"As therefore so much of the comfort and conveniency of our lives,

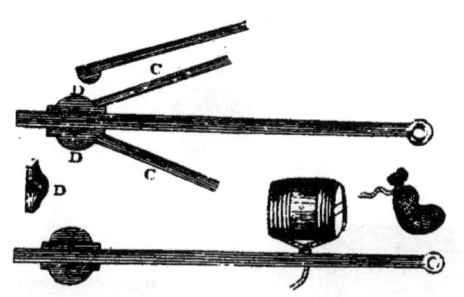

Fig. 14. Franklin's Swimming Anchor, Details of.

for so great a part of the year, depends on the article of fire; since fuel is become so expensive, and (as the country is more cleared and settled) will of course grow scarcer and dearer, any new proposal for saving the wood, and for lessening the charge, and augmenting the benefit of fire by some particular method of making and managing it, may at least be thought worth consideration.

*   *   *   *   *   *   *

"To avoid the several inconveniences, and at the same time retain all the advantages of other fire-places, was contrived the Pennsylvania Fire-Place, now to be described.

"This machine consists of
"A bottom plate (i) (See Plate annexed. figure 15.)
"A back plate, (ii)
"Two side plates, (iii iii)
"Two middle plates, (iv iv) which joined together, form a tight box, with winding passages in it for warming the air.
"A front plate, (v)
"A top plate, (vi)
"These are all cast of iron, with mouldings or ledges where the plates

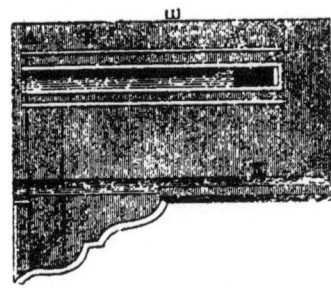

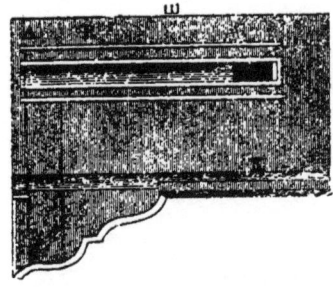

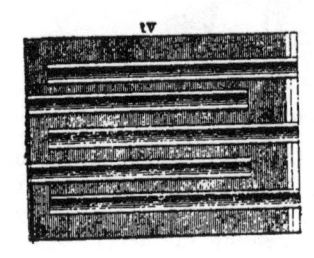

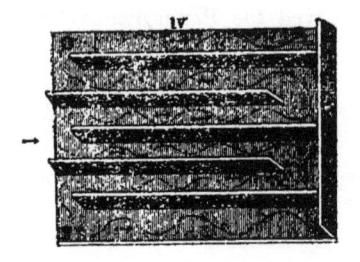

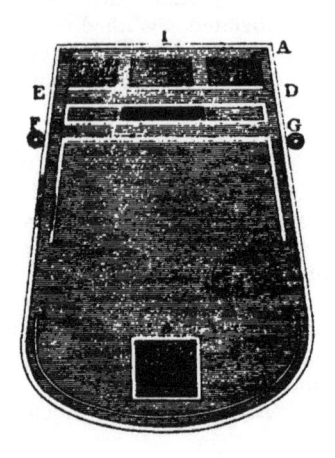

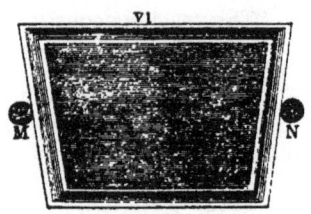

Fig. 15. The Pennsylvania

Plate & Joint of the proper size.
Fire-place.

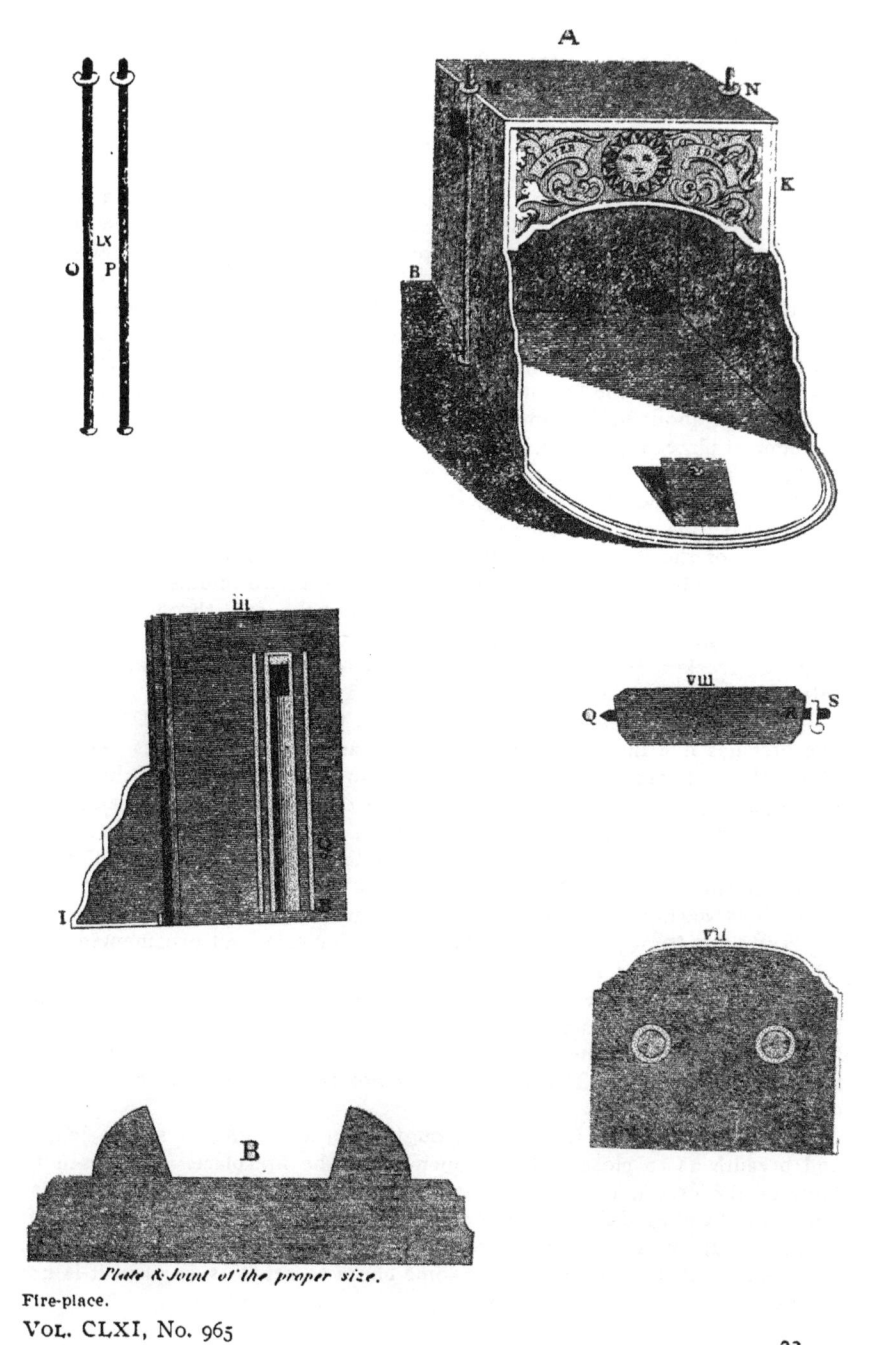

come together, to hold them fast, and retain the mortar used for pointing to make tight joints. When the plates are all in their places, a pair of slender rods with screws, are sufficient to bind the whole very firmly together, as it appears in A.

"There are, moreover, two thin plates of wrought iron, viz.: the sutter, (vii) and the register, (viii); besides the screw-rods O P, all of which we shall explain in their order.

"(i) The bottom plate or hearth-piece, is round before, with a rising moulding that serves as a fender to keep coals and ashes from coming to the floor, &c. It has two ears, F G, perforated to receive the screw-rods O P: a long air-hole, a a, through which the fresh outward air passes up into the air-box; and three smoke-holes B C through which the smoke descends and passes away; all represented by dark squares. It has also double ledges to receive between them the bottom edges of the plate, the two side plates and the two middle plates. These ledges are about an inch asunder, and about half an inch high; a profile of two of them joined to a fragment of plate, appears in B.

"(ii) The back plate is without holes, having only a pair of ledges on each side, to receive the back edge of the two.

"(iii iii) Side plates: These have each a pair of ledges to receive the side-edges of the front plate, and a little shoulder for it to rest on; also two pair of ledges to receive the side edges of the two middle plates which form the air-box; and an oblong air-hole near the top, through which is discharged into the room the air warmed in the air-box. Each has also a wing or bracket, H and I, to keep in falling brands, coals, &c., and a small hole, Q and R, for the axis of the register to turn in.

"(iv iv) The air-box is composed of the two middle plates DE and FG. The first has five thin ledges or partitions cast on it, two inches deep, the edges of which are received in so many pair of ledges cast in the other. The tops of all the cavities formed by these thin deep ledges, are also covered by a ledge of the same form and depth, cast with them; so that when the plates are put together and the joints luted, there is no communication bewtween the air-box and the smoke. In the winding passages of this box, fresh air is warm'd as it passes into the room.

"(v) The front plate is arched on the under side, and ornamented with foliages, &c. It has no ledges.

"(vi) The top plate has a pair of ears, M N, answerable to those in the bottom plate, and perforated for the same purpose: It has also a pair of ledges running round the under side, to receive the top edges of the front, back, and side plates. The air-box does not reach up to the top plate by two inches and half.

"(vii) The shutter is of thin wrought iron and light, of such a length and breadth as to close well the opening of the fire-place. It is used to blow up the fire, and to shut up and secure it at nights. It has two brass knobs for handles, *d d*, and commonly slides up and down in a groove, left in putting up the fire-place, between the foremost ledge of the side-plates, and the face of the front plate; but some chuse to set it aside when it is not in use, and apply it on occasion.

"(viii) The register is also of thin wrought iron. It is placed between the back plate and air-box, and can, by means of the key S, be turned on its axis, so as to lie in any position between level and upright.

"The screw-rods O P are of wrought iron, about a third of an inch thick, with a button at bottom, and a screw and nut at top, and may be ornamented with two small brasses screwed on above the nuts.

"To put this Machine to work,

"1. A false back of four inch (or, in shallow small chimneys, two inch) brick work is to be made in the chimney, four inches or more from the true back: From the top of this false back a closing is to be made over to the breast of the chimney, that no air may pass into the chimney, but what goes under the false back, and up behind it.

"2. Some bricks of the hearth are to be taken up, to form a hollow under the bottom plate across which hollow runs a thin tight partition, to keep apart the air entering the hollow and the smoke; and is therefore placed between the air-hole and smoke-holes.

"3. A passage is made, communicating with the outer air, to introduce that air into the fore part of the hollow under the bottom-plate, whence it may rise thro' the air-hole into the air-box.

"4. A passage is made from the back part of the hollow, communicating with the flue behind the false back: Through this passage the smoke is to pass.

"The fire-place is to be erected upon these hollows, by putting all the plates in their places, and screwing them together.

"Its operation may be conceived by observing the plate entitled, *Profile of the Chimney and Fire-place.* (Fig. 16.)

"M The mantel-piece or breast of the chimney.
"C The funnel.
"B The false back and closing.
"E True back of the chimney.
"T Top of the fire-place.
"F The front of it.
"A The place where the fire is made.
"D The air-box.
"K The hole in the side plate, through which the warmed air is discharged out of the air-box into the room.
"H The hollow filled with fresh air, entering at the passage I, and ascending into the air-box through the air-hole in the bottom plate, near
"G The partition in the hollow to keep the air and smoke apart.
"P The passage under the false back and part of the hearth for the smoke.

"The arrows show the course of the smoke.

"The fire being made at A, the flame and smoke will ascend and strike the top T, which will thereby receive a considerable heat. The smoke, finding no passage upwards, turns over the top of the air-box, and descends between it and the back plate to the holes at B, in the bottom plate, heating, as it passes, both plates of the air-box, and the said back-plate; the front plate, bottom and side plates, are also all heated at the same time.

The smoke proceeds in the passage that leads it under and behind the false back, and so rises into the chimney. The air of the room, warmed behind the back plate, and by the sides, front, and top plates, becoming specifically lighter than the other air in the room, is obliged to rise; but the closure over the fire-place hindering it from going up the chimney, it is forced out into the room, rises by the mantle-piece to the ceiling, and spreads all over the top of the room, whence, being crouded down gradually by the stream newly-warm'd air that follows and rises above it, the whole room becomes in a short time equally warmed.

"At the same time the air, warmed under the bottom-plate, and in the

Fig. 16. Profile of the Pennsylvania Chimney and Fire Place.

air-box, rises and comes out of the holes in the side-plates, very swiftly if the door of the room be shut, and joins its current with the stream before mentioned, rising from the side, back, and top plates.

"The air that enters the room through the air-box is fresh, though warm; and, computing the swiftness of its motion with the areas of the holes, it is found that near ten barrels of fresh air are hourly introduced by the air-box; and by this means the air in the room is continually changed, and kept, at the same time, sweet and warm.

"It is to be observed, that the entering air will not be warm at first lighting the fire, but heats gradually, as the fire increases.

"A square opening for a trap-door should be left in the closing of the chimney, for the sweeper to go up: The door may be made of slate or tin, and commonly kept close shut, but so placed as that turning up against the back of the chimney when open, it closes the vacancy behind the false back, and shoots the soot, that falls in sweeping, out upon the hearth. This trap-door is a very convenient thing.

"In rooms where much smoking of tobacco is used, it is also convenient to have a small hole, about five or six inches square, cut near the ceiling through into the funnel: This hole must have a shutter, by which it may be clos'd or open'd at pleasure. When open, there will be a strong draught of air thro' it into the chimney, which will presently carry off a cloud of smoke, and keep the room clear: If the room be too hot likewise, it will carry off as much of the warm air as you please, and then you may stop it entirely, or in part, as you think fit. By this means it is, that the tobacco smoke does not descend among the heads of the company near the fire, as it must do before it can get into common chimneys. ?

"THE MANNER OF USING THIS FIRE-PLACE.

"Your cord-wood must be cut into three lengths; or else a short piece, fit for the fire-place, cut off, and the longer left for the kitchen or other fires. Dry hickory, or ash, or any woods that burn with a clear flame are rather to be chosen, because such are less apt to foul the smoke-passages with soot; and flame communicates with its light, as well as by contact, greater heat to the plates and room. But where more ordinary wood is used, half a dry faggot of brush-wood, burnt at the first making of fire in the morning is very advantageous, as it immediately, by its sudden blaze, heats the plates and warms the room (which with bad wood slowly kindling would not be done so soon) and at the same time, by the length of its flame, turning in the passages, consumes and cleanses away the soot that such bad smoky wood had produced therein the preceding day, and so keeps them always free and clean.—When you have laid a little back log, and placed your billets on small dogs, as in common chimneys, and put some fire to them, then slide down your shutter as low as the dogs, and the opening being by that means contracted, the air rushes in briskly, and presently blows up the flames. When the fire is sufficiently kindled, slide it up again. In some of these fire-places there is a little six-inch square trap-door of thin wrought iron or brass, covering a hole of like dimensions near the fore-part of the bottom-plate, which being by a ring lifted up towards the fire, about an inch, where it will be retained by two springing sides fixed to it perpendicularly, (See fig. 16. C. as above). the air rushes in from the hollow under the bottom plate, and blows the fire. Where this is used, the shutter serves only to close the fire at nights. The more forward you can make your fire on the hearth-plate, not to be incommoded by the smoke, the sooner and more will the room be warmed. At night when you go to bed, cover the coal or brands with ashes, as usual; then take away the dogs, and slide down the shutter close to the bottom-plate, sweeping a little ashes against it, that no air may pass under it; then turn the register, so as very near to stop the flue behind. If no smoke then comes out at crevices into the room, it is right: If any smoke is perceived to come out, move the register so as to give a little draft, and it will go the right way. Thus the room will be kept warm all night; for the chimney being almost entirely stopt, very little cold air, if any, will enter the room at any crevice. When you come to re-kindle the fire in the morning, turn open the register before you lift up the slider, otherwise, if there be any smoke in the fire-place, it will come out into the room. By the same use of the shutter and register, a blazing fire may be presently stifled as well as secured, when you have occasion to leave it for any time: and at your return you will find the brands warm, and ready for a speedy re-kindling. The shutter alone will not stifle a fire, for it cannot well be made to fit so exactly but that air will enter, and that in a violent stream, so as to blow up and keep alive the flames, and consume the wood, if the draught be not checked by turning the register to shut the flue behind The register has also two other uses. If you observe the draught of air into your fire-place to be stronger than is necessary, (as in extreme cold weather it often is), so that the wood is consumed faster than usual; in that case, a quarter, half, or two-thirds turn of the register will check the violence of the draught, and let your fire burn with the moderation you desire: And at the same time both the fire-place and the room will be warmer, because less cold air will enter and pass through them. And if the chimney should happen to take fire, which indeed there is very little danger of, if the preceding directions be observed in making fires, and it be well swept once a year; for, much less wood being burnt, less soot is proportionately made; and the fuel being blown into flame by the shutter (or the trap door bellows) there is consequently less smoke from the fuel to make soot; then, though the funnel should be foul, yet the sparks have such a crooked up and down round about way to go, that they are out before they get at it. I say, if ever it should be on fire, a turn of the register shuts all close, and prevents any air going into the chimney, and so the fire may be easily stifled and mastered."

THE ADVANTAGES OF THIS FIRE-PLACE.

"Its advantages above the common fire-places are,

"1. That your whole room is equally warmed, so that people need not close round the fire, but may sit near the window, and have the benefit of the light for reading, writing, needlework, etc. They may sit with comfort in any part of the room, which is a very considerable advantage in a large family, where there must often be two fires kept, because all cannot conveniently come at one.

"2. If you sit near the fire, you have not that cold draft of uncomfortable air nipping your back and heels, as when before common fires. by which you may catch cold, being scorched before, and, as it were, froze behind.

"3. If you sit against a crevice, there is not that sharp draft of cold air playing on you, as in rooms where there are fires in the common way; by which many catch cold, whence proceed coughs, catarrhs, tooth-aches, fevers, pleurisies, and many other diseases.

"4. In case of sickness, they make most excellent nursing-rooms; as they constantly supply a sufficiency of fresh air, so warmed at the same time as to be no way inconvenient or dangerous. A small one does well in a chamber; and, the chimneys being fitted for it, it may be removed from one room to another, as occasion requires, and fixed in half an hour. The equal temper, too, and warmth of the air of the room is thought to be particularly advantageous in some distempers; for it was observed in the winters of 1730 and 1736, when the small-pox spread in Pennsylvania, that very few children of the Germans died of that distemper in proportion to those of the English; which was ascribed, by some, to the warmth and equal temper in their stove rooms, which made the disease as favorable as it commonly is in the West Indies. But this conjecture we submit to the judgment of physicians.

"5. In common chimneys, the strongest heat from the fire, which is upwards, goes directly up the chimney, and is lost; and there is such a strong draft into the chimney, that not only the upright heat, but also the back, sides, and downward heats are carried up the chimney by that draft of air; and the warmth given before the fire, by the rays that strike out towards the room, is continually driven back, crowded into the chimney, and carried up by the same draft of air. But here the upright heat strikes and heats the top plate, which warms the air above it, and that comes into the room. The heat likewise, which the fire communicates to the sides, back, bottom, and air-box, is all brought into the room, for you will find a constant amount of warm air coming out of the chimney corner into the room. Hold a candle just under the mantle-piece, or breast of your chimney, and you will see the flame bend outwards; by laying a piece of smoking paper on the hearth, on either side, you may see how the current of air moves, and where it tends, for it will turn and carry the smoke with it.

"6. Thus, as very little of the heat is lost, when this fire-place is used, *much less wood* will serve you, which is a considerable advantage where wood is dear.

"7. When you burn candles near this fire-place, you will find that the flame burns quite upright, and does not blare and run the tallow down, by drawing towards the chimney, as against common fires.

"8. This fire-place cures most smoky chimneys, and thereby preserves both the eyes and furniture.

"9. It prevents the fouling of chimneys; much of the lint and dust that contributes to foul a chimney being, by the low arch, obliged to pass through the flame, where it is consumed. Then, less wood being burnt, there is less smoke made. Again, the shutter, or trap-bellows, soon blowing the wood into a flame, the same wood does not yield so much smoke as if burnt in a common chimney; for, as soon as the flame begins, smoke in proportion ceases.

"10. And, if a chimney should be foul, it is much less likely to take fire. If it should take fire, it is easily stifled and extinguished.

"11. A fire may be very speedily made in this fire-place by the help of the shutter, or trap-bellows, as aforesaid.

"12. A fire may be soon extinguished by closing it with the shutter before, and turning the register behind, which will stifle it, and the brands will remain ready to rekindle.

"13. The room being once warm, the warmth may be retained in it all night.

"14. And lastly, the fire is so secured at night, that not one spark can fly out into the room to do damage.

"With all these conveniences, you do not lose the pleasing sight nor use of the fire, as in the Dutch ovens, but may boil the tea kettle, warm the flat irons, heat heaters, keep warm a dish of victuals by setting it on the top, etc.

Concerning the operation of the Pennsylvania Fire-Place, Franklin remarks:

"Having in 1742, invented an open stove for the better warming of rooms, and at the same time saving fuel, as the fresh air admitted was warmed in entering, I made a present of the model to Mr. Robert Grace, one of my early friends, who, having an iron furnace, found the casting of the plates for these stoves a profitable thing, as they were growing in demand. To promote that demand, I wrote and published a pamphlet, intitled "An Account of the new-invented Pennsylvania Fire-Places," &c. This pamphlet had a good effect. Governor Thomas was so pleased with the construction of this stove, as described in it, that he offered to give me a patent for the sole vending of them for a term of years; but I declined it from a principle, which has ever weighed with me on such occasions, viz: *That, as we enjoy great advantages from the inventions of others, we should be glad of an opportunity to serve others by any invention of ours; and this we should do freely and generously.*

"An ironmonger in London, however, assuming a good deal of my pamphlet, and working it up into his own, and making some small changes in the machine, which rather hurt its operation, got a patent for it there, and made, as I was told, a little fortune by it. And this is not the only instance of patents taken out of my inventions by others, though not always with the same success; which I never contested, as having no desire of profiting by patents myself, and hating disputes. The use of these fire-places in very many houses, both here and in Pennsylvania, and the neighboring states, has been, and is, a great saving of wood to the inhabitants."

In connection with his invention of stoves, Franklin gave considerable attention to the construction of chimneys. In a letter to John Ingenhousz, at Vienna, August 28, 1785, Franklin describes at length the different ways in which chimneys may be caused to smoke, and also points out the means by which this smoking may best be avoided.

It will be impracticable, on account of lack of space, to discuss this letter at length. It will suffice if some of the more important of the causes of smoky chimneys are pointed out. These are given by Franklin as follows:

(1) Smoky chimneys are frequently produced in new houses by the mere want of air.

(2) Smoky chimneys are frequently caused by the opening into the room being too large; that is, too wide, too high, or both.

(3) Smoky chimneys may be caused by too short a chimney funnel. This happens necessarily where a chimney is required in a low building, since, if the funnel is raised high above the roof in order to strengthen the draught, it is in danger of being blown down.

(4) Smoky chimneys may be caused by one chimney overpowering another, where, for example, two chimneys exist in one large room and fires are made in both, the doors and windows being shut, the greater and stronger fire will overpower the weaker, drawing the air down its funnel to supply its demands, while air descending the funnel connected with the weaker fire will drive its smoke into the room.

(5) Smoky chimneys may also be caused by the tops of the chimneys being commanded by higher buildings, or by a hill, so that the wind, blowing over such eminences, falls like water over a dam, thus beating the smoke down the chimney.

(6) Smoky chimneys are sometimes caused by the chimney being lower than the top of a near-by house, the wind being deflected from the house and forced down the chimney.

Other causes are given, but these are the most important.

At a later date, Franklin describes another form of stove suitable for burning pit coal. In this form of stove the construction is such that the smoke itself is consumed. A paper describing this stove was read before the American Philosophical Society on the 28th of January, 1786. As will be seen, this stove is based on the principle of an inverted draught down through the burning material:

"Towards the end of the last century, an ingenious French philosopher, whose name I am sorry I cannot recollect, exhibited an experiment to show, that very offensive things might be burnt in the middle of a chamber, such as woolen rags, feathers, &c., without creating the least smoke or smell. The machine in which it was made, if I remember right, was of this form, (Plate XV, Fig. 17, No. 1), made of plate iron. Some clear

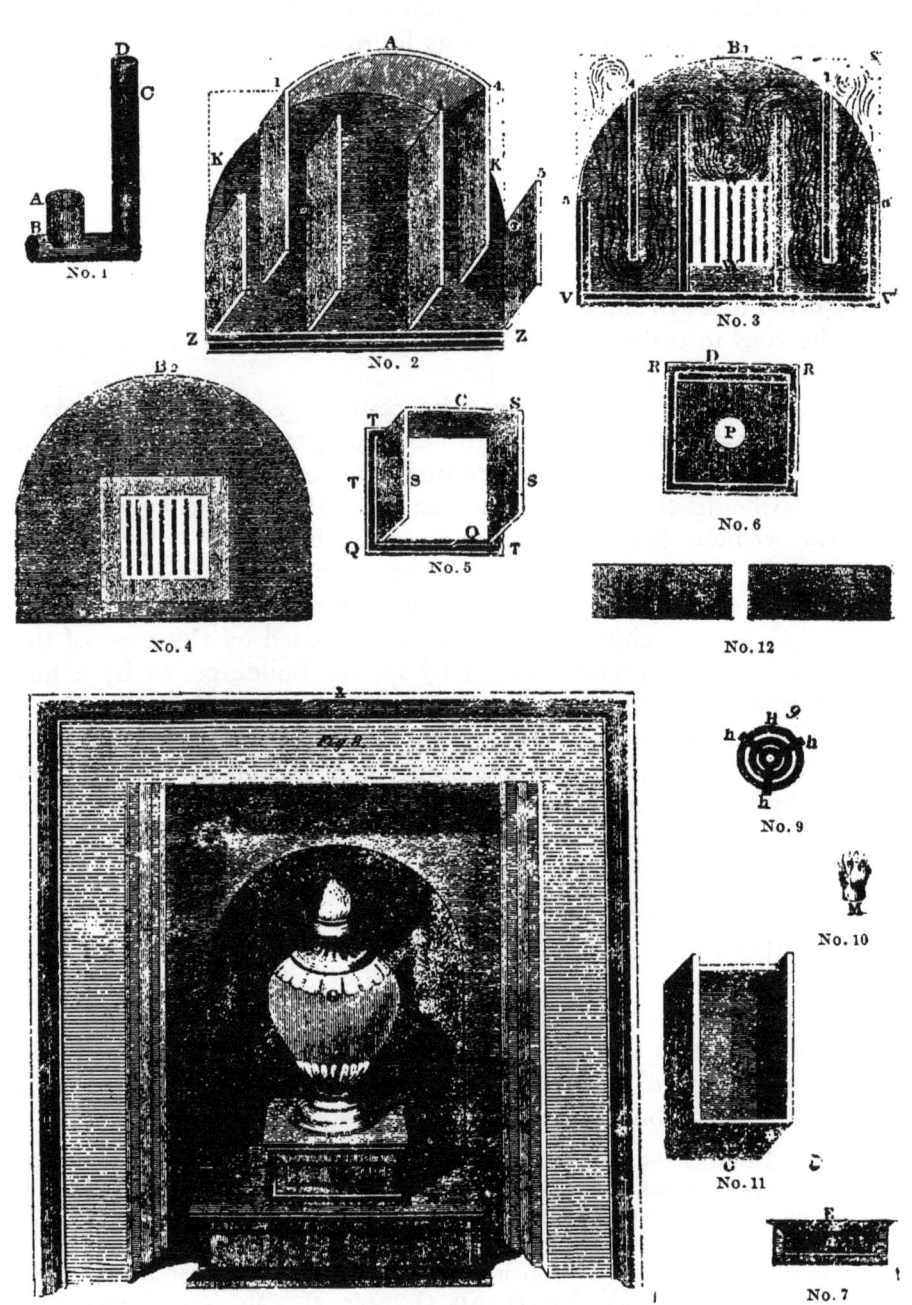

Fig. 17. Details of the Pennsylvania Fire-place.

burning charcoals were put into the opening of the short tube A, and supported there by the grate B. The air, as soon as the tubes grew warm, would ascend in the longer leg C, and go out at D, consequently air must enter at A, descending to B. In this course it must be heated by the burning coals through which it passed, and rise more forcibly in the longer tube, in proportion to its degree of heat or rarefaction, and length of that tube. For such a machine is a kind of inverted siphon; and, as the greater weight of water in the longer leg of a common siphon in descending is accompanied by an ascent of the same fluid, in the shorter; so, in this inverted siphon, the greater quantity of levity of air in the longer leg, in rising is accompanied by the descent of air in the shorter. The things to be burned being laid on the hot coals at A, the smoke must descend through those coals, and be converted into flame, which, after destroying the offensive smell, came out at the end of the longer tube as mere heated air.

"Whoever would repeat this experiment with success, must take care that the part A B, of the short tube, be quite full of burning coals, so that no part of the smoke may descend and pass by them without going through them, and being converted into flame; and that the longer tube be so heated as that the current of ascending hot air is esablished in it before the things to be burnt are laid on the coals; otherwise there will be a disappointment."

* * * * * * *

"The stove I am about to describe was also formed on the idea given by the French experiment, and completely carried into execution before I had any knowledge of the German invention; which I wonder should remain so many years in a country, where men are so ingenious in the management of fire, without receiving long since the improvements I have given it.

Description of the Parts.

"A, the bottom plate which lies flat upon the hearth, with its partitions, 1, 2, 3, 4, 5, 6, (Plate, our fig. 17, No. 2) to slide the bottom edges of the small plates Y, Y, No. 12; which plates meeting at X, close the front.

"B 1, No. 3, is the cover plate showing its under side, with the grooves 1, 2, 3, 4, 5, 6, to receive the top edges of the partitions that are fixed to the bottom plate. It shows also the grate W W, the bars of which are cast in the plate, and a groove V V, which comes right over the groove Z Z, No. 2, receiving the upper edges of the small sliding plates Y Y, No. 12.

"B 2, No. 4, shows the upper side of the same plate with a square impression or groove for receiving the bottom mouldings T T T T of the three-sided box C, No. 5, which is cast in one piece.

"D, No. 6, its cover, showing its under side with grooves to receive the upper edges S S S of the sides of C, No. 5, also a groove R R, which, when the cover is put on, comes right over another, Q Q in C, No. 5, between which is to slide.

"E, No. 7, the front plate of the box.

"P, a hole three inches diameter through the cover D, No. 6, over which hole stands the vase F, No. 8, which has a corresponding hole two inches diameter, through its bottom.

"The top of the vase opens at O O O, No. 8, and turns back upon a hinge behind, when coals are to be put in; the vase has a grate within at N N of cast iron H, No. 9, and a hole in the top one and a half inches in diameter, to admit air, and to receive the ornamental brass gilt flame M, No. 10, which stands in that hole, and being itself hollow and open, suffers air to pass through it to the fire.

"G, No. 11, is a drawer of plate iron, that slips in between the partitions 2 and 3, No. 2, to receive the falling ashes. It is concealed when the small sliding plates Y, Y, No. 12, are shut together.

"I I I I, No. 8, is a niche built of brick in the chimney, and plastered. It closes the chimney over the vase, but leaves two funnels, one in each corner, communicating with the bottom box K K, No. 2."

As we have repeatedly seen, Franklin was essentially a utilitarian. Wherever possible, he set himself the task of directly applying the principles of any great natural law he had discovered to some useful purpose. We see this in his great invention of the lightning rod. It is also to be observed in his invention of the Pennsylvania Fire-place, as well as the stove with the downward draught, that was capable of burning in a closed room, not only pit coal, but even substances that would, in an ordinary stove, give out either exceedingly disagreeable or noxious odors, or both.

To the same type of investigation is Franklin's study of the Gulf Stream. No sooner does he hear of the existence of this stream of heated water, flowing as it does along the eastern coast of the United States, in the direct path of vessels sailing between America and Europe, than he sets himself the task of having a chart drawn of this body of water, which he has engraved, sending copies to such parts of the world as would be most likely to be of benefit, thus enabling a navigator to turn what had been an evil, so far as it retarded the speed of his vessel when sailing in certain directions, into an advantage when sailing in opposite directions. Again, while this practical philosopher is employing the thermometer in determining the limits of the Gulf Stream, he points out how such use may possibly be of value to other navigators in teaching them how to avoid cold currents in the Northern Hemisphere, which, as is well known, generally tend to move from the north towards the south.

It was in this line of applied physics that Franklin made the study of the effects produced on the comfort of the individual ensured by wearing woolen clothes in cold wintry weather, and linen clothes in hot summer weather. In a letter to John Lining, of Charlestown, South Carolina, dated New York, April 14, 1757, he treats of a variety of topics, mainly, however, relating to the cold produced by evaporation. It is in this letter that he thus speaks of the effect produced by clothing:

"Thus, as by a constant supply of fuel in a chimney, you keep a room warm, so, by a constant supply of food in the stomach, you keep a warm body; only where little exercise is used, the heat may possibly be conducted away too fast; in which case such materials are to be used for clothing and bedding, against the effects of an immediate contact of the air, as are, in themselves, bad conductors of heat, and, consequently, prevent its being communicated thro' their substance to the air. Hence what is called warmth in wool, and its preference, on that account, to linnen; wool not being so good a conductor: And hence all the natural coverings of animals, to keep them warm are such as to retain and confine the natural heat in the body, by being bad conductors, such as wool, hair, feathers, and the silk by which the silk worm, in its tender embrio state, is first cloathed. Cloathing thus considered does not make a man warm by giving warmth, but by preventing the too quick dissipation of the heat produced in his body, and so occasioning an accumulation."

Franklin here gives the true explanation of the source of animal heat as the food consumed by the animal. He points out the fact that clothing does not itself supply heat to the body, but merely prevents the too rapid loss of the heat produced by the animal's food. By employing as clothing such poor conductors as wool or fur, the heat of the body is prevented from being rapidly passed into the surrounding air. Wool may properly be called warm because it is a poor conductor of heat, but not because it possesses any heat in itself. Linen, on the contrary, produces a cooling effect in that it permits the heat of the body to be rapidly passed or conducted through it to the surrounding air.

Franklin gave considerable thought to the effect of external heat on the temperature of the bodies of healthy animals. He notices the fact that it is possible for a healthy living animal to be exposed to a very high temperature without any notable increase in its temperature. In another letter to John Lining, of Charlestown, dated London, June 17, 1758, he gives an account of an exceedingly hot Sunday that occurred in Philadelphia in June, 1750, with the thermometer at 100° in the shade:

"May not several phenomena hitherto unconsidered, or unaccounted for be explained by this property? (a)During the hot Sunday at Philadelphia, in June, 1750, when the thermometer was up at 100 in the shade, I sat in my chamber without exercise, only reading or writing, with no other cloathes on than a shirt, and a pair of long linen drawers, the windows all open, and a brisk wind blowing through the house, the sweat ran off the backs of my hands, and my shirt was often so wet, as to induce me to call for dry ones to put on; in this situation, one might have expected, that the natural heat of the body 96, added to the heat of the air 100, should jointly have created or produced a much greater degree of heat in the body; but the fact was, that my body never grew so hot as the air that surrounded it, or the immediate bodies immers'd in the same air. For I remember well that the desk, when I laid my arm upon it; a chair, when I sat down in it; and a dry shirt out of the drawer, when I put it on, all felt exceeding warm to me, as if they had been warmed before a fire. And I suppose a dead body would have acquired the temperature of the air, though a living one, by continual sweating, and by the evaporation of that sweat, was kept cold.

"(b) May not this be a reason why our reapers in Pennsylvania, working in the open field in the clear, hot sunshine common in our harvest time, find themselves well able to go through that labor, without being much incommoded by the heat, while they continue to sweat, and while they supply matter for keeping up that sweat, by drinking frequently of a thin, evaporable liquor, water mixed with rum; but, if the sweat stops, they drop, and sometimes die suddenly, if a sweating is not again brought on by drinking that liquor, or, as some rather choose in that case, a kind of hot punch, made with water, mixed with honey, and a considerable proportion of vinegar? May there not be in negroes a quicker evaporation of the perspirable matter from their skins and lungs, which, by cooling them more, enables them to bear the sun's heat better than whites do? (if that is a fact, as it is said to be; for the alleged necessity of having negroes, rather than whites, to work in the West India fields, is founded upon it,) though the color of their skins would otherwise make them more sensible of the sun's heat, since black cloth heats much sooner and more, in the sun, than white cloth. I am persuaded, from several instances happening within my knowledge, that they do not bear cold weather so well as the whites; they will perish when exposed to a less degree of it, and are more apt to have their limbs frost-bitten; and may not this be from the same cause?"

(a) Franklin notes correctly the fact that, although the thermometer was 100° in the shade, and his clothing was of such a nature as to readily expose the body to the external air, nevertheless the temperature of his body did not greatly exceed its natural temperature, never, in fact, growing as hot as the surrounding air; for, as he remarks, his body was evidently colder than the desk on which he laid his arm, was colder than the chair he sat on, or the dry shirt that he took out of the drawer to replace the wet one he had on. His reference to the effect that a high temperature would produce on a dead body is also in accordance with well known facts. It is possible for a man to go into an oven, the temperature of which is sufficiently high to bake a trussed fowl, which he carries in with him, and to safely remain in such an oven until the fowl is thoroughly cooked, and then to come out unharmed. Franklin's explanation is correct as to the cause; viz., that the living body protects itself by permitting the heat to evaporate the moisture it is constantly throwing out to the surface, and thus prevent it from entering the body.

(b) The explanation concerning the Pennsylvania reapers is also correct. Their safety, while under exposure to the hot sun, is ensured by drinking copious draughts of water, the rum being very likely a matter of taste, rather than of necessity.

In another letter, dated September 20th, 1761, to a Miss Mary Stevenson, Franklin refers to the effect produced by the sun's rays on clothes of different colors as follows:

"(a) As to our other subject, the different degrees of heat imbibed from the sun's rays by cloths of different colours, since I cannot find the notes of my experiment to send you, I must give it as well as I can from memory.

(b) "But first let me mention an experiment that you can easily make yourself. Walk but a quarter of an hour in your garden when the sun shines, with a part of your dress white, and a part black; then apply your hand to them alternately, and you will find a very great difference in warmth. The black will be quite hot to the touch, the white still cool.

(c) Another. Try to fire paper with a burning-glass. If it is white, you will not easily burn it; but if you bring the focus to a black spot, or upon letters, written or printed, the paper will immediately be on fire under the letters.

"Thus fullers and dyers find black cloths, of equal thickness with white ones, and hung out equally wet, dry in the sun much sooner than the white, being more readily heated by the sun's rays. It is the same before a fire; the heat of which sooner penetrates black stockings than white ones, and so is apt sooner to burn a man's shins. Also beer much sooner warms in a black mug set before the fire, than in a white one, or in a bright silver tankard.

(d) "My experiment was this. I took a number of little square pieces of broad cloth from a taylor's pattern card, of various colours. There were black, deep blue, lighter blue, green, purple, red, yellow, white, and all other colours, or shades of colours. I laid them all upon the snow in a bright, sun-shiny morning. In a few hours (I cannot now be exact as to the time) the black being warmed most by the sun was sunk so low as to be below the stroke of the sun's rays; the dark blue almost as low, the lighter blue not quite so much as the dark, the other colours less as they were lighter; and the quite white remained on the surface of the snow, not having entered it at all.

(e) "What signifies philosophy that does not apply to some use? May we not learn from hence, that black clothes are not so fit to wear in a hot sunny climate or season, as white ones; because in such clothes the body is more heated by the sun when we walk abroad, and are at the same time heated by the exercise, which double heat is apt to bring on putrid dangerous fevers? That soldiers and seamen who must march and labour in the sun, should in the East or West Indies have a uniform of white? That summer hats for men or women should be white, as repelling that heat which gives headaches to many, and to some the fatal stroke that the French call the Coup de Soleil? That the ladies Summer hats, however, should be lined with black, as not reverberating on their faces those rays which are reflected upwards from the earth or water? That the putting of a white cap of paper or linen within the crown of a black hat, as some do, will not keep out the heat, tho' it would if plac'd without. That fruit walls being blacked may receive so much heat from the sun in the daytime, as to continue warm in some degree thro' the night, and thereby preserve the fruit from frosts, or forward its growth? With sundry other particulars of less or greater importance, that will occur from time to time to inattentive minds?"

(a) It will be noticed that this investigation is not unlike a prior investigation in which Franklin pointed out the effects produced on the temperature of the human body, arising from the use of textile fabrics that varied in their power of conducting or transferring heat. Here, however, the investigation refers only to the influence of the color of fabrics on their ability to take in heat from an external source, such as the sun.

(b) A simple but crucial experiment. After an exposure of say a quarter of an hour to the hot sun, the fact that different portions of the clothing will have been differently heated according to their color, can be readily determined by simply touching the different articles of dress, when the darker colored ones will be found to be much warmer than the light colored ones.

(c) This employment of a burning glass or convex lens is

very simple, but quite convincing. Black paper can be easily fired or burned by the concentrated solar focus of the burning glass, while white paper cannot.

(d) A charming experiment, just of the character that we should expect this true philosopher to make. With no other means than simple squares of different colored cloths of the same size for the materials which are to be exposed to the sun's heat, and with no other thermometric apparatus than the ground covered with a layer of snow, an excellent test is made. Franklin places the squares of cloth on the snow, and exposes them for a few hours to the radiation of the sun. The fact that different temperatures have been attained by the cloths varying with their color, is evident by some of them having sunk a considerable distance into the snow, while others remain apparently unaffected. It is needless to add that, in this case, it was the black cloths that sank the farthest, the dark blue

Fig. 18. Franklin's Pulse Glass.

cloths the next farthest, and then the light blue, while the pure white cloths scarcely showed any effect whatever.

(e) Here, again, the utilitarian side of the philosopher asserts itself. The application of these facts is evident. Black clothes are unsuited to wear in a hot climate; white ones should be employed.

Of a somewhat similar character, though undertaken some time before the experiments we have just referred to, are Franklin's experiments with a form of apparatus that is now known in the physical laboratory as the pulse glass. Such a device is represented in Fig. 18. This consists of two vertical glass bulbs, a and b, connected by means of a horizontal tube of small diameter. The apparatus is partially filled with water, though, in more modern forms, preferably with ether, which has a much lower boiling point. While the upper end of the glass tube, b, is still open, the water in the bulb is vigorously boiled, so as to expel the air, when the tube is hermetically sealed by the fusion of the glass. By reason of the vacuum thus produced in the tube, the temperature of the boiling point of the liquid is considerably lowered, so that the heat of the hand placed at a, as shown in the figure, is sufficient to cause a brisk boiling of the liquid.

It would appear from a letter written by Franklin to John Winthrop, dated London, July 2, 1768, that Franklin obtained a tube of this character in Germany, during the preceding year. Franklin refers to this matter as follows:

"I have nothing new in the philosophical way to communicate to you, except what follows. When I was last year in Germany I met with a singular kind of glass, being a tube about eight inches long, half an inch in diameter, with a hollow ball of near an inch diameter at one end, and one of an inch and a half at the other, hermetically sealed, and half filed with water. If one end is held in the hand, and the other a little elevated above the level, a constant succession of large bubbles proceeds from the end in the hand to the other end, making an appearance that puzzled me much, till I found that the space not filled with water was also free from air, and either filled with a subtle, invisible vapor, continually rising from the water, and extremely rarefiable by the least heat at one end, and condensable again by the least coolness at the other; or it is the very fluid of fire itself, which parting from the hand pervades the glass, and by its expansive force depresses the water till it can pass between the glass and escapes to the other end, where it gets through the glass again into the air. I am rather inclined to the first opinion, but doubtful between the two.

"An ingenious artist here, Mr. Nairne, mathematical instrument maker, has made a number of them from mine, and improved them; for his are much more sensible than those I brought from Germany. I bored a very small hole through the wainscot in the seat of my window, through which a little cold air constantly entered, while the air in the room was kept warmer by fires made daily in it, being winter time. I placed one of his glasses, with the elevated end against this hole: and the bubbles from the other end, which was in a warmer situation, were continually passing day and night, to the no small surprise of even philosophical spectators. Each bubble discharged is larger than that from which it proceeds, and yet that is not diminished; and by adding itself to the bubble at the other end, that bubble is not increased, which seems very paradoxical.

"When the balls at each end are made large and the connecting tube very small, and bent at right angles, so that the balls, instead of being at the ends, are brought on the side of the tube, and the tube is held so that the balls are above it, the water will be depressed in that which is held in the hand, and rise in the other as a jet or fountain; when it is all in the other, it begins to boil, as it were, by the vapor passing up through it; and the instant it begins to boil, a sudden coldness is felt in the ball held; a curious experiment this, first observed and shown me by Mr. Nairne. There is something in it similar to the old observation, I think, mentioned by Aristotle, that the bottom of a boiling pot is not warm; and perhaps it may help to explain that fact, if indeed it be a fact.

"When the water stands at an equal height in both these balls, and all at rest, if you wet one of the balls by means of a feather dipped in spirit, though that spirit is of the same temperament as to heat and cold with the water in the glasses, yet the cold occasioned by the evaporaion of the spirit from the wetted ball, will so condense the vapor over the water contained in that ball, as that the water of the other ball will be pressed up into it, followed by a succession of bubbles, till the spirit is all dried away. Perhaps the observations on these little instruments may suggest and be applied to some beneficial uses. It has been thought, that water reduced to vapor by heat was rarefied only fourteen thousand times, and on this principle our engines for raising water for fire are said to be constructed; but, if the vapor so much rarefied from water is capable of being itself still further rarefied to a boundless degree, by the application of heat to the vessels or parts of vessels containing the vapor (as at first it is applied to those containing the water), perhaps a much greater power may be obtained, with little additional expense. Possibly, too, the power of easily moving water from one end to the other of a moveable beam (suspended in the middle like a scale-beam) by a small degree of heat, may be applied advantageously to some other mechanical purposes."

Franklin's explanation of the phenomena of the pulse glass was, generally speaking, correct. His manner of increasing the difference of temperature between the two bulbs was ingenious, though simple; i. e., by subjecting these two bulbs to the temperature of the outer air, and of a heated room, respectively, by placing one bulb near a small hole bored through the wainscot in a window seat, through which a small quantity of cold air entered, and leaving the other bulb exposed to the warm air of the room.

During the latter part of his life, Franklin suffered from failing sight. When seventy-nine years of age, he refers to this matter, in a letter to George Whatley, dated Passy, August 21, 1784, as follows:

"Your eyes must continue very good since you can write so small a hand without spectales. I cannot distinguish a letter even of large print; but am happy in the invention of double spectacles, which, serving for distant objects as well as near ones, make my eyes as useful to me as ever they were. If all the other defects and infirmities were as easily and cheaply remedied, it would be worth while for friends to live a great deal longer, but I look upon death to be as necessary to our constitution as sleep. We shall rise refreshed in the morning."

The double spectacles to which he refers as having invented are what are now generally called bifocals. They were made for Franklin under his direction, in Paris, by a French optician. Franklin thus describes these spectacles in a letter to George Whatley, dated Passy, May 23, 1785:

"By Mr. Dollond's saying that my double spectacles can only serve particular eyes, I doubt he has not been rightly informed of their construction. I imagine it will be found pretty generally true, that the same convexity of glass, through which a man sees clearest and best at the distance proper for reading, is not the best for greater distances. I therefore had formerly two pairs of spectacles, which I shifted occasionally, as in travelling I sometimes read, and often wanted to regard the prospects. Finding this change troublesome, and not always sufficiently ready, I had the glasses cut, and half of each kind associated in the same circle, thus—

Fig. 19. Franklin's Double Spectacles or Bifocals.

By this means, as I wear my spectacles constantly, I have only to move my eyes up or down, as I want to see distinctly far or near, the proper glasses being always ready. This I find more particularly convenient since my being in France; the glasses that serve me best at the table to see what I eat, not being the best to see the faces of those on the other side of the table who speak to me; and when one's ears are not well accustomed to the sound of a language, a sight of the movements in the features of him that speaks, helps to explain; (a) so that I understand French better by the help of my spectacles."

(a) By this witty remark, Franklin of course refers to the fact that he made up, by the use of his eyes, for what his ears lacked in thoroughly appreciating the sounds of a language to which he was not accustomed.

While in London, on one of his many visits to England, Franklin saw a musical instrument called an armonica. This instrument consists of glass vessels of varying sizes, which when rubbed by passing the moistened finger around their rims, emits the notes of the musical scale or gamut. Being much pleased with the sweetness of its notes, Franklin greatly improved the instrument, so as to increase the number of tones it was capable of producing, while, at the same time, placing it in a much smaller space.

In Franklin's improved form of armonica, the different glasses were made in the form of hemispheres, provided with an open neck and a socket in the middle for fixing them to an iron spindle. The spindle, holding the glasses, was placed horizontally and moved by a treadle in a manner similar to that of the spinning wheel. The performer sat in front of the instrument and brought out the required tones by placing the moistened fingers on the rims of the rotating glass.

Franklin describes his improved instrument in a letter written to the Rev. John Baptist Becearia, dated London, July 13, 1762:

"I once promised myself the pleasure of seeing you at Turin; but as that is not now likely to happen, being just about returning to my native country, America, I sit down to take leave of you (among other of my European friends that I cannot see) by writing.

"I thank you for the honorable mention you have so frequently made of me in your letters to Mr. Collinson and others, for the generous defence you undertook and executed with so much success, of my electrical opinions; and for the valuable present you have made me of your new work, from which I have received great information and pleasure. I wish I could in return entertain you with anything new of mine on that subject; but I have not lately pursued it. Nor do I know of any one here, that is at present much engaged in it.

"Perhaps, however, it may be agreeable to you, as you live in a musical country, to have an account of the new instrument lately added here to the great number that charming science was before possessed of. As it is an instrument that seems peculiarly adapted to Italian music, especially that of the soft and plaintive kind, I will endeavour to give you a description of it, and of the manner of constructing it, that you or any of your friends may be able to imitate it, if you incline to do so, without being at the expense add trouble of the many experiments I have made in endeavouring to bring it to its present perfection.

"You have doubtless heard the sweet tone that is drawn from a drinking-glass by passing a wet finger around its brim. One Mr. Puckeridge, a gentleman from Ireland, was the first who thought of playing tunes, formed of these tones. He collected a number of glasses of different sizes, fixed them near each other on a table, and tuned them by putting into them water more or less, as each note required. The tones were brought out by passing his fingers round their brims. He was unfortunately burned here, with his instrument, in a fire which consumed the house he lived in. Mr. E. Delaval, a most ingenious member of our Royal Society, made one in imitation of it, with a better choice and form of glasses, which was the first I saw or heard. Being charmed by the sweetness of its tones, and the music he produced from it, I wished only to see the glasses disposed in a more convenient form, and brought together in a narrower compass, so as to admit of a greater number of tones, and all within reach of hand to a person sitting in front of the instrument, which I accomplished, after various intermediate trials and less commodious forms, both of glasses and construction in the following manner:

"The glasses are blown as near as possible in the form of hemispheres, having each an open neck or socket in the middle. (Fig. 20.) The thickness of the glass near the brim about a tenth of an inch, or hardly quite so much, but thicker as it comes nearer the neck, which in the largest glasses is about an inch deep, and an inch and a half wide within, these dimensions lessening, as the glasses themselves diminish in size, except that the neck of the smallest ought not to be shorter than half an inch. The largest glass is nine inches diameter, and the smallest three inches. Between these two are twenty-three different sizes, differing from each other a quarter of an inch in diameter. To make a single

Fig. 20

instrument there should be at least six glasses blown of each size; and out of this number one may probably pick thirty-seven glasses (which are sufficient for three octaves with all the semi-tones) that will be each either the note one wants or a little sharper than that note, and all fitting so well into each other as to taper pretty regularly from the largest to the smallest. It is true that there are not thirty-seven sizes, but it often happens that two of the same size differ a note or half note in tone, by reason of a difference in thickness, and these may be placed one in the other without sensibly hurting the regularity of the taper form.

"The glasses being chosen and every one marked with a diamond the note you intend for it, they are to be tuned by diminishing the thickness of those that are too sharp. This is done by grinding them round from the neck towards the brim, the breadth of one or two inches, as may be required; often trying the glass by a well-tuned harpischord, comparing the tone drawn from the glass by your finger, with the note you want, as sounded by that string of the harpischord. When you come nearer the matter, be careful to wipe the glass clean and dry before each trial, because the tone is somewhat flatter when the glass is wet, than it will be when dry; and, grinding a very little between each trial, you will thereby tune to great exactness. The more care is necessary in this, because, if you go below your required tone, there is no sharpening it again but by grinding somewhat off the brim, which will afterwards require polishing, and thus increase the trouble.

"The glasses being thus tuned, you are to be provided with a case for them, and a spindle on which they are to be fixed. (Fig. 21.) My case is about three feet long, eleven inches every way wide within at the biggest end, and five inches at the smallest end; for it tapers all the way, to adapt it better to the conical figure of the set of glasses. This case opens in the middle of its height, and the upper

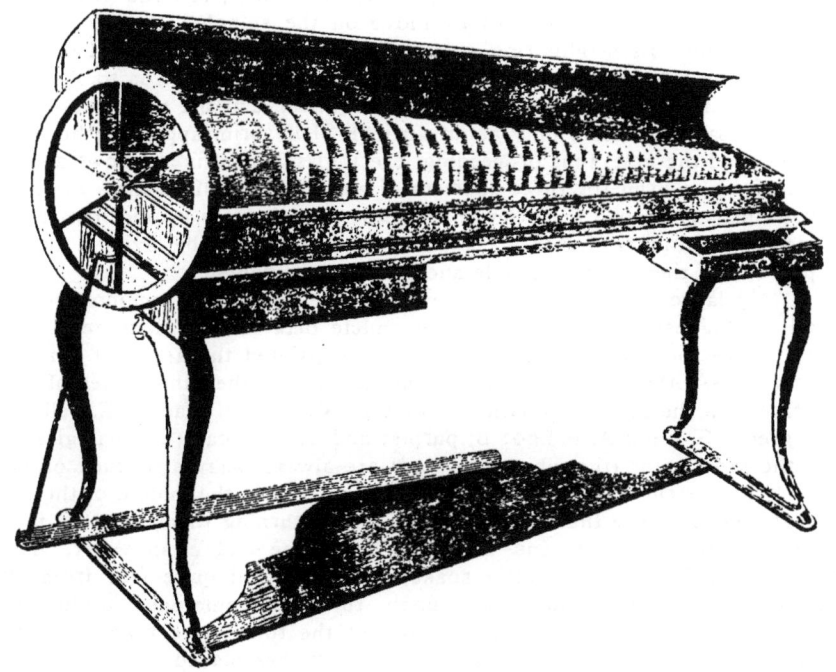

Fig. 21.

part turns up by hinges fixed behind. The spindle, which is of hard iron, lies horizontally from end to end of the box within, exactly in the middle, and is made to turn on brass gudgeons at each end. It is round, an inch diameter at the thickest end, and tapering to a quarter of an inch at the smallest. A square shank comes from its thickest end through the box, on which shank a wheel is fixed by a screw. This wheel serves as a fly to make the motion equable, when the spindle, with the glasses, is turned by the foot like a spinning wheel. My wheel is of mahogany, eighteen inches diameter, and pretty thick, so as to conceal near its circumference about twenty-five pounds of lead. An ivory pin is fixed on the face of this wheel, and about four inches from the axis. Over the neck of this pin is put the loop of the string that comes up from the movable step to give it motion. The case stands on a neat frame with four legs.

"To fix the glasses on the spindle, a cork is first to be fitted in each neck pretty tight, and projecting a little without the neck, that the neck of one may not touch the inside of another when put together, for that would make a jarring. These corks are to be perforated with holes of different diameters, so as to suit that part of the spindle on which they are to be fixed. When a glass is put on, by holding it stiffly between both hands, while another turns the spindle, it may be gradually brought to its place. But care must be taken that the hole is not too small, lest, in forcing it up, the neck should be split; nor too large, lest the glass, not being firmly fixed, should turn or move on the spindle, so as to touch and jar against its neighboring glass. The glasses are thus placed one in another, the largest on the biggest end of the spindle, which is to the left hand; the neck of this glass is towards the wheel, and the next goes into it in the same position, only about an inch of its brim appearing beyond the brim of the first; thus proceeding, every glass when fixed, shows about an inch of its brim (or three-quarters of an inch, or half an inch, as they grow smaller) beyond the brim of the glass that contains it; and it is from these exposed parts of each glass that the tone is drawn, by laying a finger upon one of them as the spindle and glasses turn round.

"My largest glass is G, a little below the reach of a common voice, and my highest G, including three complete octaves. To distinguish the glasses the more readily to the eye, I have painted the apparent part of the glasses within side, every semi-tone. white, and the other notes of the octave with the seven prismatic colors, viz: C, red; D, orange; E, yellow; F, green; G, blue; A, indigo; B, purple; and C, red again; so that glasses of the same color (the white excepted) are always octaves to each other.

"This instrument is played upon by sitting before the middle of the set of glasses as before the keys of a harpischord, turning them with the foot and wetting them now and then with a sponge and clean water. The fingers should be first a little soaked in water, and quite free from all greasiness; a little fine chalk upon them is sometimes useful. to make them catch the glass and bring out the tone more readily. Both hands are used, by which means different parts are played together. Observe, that the tones are best drawn out when the glasses turn from the ends of the fingers, and not when they turn to them.

"The advantages of this instrument are, that its tones are incomparably sweet, beyond those of any other; that they may be swelled and softened at pleasure by stronger or weaker pressure of the finger, and continued to any length; and that the instrument, being once well tuned, never again wants tuning.

"In honor of your musical language, I have borrowed from it the name of this instrument, calling it the Armonica.

"With great esteem and respect, I am, &c.,   B. Franklin."

Franklin gives the following directions as to the best manner of obtaining the sweet notes this instrument is capable of producing:

"Before you sit down to play, the fingers should be well washed with soap and water, and the soap well rinsed off.

"The glasses must always be kept perfectly clean from the least greasiness; therefore suffer nobody to touch them with unwashed hands, for even the common, slight, natural greasiness of the skin rubbed on them will prevent their sounding for a long time.

"You must be provided with a little bottle of rain water, (spring water is generally too hard, and produces a harsh tone,) and a middling sponge

Fig. 22. Franklin's Improvement on the Armonica.

in a little slop-bowl, in which you must keep so much of the water that the sponge may always be very wet.

"In a teacup keep also ready some fine scraped chalk, free from grit, to be used on occasion.

"The fingers when you begin to play should not only be wet on the surface, but the skin a little soaked, which is readily done by pressing them hard a few times in the sponge.

"The first thing after setting the glasses in motion is to pass the sponge slowly along from the biggest glass to the smallest, suffering it to rest on each glass during at least one revolution of the glasses, whereby they will all be made moderately wet. If too much water is left on them, they will not sound so readily.

"If the instrument is left near a window, let the window be shut or the curtain drawn, as wind or sunshine on the glasses dries them too fast.

"When these particulars are all attended to, and the directions observed, the tone comes forth finely with the slightest pressure of the fingers imaginable, and you swell it at pleasure by adding a little more pressure, no instrument affording more shades, if one may so speak, of the Forte piano.

"One wetting with the sponge will serve for a piece of music twice as long as Handel's Water-piece, unless the air be uncommonly drying.

"But a number of thin slices of sponge, placed side by side, their ends held fast between two strips of wood, like rulers, of a length equal to the glasses, and placed so that the loose ends of the sponges may touch the glasses behind, and by that means keep them constantly wet, is very convenient where one purposes to play for a long time. The sponges being properly wetted will supply the glasses sufficiently a whole evening, and touching the glasses lightly do not in the least hurt the sound.

"The powder of chalk is useful two ways.

"Fingers, after much playing, sometimes begin to draw out a toneless smooth and soft, and you feel as well as hear a small degree of sharpness. In this case, if you dip the ends of your wet fingers in the chalk, it will immediately recover the smoothness of tone desired. And, if the glasses have been sullied by handling, or the fingers not being just washed have some little greasiness on them, chalk so used will clean both glasses and fingers, and the sounds will come out to your wish.

"A little practice will make all this familiar; and you will also find by trials what part of the fingers most readily produces the sound from particular glasses, and whether they require to be touched on the edge chiefly, or a little more on the side; as different glasses require a different touch, some pretty full on the flat side of the brim, to bring out the best tone, others more on the edge, and some of the largest may need the touch of two fingers at once."

Fig. 22 represents a photograph of an armonica, constructed by Franklin. As will be seen, it differs in no respect from the one represented in Figs, 20 and 21, save in some unimportant details. The position of the musician operating this instrument, however, is clearly shown in the photograph.*

    Pitt Rivers Museum, University of Oxford, Sep. 9, 1904.
DEAR DR. GOLDSCMIDT:
    Thank you very much for having so kindly sent me the photograph

---

*As will be seen from the accompanying letter, dated Sep. 27, 1904, from Prof. Henry Balfour, of the University of Oxford, transmitted by Dr. V. Goldsmidt, the construction is the same as that already given above.

which is *very* interesting to me. The instrument represented is not a stringed instrument but a "Glasharmonica" and consists of 20 or more hemispherical glass bowls all fixed upon a long rod or spindle, which passes through the centres. One end of the spindles is attached to a wheel, which can be rotated by means of a pedal or foot board. When the wheel is set revolving (by pedalling with the foot) all the glasses revolve together and as they are carefully graduated in size (the base on the cleft and treble on the right) scales can be played or even chords and harmonics. The performer wets his fingers and rests them upon the revolving glasses; the friction causing them to vibrate and emit their respective notes. Your picture illustrates the method of playing excellently. I enclose a rough sketch of the instrument from the base end to show the wheel and treadle and also one to show how the bowls are fixed onto the spindle, overlapping so as to bring them all as close together as possible to allow of harmonics and chords being played. This form of the instrument was invented by the celebrated American, *Benjamin Franklin*, about the year 1760, and was an improvement upon the earlier musical glasses upon which the celebrated composer Glueck used to perform.

I have seen very few examples of the "Glasharmonica." There is one in the South Kensington Museum in London and I have seen two or three other examples. You are fortunate to have this picture, which from the costume would seem to date from 1780-1800. Can you tell me, if it is dated exactly and whom it represents in the act of playing? I assume that the picture is German.

Thanking you again for the photograph and with kindest regards.

Yours very truly,

HENRY BALFOUR.

Franklin correctly ascribes the phenomena of St. Elmo's fire, or as they are generally called by sailors, corposants or comazants, to electrical point discharges. He thus refers to these phenomena in a letter to Collinson, dated July 27, 1750:

"June 29, 1751. In Capt. Waddell's account of the effects of lightning on his ship, I could not but take notice of the large comazants (as he calls them) that settled on the spintles at the top-mast heads, and burnt like very large torches (before the stroke). According to my opinion, the electrical fire was then drawing off, as by points, from the cloud; the largeness of the flame betokening the great quantity of electricity in the cloud: and had there been a good wire communication from the spindle heads to the sea, that could have conducted more freely than tarred ropes, or masts of turpentine wood, I imagine there would either have been no stroke; or, if a stroke, the wire would have conducted it all into the sea without damage to the ship."

It will be remembered that while discussing the Franklin letter, in which he describes the construction of his movable electric clouds; *i. e.*, the cloud consisting of a movable pasteboard tube or of a charged scale pan of a balance, reference was made to the fact that Franklin had conceived an erroneous idea concerning what he called cold fusion, the reference to which was deferred to another part of the article. This we were also the more willing to do, from the fact that this article contained reference to important matters of a different character.

It would appear that Franklin was led to this change of opinion concerning the cold fusion of metals as follows: His friend Kinnersley, of London, had sent to Franklin a letter describing a form of electrical thermometer he had devised. This

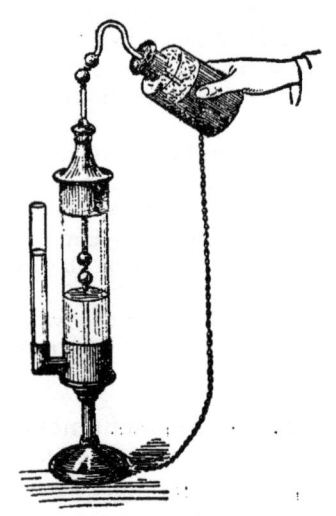

Fig. 47. Paged 17, Vol. 1, Electricity in Every-day Life.
Our Fig. 23.

form of thermometer consisted as represented in Fig. 23, of two glass stubs of unequal diameter, communicating with each other at their lower ends by means of a horizontal tube. A metallic cap closes the upper end of the larger tube. Through this cap a metallic rod extends, terminating at both of its ends in smooth metallic balls, the opening being of such dimension that the rod can be slid up and down with some little friction. Another metallic ball is attached to the upper end of a metallic rod, permanently fixed in the lower part of the enclosed tube as shown. The sliding rod is so placed as to leave a small air gap between it and the lower end of the ball connected with the fixed rod. As will be seen, the smaller tube is left open at the top. Sufficient water is placed in the tube to bring the level of the liquid below the top of the lower fixed ball. When, now, a disruptive discharge is passed between the two balls in the enclosed tube, by the discharge of a Leyden jar, the heat produced causes an expansion of the contained air, and so depresses the column of water in the enclosed tube, rising the level in the open tube. With powerful discharges, the water may even be violently thrown out from the top of the smaller tube, a part of this movement being due to the sudden blow or motion given to the air by the discharge. That, however, the depression of the column is also due to the increase in the temperature of the air consequent on the passage of the discharge, is seen by the fact that the water is maintained at a higher level in the open tube until the air in the enclosed tube has reached its former temperature.

In the same letter, Kinnersley describes another experiment which he made in accordance with a suggestion of Franklin, that also refers to what Franklin calls cold fusion. Kinnersley's description of this experiment is as follows:

"I then suspended, out of the thermometer, a piece of small harpischord wire, about twenty-four inches long, with a pound weight at the lower end, and sent the charge of the case of five and thirty bottles through it, whereby, (a) I discovered a new method of wire-drawing. The wire was red-hot the whole length, well annealed, and above an inch longer than before. A second charge melted it; it parted near the middle, and measured, when the ends were put together, four inches longer than at first. This experiment, I remember, you proposed to me before you left Philadelphia; but I never tried it till now. That I might have no doubt of the wire's being *hot* as well as red, I repeated the experiment on another piece of the same wire, encompassed with a goose-quill, filled with loose grains of gunpowder; which took fire as readily as if it had been touched with a red-hot poker. Also tinder, tied to another piece of the wire, kindled by it. I tried a wire about three times as big, but could produce no such effects as that.

"Hence it appears that the electric fire, though it has no sensible heat when in a state of rest, will, by its violent motion, and the resistance it meets with, produce heat in other bodies when passing through them, provided they be small enough. A large quantity will pass through a large wire without producing any sensible heat; when the same quantity, passing through a very small one, being more confined to a narrower passage, the particles crowding closer together and meeting with greater resistance, will make it red-hot, and even melt it.

"Hence lightning does not melt metal by a cold fusion, as we formerly supposed; but when it passes through the blade of a sword, if the quantity be not very great, it may heat the point so as to melt it, while the broadest and thickest part may not be sensibly warmer than before.

"And, when trees or houses are set on fire by the dreadful quantity which a cloud, or the earth, sometimes discharges, must not the heat by which the wood is first kindled, be generated by the lightning's violent motion through the resisting combustible matter?"

(a) "I discovered a new method of wire-drawing." Here we have a good description of a method for electrically drawing and annealing wire, made at a time when the voltaic heated the steel wire to such an extent as to permit it to be drawn out, and, on the passage of a second discharge, to be drawn out still further (at this time the wire breaking), the elongation produced by the two discharges measuring about four inches. At the same time, the wire was annealed by reason of the gradual cooling to which it was exposed. This experiment was, as Kinnersley acknowledged, suggested by Franklin.

Franklin's reply to this letter of Kinnersley was dated London, February 20th, 1762, and was as follows:

(a) "I am much pleased with your electrical thermometer, and the experiments you have made with it. I formerly satisfied myself, by an experiment with my phial and siphon, that the elasticity of the air was not increased by the mere existence of an electric atmosphere within the phial; but I did not know, till you now inform me, that heat may be given to it by an electric explosion. The continuance of its rarefaction, for some time after the discharge of your glass jar and your case of bottles, seems to make this clear. The other experiments on wet paper, wet thread, green grass, and green wood, are not so satisfactory; as possibly the reducing part of the moisture to vapor, by the electric fluid passing through it, might occasion some expansion which would be gradually reduced by the condensation of such vapor. The fine silver thread, the very small brass wire, and the strip of gilt paper, are also subject to a similar objection, as even metals, in such circumstances, are often partly reduced to smoke, particularly the gilding on paper.

(b) "But your subsequent beautiful experiment on the wire, which you made hot by the electric explosion, and in that state fired gunpowder with it, puts it out of all question, that heat is produced by our artificial electricity, and that the melting of metals in that way, is not what I formerly called a cold fusion. A late instance here of the melting of the bell-wire, in a house struck by lightning, and parts of the wire burning holes in the floor on which they fell, has proved the same with regard to the electricity of nature. I was too easily led into that error by accounts given even in philosophical books, and from remote ages downwards, of melting money in purses, swords in scabbards, &c., without burning the inflammable matters that were so near those melted metals. But men are, in general, such careless observers, that a philosopher cannot be too much on his guard in crediting their relations of things extraordinary, and should never build any hypothesis on anything but clear facts and experiments, or it will be in danger of soon falling, as this does, like a house of cards.

(c) "How many ways there are of kindling fire, or producing heat in bodies! By the sun's rays, by collision, by friction, by hammering, by putrefaction, by fermentation, by mixtures of solids with fluids, and by electricity. And yet the fire when produced, though in different bodies it may differ in circumstances, as in color, vehemence, &c., yet in the same bodies is generally the same. Does not this seem to indicate that the fire existed in the body, though in a quiescent state, before it was by any of these means excited, disengaged, and brought forth to action and to view? May it not constitute a part, and even a principal part, of the solid substance of bodies? If this should be the case, kindling a fire in a body would be nothing more than developing this inflammable principle, and setting it at liberty to act in separating the parts of that body, which then exhibits the appearances of scorching, melting, burning, &c. When a man lights a hundred candles from the flame of one, without diminishing that flame, can it be properly said to have *communicated* all that fire? When a single spark from a flint applied to a magazine of gunpowder, is immediately attended with this consequence, that the whole is in flame, exploding with immense violence, could all this fire exist first in the spark? We cannot conceive it. And thus we seem led to this supposition, that there is fire enough in all bodies to singe, melt, or burn them, whenever it is, by any means, set at liberty, so that it may exert itself upon them, or be disengaged from them. This liberty seems to be afforded it by the passage of electricity through them, which we know can and does, of itself, separate the parts, even of water; and, perhaps, the immediate appearances of fire are only the effects of such separations. If so, there would be no need of supposing that the electric fluid *heats itself* by the swiftness of its motion, or heats bodies by the resistance it meets with in passing through them. They would only be heated in proportion as such separation could be more easily made. Thus a melting heat cannot be given to a large wire in the flame of a candle, though it may to a small one; and this, not because the large wire resists *less* that action of the flame which tends to separate its parts, but because it resists it *more* than the smaller wire, or because the force being divided between more parts acts weaker on each."

(a) Franklin at once acknowledges that the continued depression of the water in the enclosed vessel unquestionably indicates an increase of temperature caused by the passage of the discharge through the air, and that, therefore, the experiment proves that electricity is capable of producing heat in bodies just like ordinary heat.

(b) The fusion of the metallic wire and the subsequent ignition of the gunpowder show beyond question that electricity produces heat, so that Franklin acknowledges that his former conception of the cold fusion of metals must be abandoned as erroneous. He then compares this result with the case of a lightning stroke recently brought to his attention while in London, in which the bolt melts a bell wire, the fused globules dropping from which burnt holes in the floor of the room on which they fell.

I have included this paragraph as interesting in showing the curious ideas Franklin had at this time respecting the causes of heat.

(c) We find among Franklin's writings on what may properly be regarded as theoretical physics, a comparatively short paper concerning the origin of light. It will be seen from this paper that he was a disbeliever in the Newtonian or corpuscular theory of light; *i. e.*, that light is due to exceedingly minute corpuscles shot off with almost inconceivable rapidity from the surface of the sun, being disposed to accept the Huyghenian theory, that light is produced by undulations or vibrations in an exceedingly tenuous medium that fills all space. This may be seen from the following quotation, taken from a letter written by Franklin, dated April 23, 1752:

"I thank you for communicating the illustration of the theorem concerning light. It is very curious. But I must own I am much in the dark about light. (a) I am not satisfied with the doctrine that supposes particles of matter called light. continually driven off from the sun's surface, with a swiftness so prodigious! Must not the smallest particle conceivable, have with such a motion, a force exceeding that of a twenty-four pounder, discharged from a cannon? Must not the sun diminish exceedingly by such a waste of matter; and the planets, instead of drawing near to him, as some have feared, recede to greater distances through the lessened attraction. Yet these particles, with this amazing motion, will not drive before them, or remove, the least and lightest dust they meet with: And the sun, for aught we know, continues of his ancient dimensions, and his attendants move in their ancient orbits.

(b) "May not all the phenomena of light be more conveniently solved, by supposing universal space filled with a subtle elastic fluid, which, when at rest, is not visible, but whose vibrations affect that fine sense of the eye, as those of air do those the grosser organs of the ear? We do not, in the case of sound, imagine that any sonorous particles are thrown off from a bell, for instance, and fly in strait lines to the ear; why must we believe that luminous particles leave the sun and proceed to the eye? Some diamonds, if rubbed, shine in the dark, without losing any part of their matter. I can make an electrical spark as big as the flame of a candle, much brighter, and, therefore, visible further; yet this is without fuel; and, I am persuaded, no part of the electric fluid flies off in such case, to distant places, but all goes directly, and is to be found in the place to which I destine it. May not different degrees of the vibration of the above-mentioned universal medium, occasion the appearance of different colours? I think the electric fluid is always the same; yet I find that weaker and stronger sparks differ in apparent colour. some white, blue, purple, red; the strongest, white; weak ones, red. Thus different degrees of vibration given to the air produce the seven different sounds in music, analogous to the seven colours, yet the medium, air, is the same.'

(a) Franklin refers to the tremendous power that the smallest conceivable particle of matter would have if it were shot off from the surface of the sun with the velocity of light. This he agrees with others in regarding as a most serious objection to the corpuscular theory of light. Instead of light possessing the power of demolishing all matter on which it falls, it is unable to move even the lightest dust particle that it falls on.

(b) Franklin simply states here the well known Huygnenian or undulatory theory of light.

Another question that may properly be regarded as belonging to theoretical physics is a belief mentioned by Franklin concerning action at a distance. Like many other philosophers, Franklin rejected the probability of action at a distance, thus writing concerning this matter in a letter which was afterwards read at the Royal Society, November 4th, 1756:

"I agree with you, that it seems absurd to suppose that a body can act where it is not. I have no idea of bodies at a distance attracting or repelling one another without the assistance of some medium, though I know not what that medium is, or how it operates. When I speak of attraction or repulsion, I make use of those words for want of others more proper, and intend only to express only effects which I see, and not causes of which I am ignorant. When I press a blown bladder between my knees, I find I cannot bring its sides together, but my knees feel a springy matter, pushing them back to a greater distance, or repelling them. I conclude that the air it contains is the cause. And when I operate on the air and find I cannot by pressure force its particles into contact, but they still spring back against the pressure, I conceive that there must be some medium between its particles that prevents their closing, though I cannot tell what it is.—And if I were ac-quainted with that medium, and found its particles to approach and recede from each other, according to the pressure they suffered, I should imagine there must be some finer medium between them, by which these operations were performed."

Franklin discovered at a very early date the power possessed by a Leyden jar discharge of imparting permanent magnetism to a steel needle, as well as to its power of reversing the polarity of the magnetic needle. It was known long before Franklin's time that lightning strokes, striking a ship, frequently resulted in a change in the polarity of the magnet.

Franklin thus refers to this phenomenon in a letter to Collinson dated July 27, 1750:

"His compasses lost the virtue of the load-stone, or the poles were reversed; the North point turning to the South.—By electricity we have (*here* at *Philadelphia*) frequently given polarity to needles, and reversed it at pleasure. *Mr. Wilson*, at *London*, tried it on too large masses and with too small force.

"A shock from four large jars, sent through a fine sewing needle, gives it polarity, and it will traverse when laid on water.—If the needle when struck lies East and West the end entered by the electric blast points North. —If it lies North and South the end that lay toward the North will continue to point North when placed on water, whether the fire entered at that end, or at the contrary end.

"The Polarity given is strongest when the needle is struck lying North and South, weakest when lying East and West; perhaps if the force was still greater, the South end, enter'd by the fire, (when the needle lies North and South) might become the North, otherwise it puzzles us to account for the inverting of compasses by lightning; since their needles must always be found in that situation, and by our little Experiments, whether the blast entered the North and went out at the South end of the needle, or the contrary, still the end that lay to the North should continue to point North.

"In these experiments the ends of the needles are sometimes finely blued like a watch-spring by the electric flame.—This colour given by the flash from two jars only, will wipe off, but four jars fix it and frequently melt the needles. I send you some that have had their heads and points melted off by our mimic lightning; and a pin that had its point melted off, and some part of its head and neck run. Sometimes the surface of the body of the needle is also run, and appears blister'd when examined by a magnifying glass: the jars I made use of hold 7 or 8 gallons, and are coated and lined with tin foil; each of them takes a thousand turns of a globe nine inches diameter to charge it."

It is pleasing to record that Franklin did not permit his enthusiasm for the great science of electricity to lead him to adopt all kinds of foolish theories concerning its therapeutic and curative powers. There are so many apparently mysterious properties possesed by electricity, that one not possessing a philosophic mind might readily be led to believe almost any statement made concerning the powers of this strange force. It is for this reason that electricity has always been a favorite remedy with quacks and charlatans, possibly in the hopes that, since so little is really known concerning this force, they could the more safely draw on their imagination as regards its curative powers.

It is true that at the earnest request of many people suffering from paralysis, Franklin consented to treat them by the discharge of the Leyden jar; but, as he afterwards acknowledged, this treatment did not appear to produce any permanent benefits. He thus refers to the matter in a letter written to John Pringle, dated December 21, 1757:

"In compliance with your request, I send you the following account of what I can at present recollect relating to the effects of electricity in paralytic cases, which have fallen under my observation.

"Some years since, when the newspapers made mention of great cures performed in Italy and Germany, by means of electricity, a number of paralytics were brought to me from different parts of Pennsylvania, and the neighbouring provinces, to be electrised, which I did for them at their request. My method was, to place the patient first in a chair, on an electric stool, and draw a number of large, strong sparks from all parts of the affected limb or side. Then I fully charged two six-gallon glass jars, each of which had about three square feet of surface coated; and I sent the united shock of these through the affected limb or limbs, repeating the stroke commonly three times each day. The first thing observed was an immediate greater sensible warmth in the lame limbs that had received the stroke, than in the others; and the next morning the patients usually related, that they had in the night felt a pricking sensation in the flesh of the paralytic limbs; and would sometimes shew a number of small red spots which they supposed were occasioned by those prickings. The limbs, too, were found more capable of voluntary motion, and seemed to receive strength. A man, for instance, who could not the first day lift the lame hand from off his knee, would the next day raise it four or five inches, the third day higher, and the fifth day was able, but with a feeble, languid motion, take off his hat. These appearances gave great spirit to the patients, and made them hope a perfect cure; but I do not remember that I ever saw an amendment after the fifth day; which the patients perceiving, and finding the shocks pretty severe, they became discouraged, went home, and in a short time relapsed; so that I never knew any advantage from electricity in palsies that was permanent. And how far the apparent temporary advantage might arise from the exercise in the patients journey, and coming daily to my house, or from the spirits given by the hope of success, enabling them to exert more strength in moving their limbs, I will not pretend to say.

"Perhaps some permanent advantage might have been obtained, if the electric shocks had been accompanied with proper medicine and regimen, under the direction of a skillful physician. It may be, too, that a few great strokes as given in my method, may not be so proper as many small ones; since, by the account from Scotland, of a case, in which two hundred shocks from a phial were given daily, it seems, that a perfect cure has been made. As to any uncommon strength supposed to be in the machine used in that case, I imagine it could have no share in the effect produced; since the strength of the shock from the charged glass, is in proportion to the quantity of the glass coated; so that my shocks from those large jars, must have been much greater than any that could be received from a phial held in the hand."

As to the probability of so-called electrical medication; *i. e.*, of causing electrified drugs, placed inside of glass vessels, to be passed through the walls of the vessel into bodies of patients in the shape of medicated effluvia, Franklin expresses himself very decidedly as follows:

"Hence we see the impossibility of success in the experiments proposed, to draw out the effluvial virtues of a non-electric, as cinnamon, for instance, and mixing them with the electric fluid, to convey them with that into the body, by including it in the globe, and then applying friction, &c. For though the effluvia of cinnamon and the electric fluid should mix within the globe, they would never come out together through the pores of the glass, and so go to the prime conductor; for the electric fluid itself cannot come through; and the prime conductor is always supply'd from the cushion, and that from the floor. And besides, when the globe is filled with cinnamon, or other non-electric, no electric fluid can be obtained from its outer surface, for the reason before-mentioned. I have tried another way, which I thought more likely to obtain a mixture of the electric and other effluvia together, if such a mixture had been possible. I placed a glass plate under my cushion, to cut off the communication between the cushion and floor; then brought a small chain from the cushion into a glass of oil of turpentine, and carried another chain from the oil of turpentine to the floor, taking care that the chain from the cushion to the glass, touch'd no part of the frame of the machine. Another chain was fixed to the prime conductor, and held in the hand of a person to be electrised. The ends of the two chains in the glass were near an inch distant from each other, the oil of turpentine between. Now the globe being turned, could draw no fire from the floor through the machine, the communication that way being cut off by the thick glass plate under the cushion: it must then draw it through the chains whose ends were dipped in the oil of turpentine. And so the oil of turpentine, being an electric per se, could not conduct, what came up from the floor was obliged to jump from the end of one chain to the end of the other, through the substance of that oil, which we could see in large sparks, and so it had a fair opportunity of siezing some of the finest particles of the oil in its passage, and carrying them off with it: but no such effect followed, nor could I perceive the least difference in the smell of the electric effluvia thus collected, from what it has when collected otherwise, nor does it otherwise affect the body of a person electrised. I likewise put into a phial, instead of water, a strong purgative liquid, and then charged the phial, and took repeated shocks from it, in which case every particle of the electric fluid must, before it went through my body, have first gone through the liquid when the phial is charging, and returned through it when discharging, yet no other effect followed than if it had been charged with water."

Franklin's interest in educational matters was not only evidenced by the valuable papers he wrote on such topics, but also by the practical results which flowed from these papers, among which are the establishment of the great University of Pennsylvania, and the American Philosophical Society. I will not stop to discuss these foundations of Franklin, since I have already described them in a paper prepared for publication in the *Journal* of the Franklin Institute. It is interesting, however, to note, in this connection, that there is another character of work for which Franklin was eminently fitted, and in which he might fairly have claimed a place in the first rank; *i. e.*, as a teacher of natural science.

Franklin not only possessed the rare ability of being able to thoroughly understand the causes of natural phenomena, but he also possessed the still rarer ability of being able to impart his ideas concerning such phenomena to others. This, I think, was the natural result of the logical methods employed by Franklin in his scientific investigations. The successive steps taken by him in his experimental researches were never made in a haphazard manner, but invariably followed one another in strict logical sequence, and when he endeavored to impart these principles to others, he always so presented them that the successive steps of his explanation followed the same strict logical order of sequence.

To a certain extent, all of Franklin's letters and papers are properly to be regarded as a species of scientific instruction to the public. The many quotations I have given from some of the more important of these papers will enable one readily to form an idea as to the clear manner in which they were written. The style of these papers may still be regarded as an admirable example of scientific teaching.

In this connection, we should not lose sight of the important role which the Junto, the select scientific debating and literary society formed by a few of Franklin's immediate friends, played in the development of this side of Franklin's character. The necessity imposed on each of the members of this Society to prepare papers on various topics at more or less regular intervals; to read such papers before the Society; and to take part in the discussion of similar papers prepared by other members, could not have failed to develop any latent powers of teaching that may have been possessed by the different individual members. I think there can be no doubt but that this phase of the Junto greatly improved Franklin's powers in this respect.

An admirable example of Franklin's rare ability as a teacher is to be had in a series of letters on scientific subjects written to a little girl, a Miss S—n, at Wanstead. I will give only a brief quotation from one of these letters, dated May 17, 1760:

"I send my dear, good girl the books I mentioned to her last night. I beg her to accept them as a small mark of my esteem and friendship. They are written in the familiar, easy manner for which the French are so remarkable, and afford a great deal of philosophic and practical knowledge, unembarres'd with the dry mathematics used by more exact reasoners, but which is apt to discourage young beginners.—I would advise you to read with a pen in your hand, and enter in a little book short hints of what you find there that is curious, or that may be useful; for this will be the best method of imprinting such particulars in your memory, where they will be ready, either for practice on some future occasion, if they are matters of utility; or at least to adorn and improve your conversation, if they are rather points of curiosity.—And, as many of the terms of science are such as you cannot have met with in your common reading, and may, therefore be unacquainted with, I think it would be well for you to have a good dictionary at hand, to consult immediately when you meet with a word you do not comprehend the precise meaning of. This may at first seem troublesome and interrupting; but 'tis a trouble that will daily diminish, as you will daily find less and less occasion for your Dictionary as you become more acquainted with the terms; and in the meantime you will read with more satisfaction because with more understanding.—When any point occurs in which you would be glad to have farther information than your book affords you, I beg you would not in the least apprehend that I should think it a trouble to receive and answer your questions. It will be a pleasure, and no trouble. For though I may not be able, out of my own little stock of knowledge to afford you what you require, I can easily direct you to the books where it may most readily be found."

In this series of letetrs will be found some charming teaching concerning a variety of scientific matters, to which it will be impossible here to do more than briefly refer; for example, why barometers, although placed in rooms with the windows and doors shut, can give the pressure of the outer air; or, even if the windows and doors be open, how can the pressure of the air act on the mercury in an apparently tightly closed vessel; besides other interesting matter which will amply repay a close perusal, on account of the excellent method employed in imparting the information. In one of those letters, Franklin gives his explanation as to the reason of a well-known fact; viz., that it is possible to utilize the power possesed by the skin of abstracting fresh water from the very salt water of the ocean. He calls attention to this matter as follows:

"I have a singular opinion on this subject, which I will venture to communicate to you, though I doubt you will rank it among my whims.—It is certain that the skin has *imbibing* as well as *discharging* pores; witness the effects of a blistering plaister, &c.. I have read that a man hired by a physician to stand by way of experiment in the open air naked during a moist night weighed near three pounds heavier in the morning. I have often observed myself that however thirsty I may have been before going into the water to swim, I am never long so in the water. These imbibing pores, however, are very fine, perhaps fine enough in filtering to separate salt from water; for though I have soaked by swimming, when a boy, several hours a day for several days successively in salt-water, I never found my blood and juices salted by that means, so as to make me thirsty or feet a salt taste in my mouth: And it is remarkable that the flesh of sea fish, though bred in salt water is not salt.—Hence I imagine, that if people at sea, distressed by thirst when their fresh water is unfortunately spent, would make bathing tubs of their empty water casks, and filling them with sea water sit in them an hour or two each day, they might be greatly relieved. Perhaps keeping their cloaths constantly wet might have an almost equal effect, and this without danger of catching cold. Men do not catch cold by wet cloaths at sea. Damp but not wet linen may possibly give colds; but no one catches cold by bathing, and no cloaths can be wetter than water itself. Why damp cloaths should then occasion colds, is a curious question, the discussion of which I reserve for a future letter, or some future conversation."

In his autobiography, Franklin speaks of his fondness for swimming, as well as the fact that, when about twenty years of age, while at Watt's Printing House, he taught an acquaintance the art of swimming in two lessons. He thus refers to this fact in his autobiography as follows:

"At Watt's printing-house I entracted an acquaintance with an ingenious young man, one Wygate, who, having wealthy relations, had been better educated than most printers; was a tolerable Latinist, spoke French and loved reading. I taught him and a friend of his to swim, at twice going into the river, and they soon became good swimmers. They introduced me to some gentlemen from the country, who went to Chelsea by water, to see the College and Don Saltero's curiosities. In our return, at the request of the company, whose curiosity Wygate had excited, I stripped and leaped into the water and swam from near Chelsea to Blackfriars; performing in the way many feats of activity, both upon and under the water, that surprised and pleased those to whom they were novelties.

"I had from a child been delighted with this exercise, had studied and practised Thevenot's motions and positions, and added some of my own, aiming at the graceful and easy, as well as the useful. All these I took this occasion of exhibiting to the company, and was much flattered by their admiration; and Wygate, who was desirous of becoming a master, grew more and more attached to me on that account, as well as from the similarity of our studies. He at length proposed to me travelling all over Europe together, supporting ourselves everywhere by working at our business. I was once inclined to it, but mentioning it to my good friend Mr. Denham, with whom I often spent an hour when I had leisure, he dissuaded me from it; advising me to think only of returning to Pennsylvania, which he was now about to do."

At a later date, Franklin writes the following as regards the best manner for a person to acquire this art without a teacher:

"I cannot be of opinion with you that 'tis too late in life for you to learn to swim. The river near the bottom of your garden affords a most convenient place for the purpose. And as your new employment requires your often being on the water, of which you have such a dread, I think you would do well to make the trial; nothing being so likely to remove those apprehensions as the consciousness of an ability to swim to the shore, in case of an accident, or of supporting yourself in the water till a boat could come to take you up.

"I do not know how far corks and bladders may be useful in learning to swim, having never seen much trial of them. Possibly they may be of service in supporting the body while you are learning what is called the stroke, or that manner of drawing in and striking out the hands and feet that is necessary to produce progressive motion. But you will be no swimmer until you can place some confidence in the power of the water to support you; I would therefore advise the acquiring of that confidence in the first place; especially as I have known several who by a little of the practice necessary for that purpose, have insensibly acquired the stroke, taught as it were by nature.

"The practice I mean is this. Chusing a place where the water deepens gradually, walk cooly in until it is up to your breast, then turn round, your face to the shore, and throw an egg into the water between you and the shore. It will sink to the bottom, and be easily seen there, as your water is clear. It must lie in water so deep that you cannot reach it to take it up but by diving for it. To encourage yourself in undertaking to do this, reflect that your progress will be from deeper to shallower water, and that at any time you may, by bringing your legs under you and standing on the bottom, raise your head far above the water. Then plunge under it with your eyes open, throwing yourself towards the egg, and endeavouring by the action of your hands and feet against the water to get forward till within reach of it. In this attempt you will find, that the water buoys you up against your inclination; that it is not so easy a thing to sink as you imagined; that you cannot, by your active force, get down to the egg. Thus you feel the power of the water to support you, and learn to confide in that power; while your endeavours to overcome it and reach the egg, teach you the manner of acting on the water with your feet and hands, which action is afterwards used in swimming to support your head higher above water, or to go forward through it."

In the preceding pages, I have endeavored briefly to give an account of the many departments of physical science that have been covered by the great experimental philosopher whose work I have endeavored to trace. It is a matter of no little regret to me that the many calls of a busy professional life have necessitated my doing this work in so incomplete and hurried a manner. The character of the work that is under review is of such a type that its discussion deserves a much more careful treatment than I have been able to give it. If, however, this work has the effect of attracting a more careful attention to the scientific work of Franklin, I shall feel more than compensated for the labor which even these hurried pages have necessitated. But however imperfect the character of my work, I feel convinced that a careful perusal cannot fail to impress those who are capable of judging work of this character that Benjamin Franklin may be properly regarded,—

(1) As the first great American electrician.

(2) As a natural philosopher of the highest type, both in the domains of geographical physics and applied physics.

(3) As an inventor of the highest type.

Franklin demonstrated his right to the title of the first great American electrician, not only by his famous experiment with the kite, but also by other electrical work, such as, for example, his single-fluid electrical hypothesis, his masterful analysis of the manner in which the Leyden jar receives and parts with its charge, as well as by the numerous papers on a great variety of electrical subjects to which attention has been briefly called in the preceding pages.

He demonstrated his right to the title of a great natural philosopher by the wonderful skill manifested in investigations on a great variety of physical subjects, to which also attention has been called. The mind of the great master is clearly exhibited in these investigations, not only by the great results achieved, but especially by the logical manner in which the detailed facts were conceived and carried out.

It is especially interesting, in this connection, to bear in mind the great variety of such investigations that were embraced in the field of geographical physics alone, embracing as they did, the causes and phenomena of the aurora borealis, thunder-gusts, waterspouts, whirlwinds, the great northeast storms of the United States, the Gulf Stream, the electrification of clouds, &c.

Franklin's right to the title of one of the greatest of the early American inventors is based on a variety of important work, such, for example, as the invention of the lightning rod, of the Pennsylvania fire-place, of the stove with the inverted draught, of the proper construction of chimneys, the ventilation of buildings and mines, the bifocal spectacles, etc.

I have received through the kindness of Mr. Augustine Biesel of the American Embassy in Paris, through the hands of Mr. Joseph G. Rosengarten, of Philadelphia, a photograph of the Greuze portrait of Franklin, from which the accompanying plate was made.

Before concluding this paper on Franklin, I think it well to give without further comment, the full text of a paper, prepared in Paris, by special request, concerning his great invention of lightning rods. As the inventor of lightning rods, Franklin naturally had his attention repeatedly called to well marked lightning strokes in different parts of the world, so that he had an opportunity of ascertaining in what respects rods erected in accordance with his directions failed in their ability to protect the buildings on which they were placed. We would, consequently expect to find in this paper a much more complete account of Franklin's views on the subject than would be found in his other papers:

"Of Lightning and the Method (now used in America) of securing Buildings and Persons from its mischievous Effects.

"Experiments made in electricity first gave philosophers a suspicion that the matter of lightning was the same with electric matter. Experiments afterwards made on lightning obtained from the clouds by pointed rods, received in bottles, and subjected to every trial, have since proved this suspicion to be perfectly well founded; and that whatever properties we find in electricity, are also the properties of lightning.

"This matter of lightning, or electricity, is an extream subtle fluid, penetrating other bodies, and subsisting in them, equally diffused.

"When by any operation of art or nature, there happens to be a greater proportion of this fluid in one body than in another, the body which has most, will communicate to that which has least, till the proportion becomes equal; provided the distance between them be not too great; or, if it is too great, till there be proper conductors to convey it from one to the other.

"If the communication be through the air without any conductor, a bright light is seen between the bodies, and a sound is heard. In our small experiments we call this light and sound the electric spark and snap; but in the greater operations of nature, the light is what we call lightning, and the sound (produced at the same time, tho' generally arriving later at our ears than the light does to our eyes) is, with its echoes, called thunder.

"If the communication of this fluid is by a conductor, it may be without either light or sound, the subtle fluid passing in the substance of the conductor.

"If the conductor be good and of sufficient bigness, the fluid passes through it without hurting it. If otherwise, it is damaged or destroyed.

"All metals, and water, are good conductors.—Other bodies may become conductors by having some quantity of water in them, as wood, and other materials used in building, but not having much water in them, they are not good conductors, and therefore are often damaged in the operation.

"Glass, wax, silk, wool, feathers, and even wood, perfectly dry are nonconductors: that is, they resist instead of facilitating the passage of this subtle fluid.

"When this fluid has an opportunity of passing through two conductors, one good, and sufficient, as of metal, the other not so good, it passes in the best, and will follow it in any direction.

"The distance at which a body charged with this fluid will discharge itself suddenly, striking through the air into another body that is not charged, or not so highly charg'd, is different according to the quantity of the fluid, the dimensions and form of the bodies themselves, and the state of the air between them.—This distance, whatever it happens to be between two bodies, is called their striking distance, as till they come within that distance of each other, no stroke will be made.

"The clouds have often more of this fluid in proportion than the earth; in which case as soon as they come near enough (that is, within striking distance) or meet with a conductor, the fluid quits them and strikes into the earth. A cloud fully charged with this fluid, if so high as to be beyond the striking distance from the earth, passes quietly without making noise or giving light; unless it meets with other clouds that have less.

"Tall trees, and lofty buildings, as the towers and spires of churches, become sometimes conductors between the clouds and the earth; but not being good ones, that is, not conveying the fluid freely, they are often damaged.

"Buildings that have their roofs covered with lead, or other metal and spouts of metal continued from the roof into the ground to carry off the water, are never hurt by lightning, as whenever it falls on such a building, it passes in the metals and not in the walls.

"When other buildings happen to be within the striking distance from such clouds, the fluid passes in the walls whether of wood, brick or stone, quitting the walls only when it can find better conductors near them, as metal rods, bolts and hinges of windows or doors, gilding on wainscot, or frames of pictures; the silvering on the backs of looking-glasses; the wires for bells; and the bodies of animals, as containing watery fluids. And in passing thro' the house it follows the direction of these conductors, taking as many in it's way as can assist it in its passage, whether in a strait or crooked line, leaping from one to the other, if not far distant from each other, only rending the wall in the spaces where these partial good conductors are too distant from each other.

"An iron rod being placed on the outside of a building, from the highest part continued down into the moist earth, in any direction strait or crooked, following the form of the roof or other parts of the building, will receive the lightning at its upper end, attracting it so as to prevent its striking any other part; and, affording it a good conveyance into the earth. will prevent its damaging any part of the building.

"A small quantity of metal is found able to conduct a great quantity of this fluid. A wire no bigger than a goose quill, has been known to conduct (with safety to the building as far as the wire was continued) a quantity of lightning that did prodigious damage both above and below it; and probably larger rods are not necessary, tho' it is common in America, to make them of half an inch, some of three-quarters, or an inch in diameter.

"The rod may be fastened to the wall, chimney, &c., with staples of iron.—The lightning will not leave the rod (a good conductor) to pass into the wall (a bad conductor), through those staples.—It would rather, if any were in the wall, pass out of it into the rod to get more readily by that conductor into the earth.

"If the building be very large and expensive, two or more rods may be placed at different parts, for greater security.

"Small, ragged parts of clouds suspended in the air between the great body of clouds and the earth (like leaf gold in electrical experiments), often serve as partial conductors for the lightning, which proceeds from one of them to another, and by their help comes within the striking distance to the earth or a building. It therefore strikes through those conductors a building that would otherwise be out of striking distance.

"Long, sharp points, communicating with the earth, and presented to such parts of clouds, drawing silently from them the fluid they are charged with, they are then attracted to the cloud, and may leave the distance so great as to be beyond the reach of striking.

"It is therefore that we elevate the upper end of the rod six or eight feet above the highest part of the building, tapering it gradually to a fine, sharp point, which is gilt, to prevent its rusting.

"Thus the pointed rod either prevents a stroke from the cloud, or, if a stroke is made, conducts it to the earth with safety to the building.

"The lower end of the rod should enter the earth so deep as to come at the moist part, perhaps two or three feet; and if bent when under the surface so as to go in a horizontal line six or eight feet from the wall, and then bent again downwards three or four feet, it will prevent damage to any of the stones of the foundation.

"A person apprehensive of danger from lightning, happening in the time of thunder to be in a house not so secured, will do well to avoid sitting near the chimney, near a looking-glass or any gilt pictures or wainscot; the safest place is in the middle of the room, (so it be not under a metal lustre suspended by a chain) sitting in one chair and laying the feet up in another. It is still safer to bring two or three mattrasses or beds into the middle of the room, and folding them up double, place the chair upon them; for they not being so good conductors as the walls, the lightning will not chuse an interrupted course through the air of the room and the bedding, when it can go thro' a better continued conductor, the wall. But where it can be had, a hammock or swinging bed, suspended by silk cords equally distant from the walls on every side, and from the ceiling and floor above and below, affords the safest situation a person can have in any room whatever; and what indeed may be deemed quite free from danger by any stroke by lightning."

Franklin died on the 17th of April, 1790. The Congress of the United States, which was in session at the time of his death, passed the following resolution on the receipt of the sad news:

"The House, being informed of the decease of Benjamin Franklin, a citizen whose native genius was not more an ornament to human nature than his various exertions of it have been precious to science, to freedom, and to his country, do resolve, as a mark of the veneration due to his memory that the members wear the customary badge of mourning for one month."

A higher honor was paid to his memory by the National Assembly of France on the morning after the intelligence reached Paris, on June 11. The assembly was convened, when Mirabeau spoke as follows:

"Franklin is dead! The genius, that freed America and poured a flood of light over Europe, has returned to the bosom of the Divinity.

"The sage whom two worlds claim as their own, the man for whom the history of science and the history of empires contend with each other, held, without doubt, a high rank in the human race."

"Too long have political cabinets taken formal note of the death of those who were great only in their funeral panegyrics. Too long has the etiquette of courts prescribed hypocritical mourning. Nations should wear mourning only for their benefactors. The representatives of nations should recommend to their homage none but the heroes of humanity.

"The Congress has ordained, throughout the United States, a mourning of one month for the death of Franklin; and, at this moment, America is paying this tribute of veneration and gratitude to one of the fathers of her Constitution.

"Would it not become us, Gentlemen, to join in this religious act, to bear a part in this homage, rendered, in the face of the world, both to the rights of the man, and to the philosopher who has most contributed to extend their sway over the whole earth? Antiquity would have raised altars to this mighty genius, who, to the advantage of mankind, compassing in his mind the heavens and the earth, was able to restrain alike thunderbolts and tyrants. Europe, enlightened and free, owes at least a token of remembrance and regret to one of the greatest men who have ever been engaged in the service of philosophy and of liberty.

"I propose that it be decreed, that the National Assembly, during three days, shall wear mourning for Benjamin Franklin."

On the adoption of this motion by acclamation, the Assembly decreed that on the 14th day of June, they would go into mourning for three days, and directed the President of the Assembly to write a letter of condolence to the Congress of the United States.

It has been falsely alleged that Franklin was an irreligious man. An assertion of this kind surely could not have been made by anyone conversant with Franklin's writings. I will merely quote a letter written by Franklin, on March 9, 1790, shortly before his death, in which he says:

"You desire to know something of my religion. It is the first time I have been questioned upon it. But I cannot take your curiosity amiss, and shall endeavour in a few words to gratify it. Here is my creed: I believe in one God, the creator of the universe. That he governs it by his Providence. That he ought to be worshipped. That the most acceptable service we render to him is doing good to his other children. That the soul of man is immortal, and will be treated with justice in another life respecting its conduct in this. These I take to be the fundamental points in all sound religion, and I regard them as you do in whatever sect I meet with them."

If, however, this testimony had been wanting, or if anyone should allege that this was merely the opinion of a man who had nearly reached the end of his life, the following epitaph prepared by Franklin, when at the age of but twenty-three years, would prove the incorrectness of the conclusion:

"The Body
Of
Benjamin Franklin.
Printer,
(Like the cover of an old book,
Its contents torn out,
And stript of its lettering and gilding,)
Lies here, food for worms.
But the work shall not be lost,
For it will, as he believed, appear once more,
In a new and more elegant edition,
Revised and corrected
By
The Author."

THE ALUMINUM INDUSTRY AND THE HALL PATENT.

Of no little significance to the aluminum industry in the United States was the expiration on April 2, 1906, of the patent of Charles M. Hall, granted on April 2, 1889, and under which it has been possible for the Pittsburgh Reduction Company to maintain almost a monopoly of production in the United States. Recently the demand has been so heavy that this company has not been able to keep up with it and domestic consumers have turned to Europe, but even at the high prices paid abroad it has not been possible to obtain any considerable quantities. The large requirements of the automobile industry are responsible chiefly for the expansion in the demand in recent years. Referring to the expiration of the Hall patent and its effect on aluminum manufacture the *Metal Industry* makes the following interesting statement:

It will be remembered that this patent, which became of fundamental importance for the commercial production of aluminum, protected the use of an electrolyte composed of cryolite as a solvent for bauxite, and this electrolyte had the most important property of being easily fusible. The extension of the manufacture of aluminum followed and the three-cornered litigation about patent rights between the Pittsburgh Reduction Company, as owners of the Hall patent, and the Electric Smelting & Aluminum Company, the moving spirits of which were the well-known pioneers in the aluminum industry, the Cowles brothers, as owners of the Bradley patent, and, in the third place, Grosvenor P. Lowrey.

There is no doubt that owing to the scarcity and high price of the metal efforts will be made to start competing works after the Hall electrolyte becomes public property, and in fact there are many rumors of such proceedings already in the air. As far as the Pittsburgh Reduction Company itself is concerned, it must be remembered that though its electrolyte is public property, yet its method of operation is protected by the Bradley patent, which will not expire until February, 1909. The Bradley patent is of fundamental importance for the manufacture of aluminum, covering as it does the use of the current, as well for the purpose of keeping the electrolytic bath in a molten condition as for effecting its decomposition and setting the aluminum free at the cathode. I has been claimed variously that Mr. Hall never succeeded in producing aluminum commercially until he abandoned the method of conducting the electrolysis in a vessel heated from the outside by an exterior source of heat. If fthat holds true it seems to be very difficult for other people to get around the claims of the Bradley patent, in spite of their having an electrolyte of the proper nature. Moreover, the Pittsburgh Reduction ompany has the immense advantage over its competitors of a large, well equipped plant, trained workmen and an experience of a number of years' duration, and is making strenuous efforts to enlarge its productive capacity.

What in view of the above considerations will be the future development of the aluminum industry is difficult to say. No doubt if competition should be successful the price of the metal would fall to a greater or less extent. As far as the Pittsburgh Reduction Company is concerned it must be said that it has never abused the privilege of having a monopoly for the purpose of putting a prohibitory price on the metal.

TIN IN NEW SOUTH WALES.

At Tingha, in the Inverell district, New South Wales, there are fifteen or sixteen dredges working for tin, all doing well. They are reported to have from six to ten years' work in front of them. The development of some of the lodes and deep leads of the district is urged. According to E. F. Pittman, the geological conditions of the district are almost precisely similar to those of Cornwall.

BENJAMIN FRANKLIN
(After Greuze.)

Benjamin Franklin Trust Funds to the Cities of Boston and Philadelphia.

By Dr. Edwin James Houston.

Emeritus Professor of Physics in the Franklin Institute.

Some time before his death, in 1790, Benjamin Franklin, remembering the necessity that had existed in his own case for obtaining a loan of money from two of his friends in order to be able to set up his business as a printer, placed in a codicil to his last will and testament, dated June 23rd, 1789, a legacy to each of the cities of Philadelphia and Boston the sum of 1000 pounds sterling. In the case of Boston this sum was to be managed under the direction of the Selectmen, together with the ministers of the oldest Episcopalian, Congregational and Presbyterian Churches in that town. This sum was to be loaned to such young married artificers, less than twenty-five years of age, who had faithfully served an apprenticeship in Boston, and who could obtain, not only a good moral character from two respectable citizens, but who could induce them to become sureties in a bond with the applicant for the repayment of the money, together with interest.

The conditions of such loans were as follows: The amount of the loan was not to exceed the sum of sixty pounds to each person, nor was it to be less than fifteen pounds. Interest was to be paid at the rate of 5% per annum.

In order to render the making of the repayments easier to the borrowers, Franklin imposed the condition that each borrower should be obliged to pay with the yearly interest one-tenth part of the principal, and the sum of principal and interest so paid to be again loaned out to others.

Recognizing, as Franklin did, the rate of increase of capital when left at compound interest, he calculated that at the end of the first hundred years the £1000 would have increased to

£131,000. He left directions to those in charge of the fund that as soon as this period of time had elapsed, they were to lay out, at their discretion, £100,000 in public works, such, for example, as fortifications, bridges, aqueducts, public buildings, baths, pavements, or whatever might make living in the town more convenient to its people and render it more agreeable to strangers resorting hither for health or temporary residence. The remaining £31,000, he directed to be loaned out on interest under the original conditions for another hundred years. At the end of this second term of one hundred years, Franklin estimated that the sum would have reached the amount of £4,000,061, of which £1,000,061 were to be placed at the disposal of the inhabitants of the town of Boston, and £300,000,000 at the disposal of the Government of the State, "not presuming to carry my views further," as the will naively remarks.

The fund left to the City of Philadelphia was placed under the same general conditions as that left to the City of Boston, with, however, the exception that since the City of Philadelphia was incorporated, the fund was to be placed directly under the management of the corporation.

In looking into the conditions of the City of Philadelphia at the end of the first hundred years that were to elapse from the formation of the trust, and bearing in mind the probability of the contamination of the sources of drinking water of a city like Philadelphia, Franklin recommended that at this time the sum of £100,000 be expended for the purpose of bringing the waters of Wissahickon Creek into the town, so as to supply the inhabitants with pure water. This he thought there should be no difficulty in doing by means of a general gravity system, or, if such was insufficient for any portion of the city, a reservoir could be built so as to raise the level of the water by means of a dam. He also recommended that the Schuylkill River should be made completely navigable with, presumably, the balance of the fund. At the end of the second hundred years, he directed that the sum of £4,000,061 be divided between the inhabitants of the City of Philadelphia and the Government of Pennsylvania, in a similar manner as he had directed as regards the similar legacy to the City of Boston.

We here publish in full the text of this part of Franklin's remarkable will, throwing as it does additional light on the character of the man, as well as on the plan that his benevolence had devised:

" * * * I have considered that, among artisans, good apprentices are most likely to make good citizens, and, having myself been bred to a manual art, printing, in my native town, and afterwards assisted to set up my business in Philadelphia by kind loans of money from two friends there, which was the foundation of my fortune, and of all the utility in life that may be ascribed to me, I wish to be useful after death, if possible, in forming and advancing other young men, that may be serviceable to their country in both those towns. To this end, I devoted two thousand pounds sterling, of which I give one thousand thereof to the inhabitants of the town of Boston, in Massachusetts, and the other thousand to the inhabitants of the City of Philadelphia, in trust, to and for the uses, intents and purposes hereinafter mentioned and declared.

"The said sum of one thousand pounds sterling, if accepted by the inhabitants of the town of Boston, shall be managed under the direction of the selectmen, united with ministers of the oldest Episcopalian, Congregational and Presbyterian churches in that town, who are to let out the same upon interest, at five per cent. per annum, to such young married artificers, under the age of twenty-five years, as have served an apprenticeship in the said town, and faithfully fulfilled the duties required in their indentures, so as to obtain a good moral character from at least two respectable citizens, who are willing to become their sureties, in a bond with the applicants, for the repayment of the moneys so lent, with interest, according to the terms hereinafter prescribed; all of which bonds are to be taken for Spanish milled dollars, or the value thereof in current gold coin; and the managers shall keep a bound book or books, wherein shall be entered the names of those who shall apply for and receive benefits of this institution, and of their sureties, together with the sums lent, the dates, and other necessary and proper record respecting the business and concerns of this institution. And, as these loans are intended to assist young married artificers in setting up their business, they are to be proportioned by the discretion of the managers, so as not to exceed sixty pounds sterling to one person, nor to be less than fifteen pounds, and, if the number of appliers so entitled should be so large that the sum will suffice to afford to each as much as might otherwise not be improper, the proportion to each shall be diminished so as to afford to every one some assistance. These aids may, therefore be small at first, but, as the capital increases by the accumulated interest, they will be more ample. And, in order to serve as many as possible in their turn, as well as to make the payment of the principal borrowed more easy, each borrower shall be obliged to pay, with the yearly interest, one-tenth part of the principal, which sum of principal and interest so paid in, shall be again let out to fresh borrowers.

"And, as it is presumed that there will always be found in Boston virtuous and benevolent citizens willing to bestow a part of their time in doing good to the rising generation, by superintending and managing this institution gratis, it is hoped, that no part of the money will at any time be dead, or be diverted to other purposes, but be continually augmenting by the interest, in which case there may, in time, be more than the occasions in Boston shall require, and then some may be spared to the neighboring or other towns, in the said State of Massachusetts, who may desire to have it; such town engageing to pay punctually the interest and the portions of the principal, annually, to the inhabitants of the town of Boston.

"If this plan is executed, and succeeds as projected without interruption for one hundred years, the sum will then be one hundred and thirty thousand pounds, of which I would have the managers of the donation of the town of Boston then lay out, at their discretion, one hundred thousand pounds in public works, which may be judged of most general utility to the inhabitants, such as fortifications, bridges, aqueducts, public buildings, baths, pavements, or whatever may make living in the town more convenient to its people, and render it more agreeable to strangers resorting thither for health or a temporary residence. The remaining thirty-one thousand pounds I would have continued to be let out on interest, in the manner above directed, for another hundred years, as I hope that it will have been found that the institution has had a good effect on the conduct of youth, and been a service to many worthy characters and useful citizens. At the end of this second term, if no unfortunate accident has prevented the operation, the sum will be four millions and sixty-one thousand pounds sterling, of which I leave one million sixty-one thousand pounds to the disposition of the inhabitants of the town of Boston, and three millions to the disposition of the government of the State, not presuming to carry my views farther.

"All the directions herein given respecting the disposition and management of the donation to the inhabitants of Boston, I would have observed respecting that to the inhabitants of Philadelphia, only, as Philadelphia is incorporated, I request the corporation of that city to undertake the management agreeably to the said direction, and I do hereby vest them with full and ample powers for that purpose. And, having considered that the covering a ground plat with buildings and pavements, which carry off most of the rain, and prevent it soaking into the earth and renewing and purifying the springs, whence the waters of the wells must gradually grow worse, and in time be unfit for use, as I find has happened in all old cities, I recommend that at the end of the first hundred years, if not done before, the corporation of the city employ a part of the one hundred thousand pounds in bringing, by pipes, the water of Wissahickon Creek into the town, so as to supply the inhabitants, which I apprehend may be done without great difficulty, the level of that peak being much above that of the city, and may be made higher by a dam. I also recommend making the Schuylkill completely navigable. At the end of the second hundred years I would have the disposition of the four million and sixty-one thousand pounds divided between the inhabitants of the City of Philadelphia and the government of

Pennsylvania, in the same manner as herein directed with respect to that of the inhabitants of Boston and the government of Massachusetts.

"It is my desire that this institution should take place and begin to operate within one year after my decease, for which purpose due notice should be publicly given previous to the expiration of that year, that those for whose benefit this establishment is intended may make their respective applications. And I hereby direct my executors, the survivors or survivor of them, within six months after my decease to pay over the said sum of two thousand pounds sterling to such persons as shall be duly appointed by the selectment of Boston and the corporation of Philadelphia to receive and take charge of their respective sums, of one thousand pounds each, for the purpose aforesaid.

"Considering the accidents to which all human affairs and projects are subject in such a length of time, I have, perhaps, too much flattered myself with a vain fancy that these dispositions, if carried into execution, will be continued without interruption and have the effect proposed. I hope, however, that if the inhabitants of the two cities should not think fit to undertake the execution, they will, at least, accept the offer of these donations as a mark of my good will, a token of my gratitude and a testimony of my earnest desire to be useful to them after my departure. I wish, indeed, that they may both undertake to endeavor the execution of the project, because I think that, though unforseen difficulties may arise, expedients will be found to remove them, and the scheme be found practicable. If one of them accept the money, with the conditions, and the other refuses, my will then is that both sums be given to the inhabitants of the city acepting the whole, to be applied to the same purposes, and under the same regulations directed for the separate parts; and, if both refuse, the money, of course, remains in the mass of my estate, and is to be disposed of therewith according to my will made the seventeenth of July, 1788."

But Benjamin Franklin did not begin to appreciate the wonderful rate of increase that was destined to the growth of the United States in general, or of the City of Philadelphia in particular. During these first hundred years, the City of Philadelphia has so far exceeded its area as to actually include the parts of the very Wissahickon Creek that were to form the source of the pure water supply for the inhabitants, uncontaminated by reason of its distance from the city. The $500,000, large as it seemed to Franklin, would have been entirely too small to permit of what Franklin thought would constitute a complete water system. In point of fact, the reading of the public newspapers of the present time, would show how much money Philadelphia contemplates spending on its great filtration system, so that, as can now easily be seen, the expenditure of the entire amount would have formed but a small fractional part of the sum required for such purposes, thus leaving entirely out of consideration the ensuring of the navigability of the Schuylkill River, or even the Delaware.

Speaking now of the legacy to the City of Philadelphia alone, it will be interesting to note the growth of the original sum of £1000 above referred to. The actual value of the invested legacy at different periods, as taken from the original records of the trust, are as follows:

```
Original Legacy, 1,000 pounds sterling.........$  5,000
Value of Legacy Dec. 31, 1854,................ 20,600
   "      "    "    "    "  1870 ................ 38,900
   "      "    "    "    "  1880 ................ 62,350
   "      "    April 17, 1890................. 87,600.29
   "      "    Dec. 31, 1890.................. 89,900
   "      "    "    "    "  1900 ................136,600
   "      "    "    "    "  1905 ................159,000
```

It will be observed that on the 17th day of April, 1890, exactly one hundred years after the death of Benjamin Franklin, the value of this trust fund in the case of the City of Philadelphia reached the sum of only $87,600.29, while the amount on December 31, 1905, was only $159,000.

A number of circumstances have combined to bring about the discrepancy between the theoretical and the actual value of the sum at the end of the first hundred years. In the first place, as is well known, in order to ensure the great increase that is possible in the case of money placed at compound interest, it is of course necessary that no lapse shall exist in the earning power of the money; in other words, that no loss shall occur either in the case of the principal or of the interest. This, of course, necessitates a continuance of the best financial management, together with no failures as regards the recovery of either principal or interest.

Whatever may have been the reasons, it is a matter of fact that the number of apprenticed artisans applying for the loan was comparatively small. The last loan made from the City of Philadelphia fund was made in the year 1885, and since that time, no application for a loan has been received. The reasons for this are probably as follows:

(1) There are now comparatively few indentured apprentices in the City of Philadelphia.

(2) The loan of money on the mere recommendation of two citizens cannot be regarded as sound financial policy. Consequently, a real estate security is required. Where, however, real estate security is available, the trust companies and building associations are capable of offering equally attractive, if not, indeed, more attractive terms.

It is interesting to note that in March, 1895, the balance of the Franklin fund available for public improvements was appropriated by the proper authorities towards the erection of a Museum and Art Gallery in Fairmount Park. The following minute of this fact is taken from the 26th Annual Report of the Board of Directors of City Trusts, for the year 1895:

"The question of the expenditure of that portion of the Benjamin Franklin Fund available since the expiration of the first one hundred years named in the will, has been finally settled by the acceptance, by the Commissioners of Fairmount Park, of the amount, $84,285.02, to be expended in the erection of an Art Gallery to be located in the Park, that portion of the building so paid for to have the name of Benjamin Franklin connected with it in such a manner as to serve to perpetutate his memory."

On May 11th, 1895, this appropriation was accepted by the Park Commissioners, but up to the present date, January, 1906, nothing further has been done in this direction, no money has been expended, and the funds still remain in the hands of the Board of Directors of City Trusts, invested and drawing income.

It is interesting to note than on the 27th of September, 1890, proceedings were commenced in the Orphans' Court of the City of Philadelphia to have the Franklin Trust Fund declared invalid. These proceedings were based on a petition in behalf of Albert D. Bache, a great-great-grandchild of Benjamin Franklin. This petition was presented to the Orphans' Court of Philadelphia County, alleging that the trust created by the said codicil to the will was void,

"(a) Because an accumulation was directed for a longer period than was allowed by the common law.

"(b) Because the legacy to the City of Philadelphia and the State of Pennsylvania vested at a period after the testator's death beyond that allowed by the law of this State.

"(c) Because the use of the said funds during the first hundred years after the testator's death was not a charitable use.

"(d) Because the purpose contemplated by the testator has become impossible on account of the dereliction and negligence of the trustees in not realizing the anticipated sum.

"And praying for an account.

"On the 28th day of October, 1890, a demurrer on behalf of the City of Philadelphia, trustee, was filed, setting forth:

"(a) That petitioner was barred by lapse of time,
"(b) That the legacy was valid.
"(c) That the trust was a charity.
"(d) That the Orphans' Court had no jurisdiction."

A similar petition was made on November 15th, 1890, on behalf of Elizabeth D. Gillespie.

The decision of the Orphans' Court on these petitions was that the trust was valid and a charity, and should not be set aside, or, quoting a part of the language of the Court:

"The City of Philadelphia is something more than a mere legatee, it is a purchaser for value; or, to speak more accurately, it has, at the earnest request of the testator and upon his promise to pay a certain reward, performed arduous services, for his benefit or gratification, or the benefit, at his instance, of third person, during a stipulated time. The contract so made is binding upon the original promiser, and no less upon those succeeding to his estate.

"The petitions are dismissed."

On a subsequent appeal to the Supreme Court, the decision of the Orphans' Court was affirmed.

On December 2, 1892, a bill in equity was filed in the Court of Common Pleas, No. 4, of Philadelphia County, on behalf of Elizabeth D. Gillespie, administratrix of the estate of Benjamin Franklin and of Richard Bache, deceased, against the City of Philadelphia, Trustee. The decision of the Court was, however, against the complainant. Quoting from a part of the decision:

"On the whole case we are of opinion that Dr. Franklin has established in legal form three valid, benevolent and beneficial charities, neither one of which is vulnerable when assaulted upon any of the grounds which were argued before us.

"We therefore, sustain the demurrer of the defendants, and dismiss the plaintiff's bill with costs."

Such, in brief, is the history of the Trust Fund created by

Franklin, and placed in the control of the City of Philadelphia. While it did not attain the magnificent proportions that its great benefactor hoped, yet it can, by no means, be regarded as having entirely failed in its benevolent purposes.

It is an interesting matter in this connection to note that the Franklin Fund received an addition at the hands of John Scott, Chemist, of St. Patrick's Square, in the City of Edinburgh, Scotland, who appointed James Ronaldson, of Philadelphia, Typefounder, as his true and lawful attorney, for him and in his name, as sole executor to transfer the sum of $3000 of certain stock to the Corporation of Philadelphia, which was entrusted with the management of the late Dr. Franklin's legacy, to be applied to the same purposes as the legacy.

Moreover, the same John Scott transferred a further sum of $4000 to the City of Philadelphia, to be placed with the same management as the said Dr. Franklin's legacy, with, however, the proviso that the interest and dividends to become receivable upon such sum be laid out in premiums to be distributed among ingenious men and women who make useful inventions; that no such premium shall exceed $20, and that with the same shall be given a copper medal, with the inscription "To the most deserving." This legacy, known as the John Scott Medal Legacy, is now, so far as the recommending of the prizes is concerned, in the hands of the Franklin Institute of the State of Pennsylvania. The medal is awarded by the Board of Directors of City Trusts, but that Board depends absolutely upon a recommendation of the Franklin Institute, and has never failed to award as recommended, nor award unless so recommended.

THE METRIC SYSTEM IN THE UNITED STATES.

An informal address by Dr. A. G. Bell to the Committee on Coinage, Weights, and Measures of the U. S. House of Representatives, on February 16, giving an explanation of the reasons why the United States should abandoned its heterogeneous systems of weights and measures, is printed in the *National Geographic Magazine* for March. The committee had under consideration a bill before Congress proposing that, from July 1, 1908, all the departments of the government of the United States, in the transaction of business requiring the use of weight and measurement, shall employ and use the weights and measures of the metric system. Dr. Bell gave an exhaustive account of the anomalies of the British systems of measurements in use in the United States. He pointed out that all civilized countries, with the exception of the United States and Great Britain and her colonies, have adopted the similar and more scientific decimal system. He reminded the committee that the metric system was legalized in the United States in 1866, and that its adoption by a portion of the population had increased the present confusion. By reference to the decimal system of coinage already in use in the States, Dr. Bell proved convincing instances of the simplification possible with it in the conversion of units, and explained that the United States, when it changed from the old system of pounds, shillings and pence to the present dollars and cents, did not adopt the metric system of weights and measures because the latter, as we know it, did not appear until after the American Coinage Act of 1792. The facts that our whole system of arithmetic is decimal, that no difficulty whatever is experienced by ordinary workmen in the use of the metric system—provided there is no question of converting their measurements—and that the use of the metric system need not mean the use of new tools, were all very clearly explained. It is interesting to note, in connection with this Bill before Congress, that the committee on publicity of the Metrological Society, of which Prof. Simon Newcomb is chairman, has circulated a letter urging all persons in favor of the introduction of the metric system to write, and also secure from other friends, as many letters to representatives in Congress as possible, so that they may see that public sentiment is in the direction of the adoption of decimal weights and measures.

A THERMIT REPAIR.

An interesting thermit repair was recently accomplished by L. Heynemann, consulting engineer, at San Francisco, Cal., upon a large forged steel dredge bucket arm measuring two inches thick by twelve and one-half inches wide at the break. The fracture was complete. To facilitate the flow of theremit steel through and between the broken parts a number of holes were drilled in the crack. A sheet-iron mold box was first closely fitted to the arm and carefully lined with sand and flour core mixture and thoroughly baked. After the mold was placed in position the parts were heated by a blow torch to expel moisture and avoid chilling the thermit steel.

The thermit reaction produces a temperature of approximately 5400 degrees F., so that, as in this instance, it is possible to add to the contents of the crucible a considerable percentage of steel punchings or rivets, which somewhat lowers the temperature and increases the quantity of liquid steel. Two hundred and fifty pounds of thermit were used (yielding after combustion 125 pounds of liquid steel and 125 pounds of slag) and 40 pounds of small steel rivets. On the completed repair a reinforcing collar of thermit steel was cast around the weld to give extra strength.

Tests covering the strength and permanency of thermit welds were recently made at the works of the Fore River Shipbuilding Company, Quincy, Mass. Bars of rolled steel two inches by four and one-half inches were drilled, broken and welded with thermit, and standard test bars cut from the center of the weld. As a basis of comparison, test bars were also cut from the unwelded portions of the stock. The following gives the gist of the results obtained:

Description.	Elastic limit.	Tensile strength.
Weld	32,000	59,000
Stock	38,500	60,500
Weld	33,700	61,800
Stock	36,850	63,400

As an average of the two cases, it will be noticed that the elastic limit suffers a reduction of about 12.7 per cent. after the welding, but only about 2.5 per cent. of the tensile strength is lost.—*Iron Age.*

Sections.

SECTION OF PHOTOGRAPHY AND MICROSCOPY.—*Stated Meeting*, held Thursday, March 1, 1906. Dr. Henry Leffmann in the chair. Present, seventy-two members and visitors.

The first communication of the evening was made by Prof. Henry Kraemer, of the Philadelphia College of Pharmacy, entitled "Some Experiments in Growing and Photographing Wild Plants." The speaker's remarks were profusely illustrated with the aid of lantern pictures.

The subject was discussed by the chairman, and Messrs. Ridpath and Wilbert.

The thanks of the meeting were tendered to Prof. Kraemer. Adjourned.

M. I. WILBERT, *Secretary.*

Stated Meeting held Thursday, April 19, 1906. Dr. Henry Leffman in the chair. Present, fifteen members and visitors.

Dr. Leffmann presented the first communication of the evening on "Invisible Micro-organisms."

Mr. John Bartlett followed with a paper on "Modified Form of Reduction by Farmer's Method." (Referred for publication.)

M. I. WILBERT, *Secretary*.

ELECTRICAL SECTION.—A *stated meeting* of the Electrical Section was held on Thursday, March 22, at 8 P.M. President Thos. Spencer in the chair.

The minutes of the last regular meeting were read and approved. There being no items of business, the paper of the evening on "The Mercury Arc," by Dr. E. Weintraub, of the Research Laboratory of the General Electric Company of Schenectady, N. Y., was proceeded with.

This paper treated the subject from the standpoint of the development of the Mercury Arc, and was fully illustrated with experiments, showing both the principles employed, and the practical application of the arc.

On motion a vote of thanks was extended to Dr. Weintraub for his very instructive and interesting lecture.

There were fifty members present. On motion the meeting adjourned.

GEO. A. HOADLEY, *Secretary pro tem.*

MECHANICAL AND ENGINEERING SECTION.—*Stated Meeting*, held Thursday, April 5, 8 P.M. Mr. James Christie in the chair. Present, 110 members and visitors.

The paper of the evening was read by Mr. Chas. G. Darrach, of Philadelphia, on "The Equipment and Maintenance of the Modern Community Building."

The subject was freely illustrated with the aid of lantern photographs, and was discussed by Prof. H. W. Spangler, and Messrs. C. A. Hexamer, Francis Head and James Christie. Adjourned.

FRANCIS HEAD, *Secretary*.

Stated Meeting, held Thursday, April 26, 8 P.M. President Chas. Day in the chair. After the transaction of formal business, the chairman introduced the speaker of the evening, Mr. Strickland L. Kneass, of Wm. Sellers & Co., Philadelphia, who read a highly interesting and instructive paper on "High Pressure Steam Tests of a Locomotive Injector," which was freely illustrated with the aid of latern photographs.

The subject was discussed by Mr. Henry F. Colvin, the chairman, and the author, and the paper was referred for publication.

The thanks of the meeting were voted to the speaker. Adjourned.

WM. H. WAHL, *Sec'y pro tem.*

SECTION OF PHYSICS AND CHEMISTRY.—*Stated Meeting*, held Thursday, March 29th, at 8 P.M. Dr. Robert H. Bradbury in the chair. Present, thirty members and visitors.

The paper of the evening was read by Prof. E. L. Nichols, of Cornell University. The subject covered the phenomena of Phosphorescence and Fluorescence. The speaker's remarks were profusely illustrated.

The subject was discussed by Dr. H. H. Heyl, Dr. Geo. S. Stradling, Mr. Carl Hering and others.

The thanks of the meeting were voted to the speaker of the evening. Adjourned.

E. A. PARTRIDGE, *Secretary.*

Stated Meeting, held Thursday, April 12, 8 P.M. Dr. Edward Goldsmith in the chair. Present, twenty members and visitors.

Mr. Lyman F. Kebler, of the U. S. Dept. of Agriculture, read the paper of the evening, on "Drugs and Chemicals." The speaker called attention to the universal impurity of chemicals, even of those marked C. P., by reputable manufacturers. He referred also to a similar condition in the case of drugs, and to their gross abuse, as a public scandal. The speaker received a vote of thanks. Adjourned.

EDWD. A. PARTRIDGE, *Secretary.*

Franklin Institute.

(Proceedings of the Stated Meeting held Wednesday, March 21, 1906.)

HALL OF THE FRANKLIN INSTITUTE,
PHILADELPHIA, March 21, 1906.

PRESIDENT JOHN BIRKINBINE in the chair.

Present, 132 members and visitors.

The report of the Board of Managers exhibited the election of eight new members.

The first communication of the evening was presented by Mr. Herbert E. Ives, of Johns Hopkins University, on "An Improved Process of Diffraction Color Photography."

The speaker illustrated his remarks by means of lantern photographs and specimen color photographs made by the process.

The subject was referred for investigation to the Committee on Science and the Arts.

Mr. U. C. Wanner gave an interesting informal address on "The Photography of Flowers;" illustrated by numerous flower pictures very artistically executed and colored in imitation of nature.

The appointment of President Birkinbine by the Board of Managers to represent the Institute at the bi-centennial anniversary of the birth of Franklin, to be celebrated on April 17-20, by the American Philosophical Society, was announced and confirmed.

WM. H. WAHL, *Secretary.*

(Proceedings of the Stated Meeting held Wednesday, April 18, 1906.)

HALL OF THE FRANKLIN INSTITUTE,
PHILADELPHIA, April 18, 1906.

PRESIDENT JOHN BIRKINBINE in the chair.

Present, eighty-two members and visitors.

Addition to membership since last report, six.

The first communication of the evening was read by Mr. W. S. Ayers, Mining Engineer, of Hazleton, Pa., describing the recently-discovered Deutschman Cave, in the Silkirk Mountains of Canada. The communication was profusely illustrated with the aid of lantern pictures.

Mr. John C. Troutwine followed with an account of the Aqueduct of Philadelphia's First Water Works as uncovered by the excavation for the Subway of the Philadelphia Rapid Transit Co., on South Penn Square, opposite to the South Entrance to City Hall.

The meeting was closed by Mr. W. W. Jennings, who exhibited and described a series of photographs taken by him during a recent visit to Mt. Vesuvius.

Adjourned.

WM. H. WAHL, *Secretary.*

Committee on Science and the Arts.

(Abstract of Proceedings of Stated Meeting held Wednesday, April 11, 1906.)

MR. HUGO BILGRAM in the chair.

The following reports were adopted:

(No. 2358.) *Browning's Automatic Pistols.* Colt's Patent Fire-Arms Mfg. Co, Hartford, Conn.

ABSTRACT: A supplementary report. In the original report on this invention, approved May 5th, 1905, it was stated that the Browning weapon examined by the Committee did not function properly with soft-pointed bullets as the slide jammed and refused to close automatically.

In consequence of this objection, the inventor submitted to the Committee an improved pistol—thirty-eight calibre military model—which in the hands of the chairman functioned quite satisfactorily with 100 soft-pointed cartridges. (*Sub-Committee,* Wm. O. Gregg, Chairman.)

(No. 2361.) *System of Electric Distribution and Locomotion.* Josef H. Hallberg.

ABSTRACT: The report considers the several systems in use, viz.: The direct-current low voltage system as typified in the equipment of the N. Y. Central Railway Terminal in New York City and suburban lines; the single-phase alternating current commutator type motor, as employed on the New York, New Haven & Hartford Railroad, and which obviates the necessity of converting alternating current to direct current in the sub-stations; and the system of multiphase distribution (usually three-phase) necessitating two overhead conductors and the rail as the third conductor.

The objection to the use of the multiphase distribuion is chiefly on account of the complicated construction of frogs and cross-overs.

Mr. Hallberg proposes to use a traveling sub-station—in other words, he locates his sub-station in the locomotive, an idea first suggested by Mr. H. Ward Leonard.

In the Hallberg system it is proposed to install a motor-generator set on the locomotive, consisting of a single-phase Synchronous motor which receives current direct from the line, and to provide a three-phase A. C. generator of special construction consisting of a double set of poles, of which one or both can be energized, changing the frequency from twenty-five to fifty cycles so as to obtain the desired torque at the various rates of speed demanded by the transmitters.

In view of the careful attention the writer has given to the various steps in the operation to meet the conditions demanded in railway service, the report grants to him the award of a Certificate of Merit. (*Sub-Committee*, W. C. L. Eglin, Chairman; Richard Gilpin.)

(No. 2384.) *Moving Platforms.* Max E. Schmidt, New York.

ABSTRACT: The invention is covered by U. S. letters patent No. 440,725, Nov. 18, 1890, granted to M. E. Schmidt and I. L. Silsbee, and No. 747,090, Dec. 15, 1903, granted to applicant. The Committee granted the Longstreth award to the inventors (see report No. 1769) for the first-named invention.

The essence of the first invention consisted in the use of a flexible traveling rail in driving a high-speed platform. The low-speed platform was practically the platform of our ordinary car carried on two axles and four flanged wheels, with the difference that the platform or body is carried outside of the wheels, the space between the wheels being occupied by the high-speed platform. This platform was carried upon two continuous flexible rails upon which it rested, and the rails themselves rested edgewise upon the tops of the treads of the car wheels mounted on the axles of the slow-speed platform. [An installation of this description was successfully operated at the Expositions in Chicago, Paris and Berlin.]

It was found, however, that the continuous flexible rails were unable to stand the constant bending, and this fact led to the complete change of the system in accordance with the design and construction now submitted and covered with the more recent patent (No. 747,090) of applicant.

In the new form, the idea of a car mounted on traveling axles is entirely abandoned. The platforms are six feet long and are supported every two feet nine inches by wheels mounted on axles running in stationary bearings, the wheels being proportioned to give the desired speed to the respective platforms. These axles are driven by electric motors placed about seventy-five feet apart. The platforms are linked together by pivotal joints and are guided by a central fixed rail and two sets of horizontal guide wheels stradling the rails, attached to each platform.

The report finds that from an engineering standpoint the patent of Schmidt, here under consideration, represents a substantial improvement over the original invention of Schmidt and Silsbee. The construction is more substantial; the flexible rail is altogether eliminated; the greater part of the machinery is stationary and will permit of inspection at any time; much sharper curves are admissible; and the possibility of serious accidents from derailment or other cause appears to be remote.

The report further states that these recent improvements have been made with the view of bringing the moving platform into daily practical use as a means of transportation in large cities, and not merely for amusement purposes. It is noted, also, in passing, that no less than 12,000,000 passengers have been carried on the three platforms already built without injury to a single passenger.

The report recommends the award to the inventor of the John Scott Legacy Premium and Medal. *(Sub-Committee,* Arthur Falkenau, Chairman; Chas. E. Ronaldson, James Christie.)

(No. 2385.) *Historic Collection of Incandescent Electric Lamps.* Wm. J. Hammer, New York.

ABSTRACT: This collection, which was honored at the recent world's fair in St. Louis by the award of a "Grand Prize," is the result of continuous and vigilant efforts on the part of Mr. Hammer covering the past twenty-five years, and includes over 1000 specimens, illustrating every step of the development of this type of lamp.

Mr. Hammer's application is accompanied by a photograph of his collection, and its priceless value from the historical standpoint is approved and endorsed by most eulogistic letters from the most noted electrical engineers of both hemispheres. The award of the Elliott Cresson Medal is made to Mr Hammer. *(Sub-Committee,* Chas. J. Reed, Chairman; Thos. Spencer.)

JOURNAL

OF THE

FRANKLIN INSTITUTE

OF THE STATE OF PENNSYLVANIA

FOR THE PROMOTION OF THE MECHANIC ARTS

VOL. CLXI, No. 6   81ST YEAR   JUNE, 1906

The Franklin Institute is not responsible for the statements and opinions advanced by contributors to the *Journal*.

THE FRANKLIN INSTITUTE.

(*Stated Meeting, held Wednesday, March 15, 1905.*)

Notes on Great Tunnels.*

BY LEWIS M. HAUPT.

Prof. of Civil Engineering, Franklin Institute.

Mr. President and Members of the Institute:—

Tunnels, like bridges, are engineering devices for overcoming physical obstacles to traffic.

Although of ancient origin their recent great development is the outcome of the demands of modern civilization resulting from the increase in population and commerce.

The advent of railroads in 1827 gave an impetus to the building of tunnels to save time, distance and grade in the interchange of traffic. These "horizontal wells," as they were called, were excavated by hand labor, as they were comparatively short, but with the increase of tonnage and demands for greater economy in cost and time, more pretentious enter-

---

*Revised for publication.

prises were undertaken, requiring the use of machinery and power for their completion. Thus the ingenuity of man was impressed to meet the difficult requirements of providing an unobstructed passageway through mountains or under rivers or city streets, in the least time and at a moderate cost.

It was not until tunneling machinery was reasonably perfected that long tunnels without shafts became at all practicable for transportation, although there were several constructed at enormous waste of time and labor, for drainage purposes.

The introduction of efficient drills may be said to date from the first American patent, issued to J. J. Couch, March 27, 1851, for a reciprocating, hollow piston-rod, working in a cylinder, operated by steam, but it had a constant feed and no provision for the rotation of the bit. He took out a second patent November 3, 1852, for some improvements, but in the meantime Stuart Guynn was busy on independent lines in applying the same idea in a practical machine,—also having a constant feed,—but it slumbered from 1851 until the contractor of the Hoosac Tunnel, in Massachusetts, Herman Haupt,* embodied with this hollow piston-rod an automatic feed and rotating device which produced a simple, light and effective drill capable of piercing granite at a rate of one inch per minute and weighing only about 125 pounds. In ordinary shale the speed was doubled. It was estimated that with this tool a progress of twelve feet per twenty-four hours could be made at a cost of $16.33 per lineal foot of single track (15x18), or at about half the price of hand work, and with four times the speed.

To prevent the loss of time in clearing the heading from the debris and gases incidental to blasting—as well as to facilitate the drilling of the holes—a system was devised by the contractor whereby the drills were mounted in series of three or four on columns, which served both as supports and feed pipes for power (steam or compressed air). Thus mounted the drills had a large range in altitude or azimuth, as shown in Fig. 1. The jacks for clamping the stanchions in position and the ball and socket joint at the bottom of frames enabled the bits to be readily changed without removing the gang. The plant also included a low counterpoised derrick

---

*Died December 14, 1905.

and truck, enabling the gangs of drills to be rolled to the rear of the heading, which was driven at the bottom. Prepared cartridges were placed in the holes, and held by a plug, wired in series, and discharged by a hand battery, simultaneously. A vacuum fan and conduit laid on the floor soon relieved the heading of foul gases, and the drills were rolled forward, set up, coupled and started without awaiting removal of débris. To protect men and machinery a section of heavy timbers laid on the floor was hoisted by a bell-crank lever to an

Fig. 1

upright position to serve as a screen during blasting. These several features are clearly shown in Figs 2 to 4. which also illustrate the auxiliary steam plant with tender, when that motor is found to be more expedient.

Modification of this pioneer rock drill were made by Burleigh, DeVolson Wood, and others, to adapt them to the ever varying requirements of the service, and have led to the extensive manufacturing plants of the Ingersoll-Rand, and

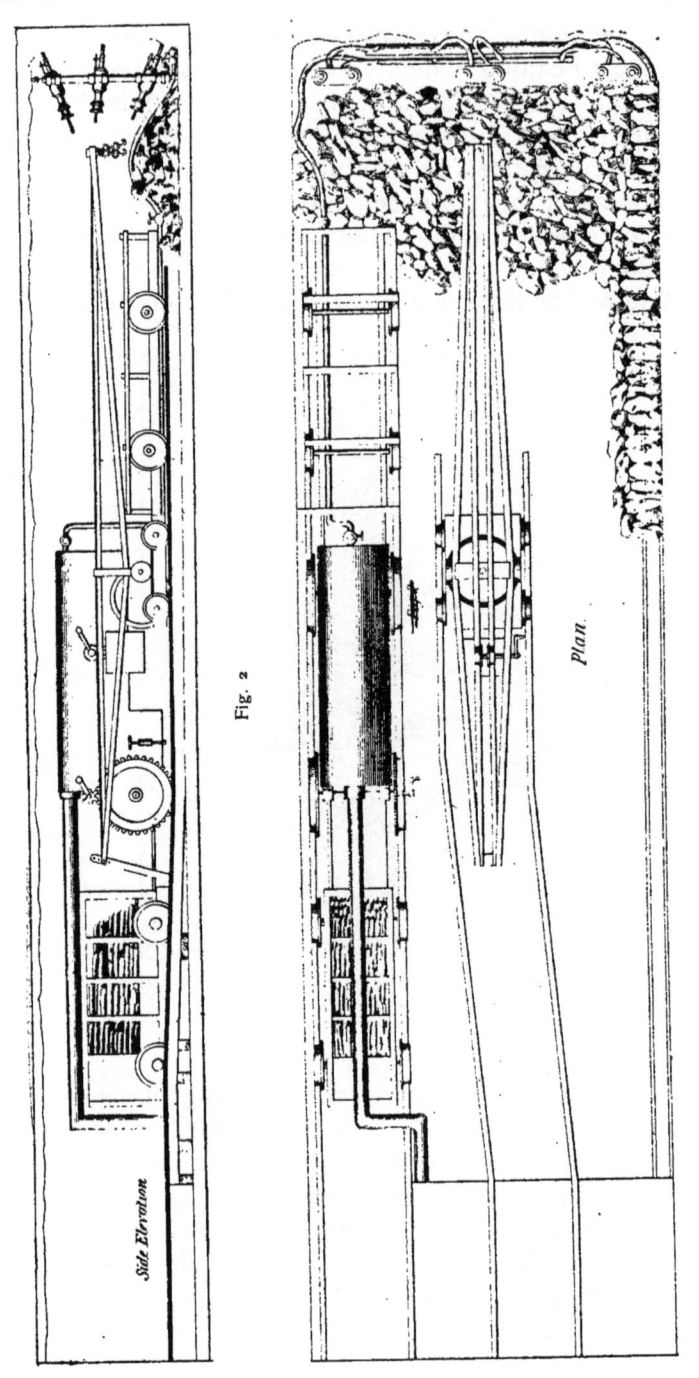

Fig. 2

Side Elevation.

Fig. 3

Plan.

other companies for supplying complete outfits of tunneling and mining machinery now in common use for all classes of excavations in hard materials.

The contract price for this tunnel was $2,000,000, while the cost was $50 per lineal foot, or for the full length nearly $1,300,000. It was the purpose of the contractor not to use a shaft but to drive from the ends on a rising gradient and ultimately, when the traffic increased sufficiently, to open a parallel tunnel by means of galleries from the first, at frequent intervals, thus greatly reducing the cost, improving the ventilation and avoiding lining. The State, however, saw great political possibilities in this enterprise, and by a change of administration, during the civil war, its control was assumed by the Commonwealth, the dimensions were changed to double track, a large dam built for power, and other radical departures made, which swelled the cost to about $10,000,000, or five times the amount of the original contract. A shaft was sunk 1028 feet in depth, which was of little use in expediting the work and was of doubtful utility. This Hoosac Tunnel of five miles in length, the longest in the United States at the time, was simultaneous with that under Mt. Cenis, where a compressed air plant, operated by water-power, was installed by Mr. Sommellier, the experience from which was a factor in the designing of the plans for the Hoosac, and this in turn, for works of the present day. The close relations of the several long railroad tunnels will become more apparent by a brief chronological statement of their statistics and time of construction.

1. HOOSAC, MASS.

Work was commenced in 1854 and prosecuted by private contractors until the State took possession, September 4, 1862. It was enlarged, and completed July 1, 1876, when it was officially accepted. The total cost of the railroad and tunnel, forty-four miles, was $17,322,019, of which $3,287,835 was interest. The progress was greatly accelerated by the introduction of power drills, 1867, and of nitro-glycerine, in 1870, which was manufactured by Geo. W. Mowbray, in the vicinity of North Adams. This gives an average progress of 5.5 feet per diem.

2. MT. CENIS.* THE ALPS.

The project was first broached by Guisseppe Medail, a Swiss peasant, in 1838, and subsequently, in 1852, M. Colladou suggested driving it by the use of compressed air drills, operated by water-power, but it was not until 1857, after a report by experts, that the Piedmontese Parliament granted a charter and agreed to pay half the estimated cost, which was $7,760,000, but it ultimately paid for the entire work.

The first blast was fired August 18, 1857, and the headings met December 25, 1870. It was opened for use September 17, 1871. It was double-track, driven with bottom heading $9\frac{1}{2}$x$8\frac{1}{2}$ feet. Average rate of progress, 8.00 feet per day. Total length, 42,158 feet, or 8.0 miles. Time, fourteen years in building. Cost, $14,498,352. Machine drills were introduced at the north end in 1861 and at the south end in 1863. The last year the rate was $14\frac{3}{4}'$ a day.

3. SUTRO TUNNEL.

The third long tunnel built was designed to open up the famous Comstock lode, in Nevada, by providing better drainage and greater accessibility to the mines. The section was 12x16 feet and length 20,351 feet, or nearly four miles. Work was begun October 19, 1869, and completed about 1878. There were four shafts designed to be used in construction, three of which were over 1000 feet deep, but two of them were abandoned before reaching grade because of the great influx of water. Progress per day (1875) in heading (10x8) was 10.24 feet.

4. ST. GOTHARD. ALPS (1872-81).

At the St. Gothard tunnel the Sommellier drills were used for a few years but were supplemented by those of McKean & Ferroux, making about 180 blows per minute. These McKean drills were modifications of the system designed for the Hoosac Tunnel and taken to Europe by Blanchard & McKean, agents for the contractor.

The contract for the St. Gothard was awarded to Mr. Louis Favre, August 7, 1872, at an estimated cost of $10,000,000, and

---

*Drinker's Tunneling, p. 266.

was completed in eight years, with a premium of $1000 per diem for every day saved or a similar penalty for each day lost. With the system of rapid firing and removal of débris great progress was made in each heading, and the work was opened in 1880—the first train passed through in 1881.

Its total length is nine and one-quarter miles; time, nine years, five months, and average daily progress, 14.6 feet, notwithstanding great difficulties from water and the caving of the lining. The early location of the approaches was placed high on the mountain slopes instead of the valley bottom, adding

Fig. 5

much to the expense, but this was afterward modified by the use of the spiral tunnel and loops. See Fig. 5.

THE ARLBERG TUNNEL. ALPS (1880-84.)

Through the Austrian Tyrol a tunnel of 6.38 miles was driven between 1880 and 1884 at a much more rapid speed than any of its predecessors, at an estimated cost of $7,000,000. The time consumed was three years, nine months, and the average rate of progress 27.8 feet per diem. Work was conducted very systematically and construction trains were run by time schedules. The headings, which were 7.5 feet high and 9.2 feet wide, were placed at the bottom instead of the top of the tunnel. In cost, therefore, this work was but $1,000,000 per mile, $208 per foot, and in speed of execution three times as fast as Mt. Cenis.

THE SIMPLON. 1893-1905. 12¼ MILES LONG.

This tunnel, which connects Brig, in the Valley of the Rhone, with Iselle, on the Diveria, in Italy, is twelve and one-quarter miles in length. The contract was let in September, 1893. The plans contemplated the use of the system proposed for Hoosac of two single-track parallel tubes 16.5 feet wide, 55.7 feet apart, but connected by oblique galleries at frequent intervals. Only one of them was to be completed to full dimensions, while the second was to be used for subsidiary purposes to provide trackage for construction and ventilation, and to await the demands of traffic for its final enlargement. The cost is reported to be $15,500,000 to February 24, 1905, when the headings met, thus opening the shortest route by eighty miles between Paris and Milan, and connecting the Atlantic and the Mediterranean. The large amount of water and high temperature (reaching 118°F.) made this work one of unusual difficulty.

Actual work was begun in 1898, so that the average ratio has been nearly two miles a year, and it is a monument to the enterprise and energy of the countries which have contributed so liberally to the colossal work which at best will save but a few hours in transit between the Italian seaports and the British Isles. (For difficulties from hot springs see Scientific American of March 18 and 25, 1905.)

SUBWAYS.

The feature of the above works is that they have generally been excavated from the ends, without intermediate shafts or slopes. Many more miles of continuous underground ways are now in existence, as in the numerous subways of Continental and American cities, but these have been built largely as open cuts or covered-ways, or with numerous shafts.

SUB-AQUEOUS.

Another important class of structures is the sub-aqueous passages connecting great centers of industries and taking the place of bridges. These involve greater difficulties and risks than are to be found in the previous classes, yet to avoid the transfers, delays and risks of ferries or the obstructions due to bridges, ice and fogs, they are found to be expedient, even at very great cost.

The latest and best practice in works of this kind is illustrated in the extensive system of tubes now completed and under construction by the Pennsylvania Railroad Company, under the direct supervision of Mr. Chas. M. Jacobs, Alfred Noble and others, in and across the rivers surrounding the Island of Manhattan, and connecting its western ramification directly with the greatest seaport of the continent, without break of bulk.

Whilst the tube-system as used underneath the North River at New York has required much ingenuity to adapt it to its purpose, the general idea was first applied in the pioneer work built under the Thames, at London, by M. I. Brunel, in 1825 to 1846. This structure is 1200 feet long, with two passageways 14 feet wide by $16\frac{1}{2}$ feet high, and is now used by the East London Railway.

The influx of mud and water were so great as to cause the invention of a shield to cover the whole face of the excavation, 38 feet wide and $22\frac{1}{2}$ feet high. An attempt was made to introduce the Beach pneumatic system in New York about the year 1863, but it was untimely, and the traffic had not at that time reached such magnificent proportions as to justify the expense, and there was an aversion to being shot through a hole in the ground by the public. Electric motors were not then available. Again, in 1868-69. W. H. Barlow used a modified form of Brunel's shield in building the "Tower" subway under the Thames. This was circular in section, eight feet in outside diameter, and was lined with ribbed cast-iron plates. It was the prototype of the present generally adopted system. In 1889, a pair of tubes, each ten feet in diameter, was laid under the Thames by Mr. Greathead for the South London Railway by means of an improved shield telescoped over the outer end and pressed forward by jacks, as in the Tower subway.

The earliest sub-aqueous aqueduct tunnels in this country were those built at Chicago in 1864-67, two miles long, at a cost of $457,844, and subsequently extended four miles further for a fresh water supply. A second conduit eight feet in diameter and four miles long was added in 1887-1892, and also at Cleveland in 1869-74, when a conduit five feet in diameter and 6,606 feet long was built.

In 1888-1902, the Grand Trunk Railroad built a single-track circular tunnel, lined with cast-iron segments, by the use of shields, under the Detroit River, at Sarnia, through soft clay, sand and gravel 6000 feet in length. To-day similar systems of sub-aqueous tunnels are being rapidly and successfully built under the North and East Rivers.

AQUEDUCTS.

Still another extensive group of tunnels is to be found in the aqueducts for supplying large communities with water. The most conspicuous example of this class is the Croton Aqueduct of thirty-three miles in length and about fourteen feet in diameter, crossing the Harlem River by means of an inverted syphon at a depth of 306 feet below the surface.

DRAINAGE TUNNELS.

Amongst the most interesting works of this class may be mentioned the ancient Desague de *H*uehuetoca, undertaken by Enrigue Martinez, a Dutch engineer, in 1607, for draining the basin of the City of Mexico. The tunnel was four miles long and the drain thirteen, but before the lining was completed a great flood caused it to cave in. As a reward for his effort the engineer was imprisoned for three years, and when released he was ordered to make an open cut, in which he spent the rest of his life. But the work was continued for 120 years, yet it was not made deep enough to relieve the basin to any great extent. In 1888, another tunnel was built six miles long and 150 square feet in section, supplemented by twenty-seven miles of large canals, this tunnel had twenty-four shafts, varying in depth from 75 to 325 feet and a discharge capacity of 450 cubic feet per second, and furnishes an excellent precedent for the problem now confronting the Isthmian Canal Commission engaged in the regulation of the floods of the Chagres River at Panama, where the success of the enterprise is made to depend upon the diversion of these torrential waters by means of tunnels from seven to ten miles in length through the Cordilleras. But in this case little is known of the geology or stratigraphy, and as to the possibility of any shafts being used.

Before closing these brief remarks it may be found expedient to glance at the attitude of the traffic problem in this city in 1888 as contrasted with present conditions. Then the necessity of additional facilities was urged by the official publications of this Institute, but was opposed by the vested interests handling the interurban traffic, because of the extreme cost of such works and terminals; of the "impossibility of satisfactorily operating the subway;" because "the subway car motor connected with the other lines of the road;" because "it would be unwholesome and unsatisfactory to the public;" because "it would forever preclude the growth of business on the lines of the company and prevent any extensions in the future to meet increased business," and lastly, because "the destruction of the railroad's terminal facilities in Philadelphia would entail a loss* impossible to estimate."

That these objections were untenable is evident from the fact that the "impossible" has vanished, and the underground and elevated roads have come to stay. Councils are now considering the necessity of abolishing all grade crossings and the steam routes are cheerfully asquiescing, while the city and railroad engineers are harmoniously working to relieve the surface from all rapid transit trains, as is best for the interests of all parties.

The Market Street subway contracts for sections 3 and 4, extending from Fifteenth Street west to the Schuylkill River, were let to E. E. Smith on April 1, 1903, and the work, which was begun April 6,. has been vigorously prosecuted so that it is expected to be open for traffic this year (1905). It is designed for four tracks, two of which will carry express trains to connect with the elevated railroad building out Market Street from the east side of the Schuylkill to Delaware County. The four track bridge across the river is now well under way. The local or

---

*See rapid Transit in Cities. Jan., 1888, Journal of F. I.. Feasibility of Underground Railroads, Dec., 1888, Journal of F. I.

outer tracks will make a loop at the eastern end, passing down Fifth to Walnut, thence to the River Delaware, to Arch, to Fifth and return to Market. The dimensions, in the clear, are 48 feet 6 inches in width and 14 feet 6 inches in height from top of rails. The roof girders are supported by three rows of steel columns. The side walls are of reinforced concrete, and the work is being carried on without seriously interrupting local traffic.

Stations are placed at Fifteenth, Nineteenth and Twenty-fourth Streets.

Chicago has already constructed some twenty-eight miles of freight and passenger subways at a rate of twenty-one feet per day from each of the fourteen headings, the material being firm clay (See Scientific American, March 11, 1905) giving the unprecedented rate of progress of twelve miles in less than a year.

Boston has built a much-needed underground transit way, which is very popular and which has to some extent relieved the congestion of the surface, and the great work in New York goes rapidly on, but it is demonstrated that by the time one system is completed the increase of traffic has reached the limit of its capacity and another is demanded, so that to-day applications are being made for new charters under additional streets.

In view of the record of the past it is reasonable to look for the construction in the near future of the long-projected tunnels under the Straits of Dover to connect the British Isles with the Continent by a continuous line of rails, and the project of M. Lobel under Behrings Straits to join Alaska and Siberia, and thus furnish an all-rail overland connection between New York and St. Petersburg or Paris, or even with Cape Town in Africa, via the Cape to Cairo route.

These are some of the transformations which the engineers of the future may effect in the traffic routes of the world, through the instrumentality of great tunnels made practical by improved drilling machinery.

For further details see "Tunneling," by Henry S. Drinker. Johnson's Encyclopœdia article Tunnels, by Wm. R. Hatboro. Tunneling by Machinery, Herman Haupt, 1867.

The Economic Future of Japan.

By Mr. Achille Viallate.

Professor in The Free School of The Political Sciences at Paris, in Conference with The Société Industrielle de Mulhouse.

Translated from the April, 1905, Bulletin of the Société, Séance of February 22d, 1905, by Chief Engineer Benjamin F. Isherwood, U. S. Navy.

Japan was made known to Europeans for the first time by the writings of the Venetian traveller, Marco Polo, in the Thirteenth Century. Regular relations, however, were only established between the Occidental Powers and the Empire of the Rising Sun three centuries later, towards the middle of the Sixteenth Century. In 1543, a Portugese adventurer navigating along the Chinese coast, was driven by a storm upon the coast of Japan, where he landed. Following the Portugese came the Spanish—masters of Manila—and then came the Hollanders and the English, each in turn seeking commerce with Japan. This traffic had, with difficulty, began to acquire some amplitude, when the Japanese Government brusquely closed its dominion to foreigners. The edict of 1638 interdicted them, with the exception of Chinese and Hollanders, from trading, and restricted these latter two to the small island of Desima, in the Bay of Nagasaki. Also, the Japanese were forbidden, under penalty of death, from leaving their country.

During two centuries Japan continued to live an isolated life. The ardent desire of Europeans for a renewal of commercial relations which they hoped would be profitable, menaced, from the commencement of the Nineteenth Century, this isolation. In 1824, the English appeared at Mito; and in 1846 an American squadron vainly attempted to open negotiations with the Japanese authorities. The Americans returned in 1853, and in the following year Commodore Perry obtained, by means of the fear which his cannon inspired, a treaty of peace and amity between Japan and the United States, together with the opening of the ports of Shimoda and Hakodate to American commerce. A little while afterwards Great Britain and the other European powers compelled the signing of treaties giving them the same advantages. In July, 1858, the United States concluded a treaty of commerce which provivded for the opening within three years of five additional ports, and limited the maximum tariff; the tariff of importation was not to exceed 20 per centum, *ad valorem*; and the tariff of exportation was not to exceed 5 per centum. By new conventions made in 1866, Japan engaged to limit to 5 per centum, *ad valorem*, the tariff of importation. By means of the insertion into all the treaties made by other powers with Japan of the clause of the most favored nation, the advantages obtained by any one of them becomes extended, *ipso facto*, to the others, and all Europeans are thus placed in an analogous situation.

The concluding of these treaties hastened a political revolution which ended in the destruction of the feudal system prevailing up to this time in Japan. The Emperor regained the power of which for more than three centuries the Shogun had dispossessed him. In 1881, an imperial ordinance established the representative system, and in 1889, the Emperor granted a constitution to his people, in which he remains the supreme authority, but delegates a part of his powers to a Diet, composed of two Chambers: the Ministers remain responsible to him alone.

Modernized Japan endured with discontent the treaties of the period between 1854 and 1866, which gave to foreigners the rights of exterritoriality in the open ports, and made them answerable to only the Consular tribunals. In 1894, England, and, afterwards, the United States; then, Russia in the following year; Germany in 1896; and, then, France; consented to the substitution for the old treaties, of new ones satisfactory to the Japanese. After the making of these treaties, which became operative on the 4th of August, 1899, and which now actually regulate the relations of Japan with foreign powers, the latter abandoned their privilege of national jurisdiction; they also surrendered to Japan the elaboration of its own tariff with reservation of the conventional rights defined by these latter treaties. On the other side, Japan has opened its entire territory to foreigners.

Japan has not limited its innovations to the Europeanizing of its political institutions. It has with great boldness undertaken a profound economic transformation that has caused serious disquiet to the rest of the world.

What, then, are these efforts now being made by modernized Japan in the evolution of its economic system. What causes exist from which their success may be inferred in the future? Has Japan truly the necessary elements for becoming a great industrial power? Up to what point has other nations to fear the advent of this new competitor? Such are the questions which are proposed to be briefly answered in this paper.

I.

The Empire cf Japan consists of a long chain of mountainous islands or volcanic origin, extending from mid-Kamtchatka to Formosa, which island was taken from China at the close of the war of 1894. It is formed, exclusive of Formosa, which is a colony, of four large islands, "Hokkaido," "Honshu," "Shikosu." and "Kiushu," and of 600 small islands. Its surface is about 147,500 square miles. Excepting "Hokkaido" and the north cf "Honshu," which, because of their latitude, are subjected to rigorous cold, the rest of the Empire has a temperate and humid climate.

The population has rapidly increased. From 33,000,000 in 1872, it rose in 1903 to 46,000,000.

The Japanese remain to the present time almost wholly an agricultural people, notwithstanding that the country is but little adapted for agriculture. The arable land is only 15.7 per centum of the whole surface of the Empire. Of this whole 45 per centum (the low flat ground) is utilized for the cultivation of rice; and 38 per centum (situated on the sides of the hills) serves principally for the tea plantations, and for the plantations of mulberry trees, the leaves of which are used as food for the silk worms.

The food of the Japanese is principally rice and fish.

The agricultural population is estimated at 23,000,000 of persons. The predominant character of the cultivation is the small scale on which it is practiced. The work is almost wholly done by hand. The implements are mediocre, and often not of metal.

The productive capacity of the worn-out soil is maintained by constant labor, a very careful irrigation, and the abundant use of manures. By these means, two, and sometimes even three, crops are obtained during the year from the same ground.

Notwithstanding all efforts, agriculture with difficulty supplies the needs of the population, the rapid increase of which is a most disquieting problem for the Japanese Government. The least deficit in the rice crop compels recourse to importation.

The agriculture of Japan, though still very backward, can hope for great improvement, to accomplish which the agriculturists, now entirely guided by tradition, must be properly instructed, and the necessary money for the desired ameliorations must be provided, because they are so ignorant and so poor.

The government is applying itself to this double task by supplying agricultural teaching, and by creating financial institutions for agricultural credit. But whatever may be the efforts made and the results obtained, the progress of agriculture will be fatally limited by the small extent of territory susceptible of cultivation. Doubtless, the sides of the hills having a moderate inclination are capable of cultivation, but they represent only 8 per centum of the total surface of the country, so that the most optimistic previsions cannot hope the Empire of Japan will ever be able to devote more than 25 per centum of its area to the production of food.

* * *

The Japanese agriculture of the present day does not differ much from what it was before the Revolution; but the case is quite different with industrial Japan, for here the progress made in thirty years has been considerable. The Japan of 1870 knew only the industries that could be practiced by the household or in small workshops, and was ignorant of all the modern methods of production. The tools were rudimentary, and the only motor employed was water-power.

Industries on the great scale are to-day established in Japan. Their progress has been rapid. In 1894 only 1808 steam-engines developing an aggregate of 32,808 horse-power were employed. In 1902 the number of steam-engines returned was 4,057, developing an aggregate of 90,778 horse-power. In the suburbs of the large cities, "Tokyo," "Kyoto," "Osaka," etc., formerly dominated by the traditional low buildings of wood, now rise the vast brick constructions and the lofty chimneys of the modern factory.

Like the political transformation, the industrial transformation is the work of the statesmen who have made modern Japan. With that country remaining completely isolated from European movement, to count upon private initiative to introduce the modern methods of production, was impossible. The government of the Revolution, conscious of the task imposed upon it, sent to Europe and to the United States missions charged to study the industrial methods of those countries, and it also obtained from them experts for organizing such of those industries as seemed the most pressing.

Mining was on an ancient system. The government employed foreign engineers to introduce into this industry the modern processes of exploitation.

The ministers of war, of the navy, of the treasury, became ministers of the corresponding industries, and thus played the part of veritable educators. The first undertook the manufacture of powder for cannon, and of the material of war. The second created a dockyard and workshops for the construction of small vessels, and for the reparation of the large ones that during several years he was obliged to buy from foreigners. To furnish the castings for these purposes a foundry was put in operation.

The minister of the treasury needed a mint, and this required an establishment for making the assays. He desired to manufacture, according to the methods of Europe and America, the material for the paper money issued by the State, and for the notes issued by the banks, and for the postage stamps. For these purposes he was obliged to create a paper making factory, a manufactory of sulphuric acid, of soda, etc. The government having undertaken the construction of railroads, had to manufacture cement.

In 1872, he established at Tourioka (Gummakin) a model spinning mill, with the view of introducing machine-spinning in the silk industry. In 1877, he created at Senju, near Tokyo, a weaving mill for wool. In 1881, he erected model spinning mills for cotton at Nukada-Gun and at Aki-Gun; and he sold, on credit, looms bought in England, to private persons living in different cities in order to spread this new industry. In 1876, he founded a glass-making factory at Shinagander.

The government having thus given the impulse, abandoned the part of initiator which circumstances had compelled it to assume, as soon as private interests in their turn commenced to follow in the way shown by it and in which it still guided them. Commencing in 1880, it sold, one after another, its model establishments to private industry; but it retained, nevertheless, the workshops necessary for the needs of the army, and of the navy, and of some other departments of the State.

At the same time that the government was creating modern industries by the aid of foreigners, it commenced the creation of technical education for its own people.

The efforts of the Japanese government have been crowned with success. Without doubt the old industrial methods still exist; the family industry has not ceased, and numerous industries are operated in an incomplete manner; but the great modern industries, using mechanical tools, and steam-engines for driving them, together with their scientific methods, have taken firm root in Japan, and are extending farther and farther every year.

In 1902, there were 2,427 manufacturing companies, with an aggregate capital of $86,500,000; and in 1898, establishments employing over thirty persons each, had an aggregate personnel of 506,912.

The following statistics enable an idea to be formed of the industrial development during the last decade, and of the actual situation.

PRODUCTION OF SOME MINERAL MATTERS.

	Coal tons.	Iron tons.	Copper lbs.	Petroleum gallons.
1892	3,200,000	18,500	248,400	4,491,400
1902	9,700,000	31,400	350,400	41,743,600

WORKSHOPS AND FACTORIES WITH MOTIVE POWER.

	Steam Power		Water Power		Steam and Water Power	
	Number	Horsepowers	Number	Horsepowers	Number	Horsepowers
1894	1098	32,858	1090	2,429	221	5,744
1902	2449	90,778	497	5,298	45	4,825

PRODUCTION OF THE PRINCIPAL INDUSTRIAL ARTICLES.

	Raw Silk. Pounds	Waste Silk. Pounds	Tissues. Dollars	For Exportation Straw Goods. Dollars	Porcelain and Faience. Dollars
1892	9,064,700	4,320,700	24,500,000	282,000	1,881,000
1902	14,827,500	6,303,800	75,500,000	2,631,000	3,455,000

	Lacquered Articles. Dollars	Beer. Gallons	Chemical matches. Dollars	Japanese paper Dollars	European paper Dollars
1892		333,500	2,500,000	2,500,000	500,000
1902	2,769,000	3,702,000	4,300,000	7,000,000	3,500,000

The greatest development has been in the spinning of cotton-thread industry, which has become a really great modern manufacture.

COTTON-THREAD.

	Number of factories	Capital engaged. Dollars	Mean daily number of employees	Raw Cotton used. Pounds	Cotton-thread manufactured. Pounds
1892	39	4,500,000	403,314	100,900,000	82,700,000
1902	80	17,200,000	1,301,118	365,500,000	317,600,000

Certain of the above industries have worked solely for the home market, but now they have commenced to work for the foreign market also, and exportation has made sensible progress.

EXPORTATION VALUE IN DOLLARS.

	1890.	1903.
Cotton goods	28,000	19,029,000
Silk goods	1,662,500	15,724,500
Coal	1,549,500	9,603,000
Chemical matches	744,500	4,236,500
Straw goods	173,500	2,325,500
Porcelain and faience	622,500	1,584,500
Cigarettes	4,000	1,023,500

* * *

The imperial government became occupied at an early date with the means of communication and of transportation.

In March, 1871, the postal system, organized in imitation of those of Western nations, was inaugurated between the cities of Tokyo, Osaka, and Kyoto, and was rapidly extended. In June, 1874, Japan joined "The International Postal Union." In 1879, it became a member of "The Universal Telegraphic System."

The first line of railroad was opened in 1872. It connected the cities of Tokyo and Yokohama, and was a governmental initiative. In 1883, private companies commenced the construction of railroads. In 1903, the extent of the system was 4,495 miles, of which 1,344 miles were built by the State. In the last decade the number of passengers has nearly quintupled, and the freight has more than decupled.

From an economic point of view, the organization of a national monetary system was not less important than the creation of the means of rapid transport. The imperial government was obliged to consider this problem several times before it reached a solution.

A law of November, 1869, decided to adopt the metric measures and with silver for the standard metal, as the basis of the imperial monetary system. The law of May 10th, 1871, modified this system by substituting the gold standard for the silver standard. The change was premature, and embarrassed the government, which was obliged to have recourse to paper money, and this both drove the metallic money out of, and prevented it from passing into, circulation. Previous to the failure of the law of 1871, the ordinance of the 11th of May, 1878, legalized the silver "yen," and without restriction of coinage. Theoretically, Japan passed into bimetalism; practically, it remained under an exclusively paper-money regime. In 1886, the return to specie payment was realized and became definitive; but the country remaining subjected, in fact, to an exclusively silver standard, suffered all the consequences of a gradual diminution in the price of that metal.

The war indemnity obtained from China in 1895, enabled Japan to adopt the gold standard. The law of the 26th of March, 1897, prescribed for the monetary unit the "yen" of the weight of two *fun* (0.75 grammes, centigrade) of pure gold, and, consequently, of an intrinsic value of 2.50 francs.

There is not in circulation any more paper money issued by the State. The sole existing paper money is the notes of the Bank of Japan, which bank was created in 1882 with the exclusive privilege of emitting them. It has an entirely paid up capital of $15,000,000, and issues notes payable in gold on demand, which notes have the quality of legal money.

Foreign commerce has had a rapid development. The aggregate of importations and exportations in 1872, was only $16,500,000; but in 1888, it surpassed $50,000,000. In 1894, it exceeded $100,000,000; and in 1903, it rose to $303,000,000.

More than 30 per centum of the total importations of Japan is food stuffs, which the insufficiency of its own agricultural productions compels it to obtain abroad; and 23 per centum is of material in the raw state; the cotton, wool and hemp necessary for its newly developed industries. The remainder is composed of articles of European manufacture, the consumption of which rapidly increases. Among these articles are prime motors and power tools to the annual amount of $2,500,000.

About 34 per centum of the exports consists of agricultural products: raw silk, tea, camphor (from Formosa); and 11 per centum of mineral products, copper ore and coal. The surplus consists of manufactured products, of which the two principal ones are articles of silk and articles of cotton (for the major part of the cotton is in the state of thread), representing altogether about 23 per centum of the total exportation.

The greatest part of the commerce of Japan is with Asiatic nations—about 50 per centum of its importations and of its exportations. Its greatest sources for supplies are as follows: Great Britain, the United States, Germany; then, but a long way after, Belgium and France. Its best clients among the western powers, are: the United States, France, then Great Britain, Italy and Germany.

The Japanese Government is strongly endeavoring to create a national merchant marine. Since 1870, it has subsidized marine construction. The laws of 1896, which regulated anew the bounties for construction and for navigation, have been largely increased. The results obtained are important. The total tonnage of the merchant fleet (steamers and sailing vessels) of the European type, has risen from 77,000 tons in 1879 to 225,000 tons in 1893, and to 919,000 tons in 1901. At this latter date, the number of steamers was 1395, measuring an aggregate of 583,000 tons, of which number 969 measured in the aggregate 577,000 tons, counting among them 67 which measured between 2000 and 5000 tons each, and 22 among them which exceeded 5000 tons each.

Following the example of the western powers, Japan desired a Postal Company, with rapid steamers, and for that purpose subventioned the "Nippon Yousen-Kaisha." This company, just previous to the war with Russia, which disorganized its service, had made six regular departures per month; once a fortnight for Europe and for America, and once every four weeks for Bombay and for Oceanica.

Up to 1883, the part taken by the Japanese flag in the foreign commerce of the country was nearly nothing. Since then, it has increased rapidly, rising to 10 per centum in 1890. and to 30 per centum in 1900.

II.

The progress made by Japan in the short time of thirty years, has indeed been very considerable. She surprises us simultaneously by the importance of that progress, and by the decision and spirit with which her government commenced and continued the economic transformation of the once merely agricultural Japan into the present industrial Japan, equipped with western knowledge, and eager to compete with western nations in the markets of the world. The statesmen of the Revolution had the intelligence to perceive promptly and clearly that a pressing economic necessity compelled Japan to become, and at once, a great industrial power, and that only under this condition could she achieve the political destiny for which she was ambitious.

With its agricultural territory narrowly limited by the physical configuration of its soil, Japan already could not feed its population; and to become an industrial nation was a vital necessity. Only in the development of its manufacturing industries could be found the lacking elements of wealth. And to become a great political power; to play the role that it wished to play in the Extreme-Orient; to make itself the guide in the path of western material civilization for the Oriental peoples; to substitute its protectorate for those which European nations were attempting to impose on them; there was an equal necessity for Japan to become a great industrial power.

But has Japan truly the many indispensable elements for becoming a great industrial power? Will it not encounter obstacles that will obstruct or arrest its course? Is the competition with which it menaces the industries of the Ocident as redoubtable and as near as has been proclaimed by many European publicists?

To form an opinion on that point, an examination must be made of the conditions of Japan relatively to the abundance of its prime motors and raw materials, and to the facility with which it could find the necessary capital for the erection of so vast an edifice. Finally, there must be known the personal qualities and defects of the captains and of the troops of its industrial army.

The subsoil of the Japanese Empire is not yet sufficiently

known to appreciate with any degree of exactness its deposits of coal; but the basins of Kyushu, of Hokkaido, and of Hitachi-Iwaki, the exploitation of which is still far from being complete, assures a sufficient quantity of fuel for a long period. Should this be an error, coal from the Chinese mines can be cheaply procured, and would add considerable resources. Petroleum is another fuel which nature seems to have as liberally bestowed. There is thus a nearly inexhaustible reserve of motive power. Also, the numerous waterfalls along the coasts of Japan will become a precious advantage to its industries.

Japan is less favored in the matter of raw materials. In that respect, there is abundance only of silk; but the location of the country permits it to cheaply procure all the others it may need. Well endowed as regards copper, it appears to be less so as regards iron; but of the latter it could obtain large quantities in China.

The southern islands of Japan still cultivate cotton, but this crop, little remunerative, is diminishing instead of increasing. The Japanese industries import their cotton ordinarily and cheaply from British India and from China. Formosa will, without doubt, be in a few years the center of an important production. As to American cotton, the completion in two or three decades of the Panama Canal, will enable it to be imported under excellent commercial conditions.*

As regards wool, the difficulty of raising sheep, notwithstanding the attempts hitherto made, seems insurmountable. But, on the other hand, the Japanese are no farther than the Europeans from the great producing centers of Australia and of South America.

Thus, in what concerns motive power and raw materials, Japan will encounter no extraordinary obstacles. It has not to fear any impediments that could long arrest its progress.

The question of capital offers a more delicate problem. Japan cannot find among its own people sufficient resources for the building of factories with which to create large industrial enterprise still in embryo, much less can it progress rapidly. The paucity of capital for such purposes is a serious matter, and the money will have to be obtained abroad. Now, it will have to overcome among the capitalists of the

West, an apprehension that exists there at present with regard to such loans, and which nothing has thus far lessened.

It ought, if it does not wish to cripple its industrial progress, to facilitate the coming of capital into the country by granting to foreign capitalists guarantees which the Japanese Legislature has, up to the present time, refused to do.

The question of the personnel—captains of industry, leaders of enterprises, agents, superintendents and foremen, the mass of workmen—is the gravest that confronts Japan in what concerns its development. Herein it will encounter the most serious obstacle, if not the stumbling-block, capable of arresting, or, at the least, of sufficiently hindering, the course of its economic transformation to prevent the accomplishment of its grandiose projects.

Pure science, vast hypotheses, seem repugnant to the Japanese mind. It lacks invention, and closely follows the European lead. But if the Japanese are deficient in genius, they have, on the contrary, the talent of imitation strongly developed, and they very quickly acquire the industrial methods originated by others. And, certainly, the directing classes among them have other qualities more important at the present era than formerly from the economic point of view by reason of the already great development of mechanism, namely, a concentrated power of observation, and the faculty of never becoming discouraged neither by the minutia of details nor by the length of preparation necessary to the successful execution of a preconceived plan. They have shown these two qualities in a rare degree in the organization of their late campaigns against Russia; and their success in them was due for the most part to the attention they had given during the preceding years to the preparations for that war. That they will show the same qualities in industrial pursuits is quite probable. They seem also to have remarkable talents for organization, one of the most important things demanded by modern industry where concentration is continually on the increase, and which places under the direction of the industrial chiefs large masses of workmen whose labor in order to be made efficient and remunerative must be coördinated in the most rigorous manner.

A great defect tarnishes the reputation of the industrial and commercial classes of the Japanese; it is their want of probity.

Notwithstanding, certainly, very numerous exceptions, this defect is so general that their dishonesty in business affairs has become a proverb in the extreme Orient; and many European houses which have traded directly with them have suffered heavy losses from this cause. The defect is so inveterate, that there should be created in their commercial schools courses of commercial morality. Such education, as a necessity of foreign commerce, will probably sooner or later come.

The question of labor presents still more serious difficulties than that of the education of those who direct it. The great modern industries need abundance of labor habituated to long hours of work; regular and attentive in the workshop; capable of continued effort; and sufficiently instructed to use in an intelligent manner, the implements and tools, often delicate, intrusted to them, and to understand the importance of keeping them in the good order which they require. This class of workmen Japan has barely commenced to form, and they are yet very far from reuniting the numerous qualities possessed by the workmen of western nations where the great industries flourish.

Carelessness, love of idling, want of application, are the most frequent faults found with the Japanese workman.

This working population, recently arrived from the rural districts, and drawn to the large cities by the hope of higher pay, and of an easier and, above all, more varied life, often deceived by the fallacious promises of the agents sent from province to province to engage them, is of the most worthless kind.

In general, the Japanese workman understands badly the machine placed in his hands; it is still for him a strange thing for which he continues to feel a stupid antipathy. He has not yet acquired a comprehension of its utility or of the numerous advantages to be obtained from its use.

The many deficiencies and insufficiencies of his workmen, nullify largely for the Japanese industrial the advantage of the small pay given by him relatively to the pay received by the European workman. Further, in these latter years, there has been so great an increase in the pay of labor, that an increase of its output does not seem justified, and which, under the influence of the general causes resulting from such an economic transformation and under the pressure of the labor unions, does not appear to be yet arrested. The increase of pay during the last seven years has varied between 50 and 70 per centum.

The increase of socialistic ideas, is also a menace, increasing the difficulties to be encountered by the Japanese industries, and which might signally retard their progress. The labor agitation commenced towards 1882. In 1889, was founded the syndicate of iron workers which to-day contains nearly 3000 members. In 1897, was created the Fraternal Union of Workmen, and in two years it had 6000 members. Then, in 1898, the railroad engineers and firemen formed a union, which, in 1900, engaged in a successful struggle with the companies. The syndicate movement at present very feeble and having only precarious funds at its disposal, is surely destined to follow, as in other countries, the development of the great industries.

Constrained by its economic necessities, and obliged, in order to realize its political ambitions, to become a great industrial power, Japan has no insurmountable obstacle to fear that could prevent its continued progress in the path in which it is so resolutely moving, but it will not realize its progress without halts. Of the numerous impediments, one of the most serious will be the labor question in slackening its speed. The war with Russia, by reason of the loss of men it caused, will have an analogous effect. The European industrials should, nevertheless, endeavor to maintain in the Japanese markets a more and more active competition on the part of their own national industries.

Will Japan, after the example of the Western nations, have recourse to political tariffs for hastening the development of its industries? Up to the making of the recent Japanese treaties it has not been at liberty to inaugurate a tariff protection system. These treaties have given this liberty, but they still limit its extent as regards numerous products, which, however, are the most important ones, and which have been made the objects of a special tariff. What will happen at the expiration of these treaties made to last a dozen years? Nothing in this respect can be forseen, but if, in this case, Japan should follow the example of the Western nations, there would be no reason for astonishment.

The Japanese industries in order to render the services to their country which are expected from them, must seek foreign outlets. On the doors of what markets with the European products knock in this new competition? Naturally, on the doors of the markets of the extreme Orient, as those will be the first encountered, and the Japanese consider that these markets will always and exclusively belong to them. A Japanese writer looking at the economic future of his country, concludes as follows: "The domination of the Pacific Ocean is now the object of all the commercial nations, and China is the El Dorado of the Orient." On to China! will, therefore, be the word of command given by the politics, the manufactures, and the commerce of modern Japan.

The struggle for these markets will be difficult for the western industrials against their Japanese competitors. The latter will have the great advantages of the proximate geography, and of the similarity of race and customs. They will be more apt to divine the needs of these Asiatic populations, to stimulate them, to create new wants, and to satisfy them in respect to their habits and even to their manias. The more so, as Japan has already commenced this commercial invasion of China.

Will Japanese competition be one day found in the markets of the West? This would be a disquietude which at the present time is without cause. If the industrial development and the commercial expansion of Japan be injurious to certain European industries, they will, on the contrary, be beneficial to certain others.

During a long time yet, Japan must obtain from Europe the greater part of its machinery and large tools; and, in measure as its people, in consequence of its economic development, become richer, they will consume a greater quantity of European products of superior quality to what the Japanese industries will be capable of manufacturing.

Examined closer, the Japanese peril singularly diminishes in importance. The Japanese industries could not progress with the rapidity assumed in the attempts that are sometimes made to frighten us with it. They will encounter more than one formidable obstacle which will diminish their speed. Their chiefs will have in the education of the different classes of their workmen a laborious and extremely difficult task, and they will not be able, in the course of their growth, to avoid economic and financial crises which will cause them serious perturbations. On the other hand, the field of their competition with European industries must necessarily be a limited one.

EXPLORATION OF THE CANADIAN ROCKIES.

The work of the expedition dispatched by the Smithsonian Institution of Washington, D. C., to the Canadian Rockies and Selkirks, under the direction of Prof. W. H. Sherzer, of the Michigan State Normal School, is described in the report of the late Dr. S. P. Langley for the year ending June 30, 1905. The expedition had a successful season's work on the glaciers along the line of the Canadian Pacific Railway. A selection was made of those five glaciers which are most accessible to the student of glacial geology, and these were found to exhibit the characteristics of glaciers throughout the world. Four or five days of comfortable railway travel places an investigator in the midst of snow-fields rivalling those of Switzerland, and the ice bodies descending from these fields may be studied from modern hotels as a base, and a horse may be ridden to the feet of the glaciers studied by the expedition. So far as is known, there is in this district the most magnificent development of glaciers of the Alpine type on the American continent, and the purpose of the survey was to gather as much information as possible concerning them. Many photographs illustrating the details of glacial structure were obtained, and a full report of the expedition may be expected later.—Nature.

SMOKE SUPPRESSION IN NEW YORK.

The Anti-smoke League recently formed in New York City, is not designed, its patrons say, to prevent the burning of soft coal, but to insist on proper firing and to abolish smoke as far as possible. One firm affiliated with the league says that formerly it was able to manufacture in New York City the most delicate silk fabrics, with a very small percentage of damaged goods. For the last three years, and particularly for the last two years, it has become almost impossible to manufacture delicate shades, on account of the impregnation of the air with soft coal smoke.

THE FIRST TURBINE STEAMSHIP built in the United States was launched at the Roach Shipyard, at Chester, Pa., on Saturday, April 21. The vessel is the Governor Cobb, and is intended for the Boston-New Brunswick trade of the Eastern steamship Company of Boston. Its length is 290 feet and the width 51 feet. The motive power is Parson's turbines, which were built by the W. & A. Fletcher Company, Hoboken, N. J.

AN IMMENSE DOCK nearly half a mile in length, 1000 feet in width, and covering an area of thirty-four acres, has recently been completed at Cardiff, Wales. It is designed to accommodate the largest vessels afloat. About 1500 workmen were employed for seven years in its construction.

MARCH EXPORTS of all kinds of merchandise totaled $145,522,342, being the largest for March ever reached. The imports were valued at $133,625,066, which is the largest total for any month in the country's history.

Influence of Benjamin Franklin Abroad.

By Victor Straus-Frank.

All matters, details, however minute, pertaining to Benjamin Franklin are of the greatest import and necessity to young and old, to poor and rich, to American, European, Asiatic. Franklin's life and work are as salt to humanity. Everybody must seek at some time in life either advice or solace, often both, in the perusal and meditation of his books. All, printer, statesman, economist, farmer, in fact, every citizen.

Yet Franklin's work is not sufficiently known, popularized. His writings, most of them, should be placed in the hands of each child at the same time as the Bible. In France many a scholar learns to read out of "Bonhomme Richard." Is not Franklin one of France's greatest "citoyens?" The "demi-frère" of Voltaire.

The famous French National Library (Bibliothèque Nationale) is particularly proud of its documents and books by and concerning Benjamin Franklin. Its catalogue boasts of no less than 250 manuscripts, documents, books, referring to America's Solon.

The Franklin bi-centennary, which was celebrated last April in Paris with great pomp, has resulted in a general Franklin "book-craze." This can but benefit the nation, the world at large. But a few years ago a parcel of "Bonhomme Richard" were found in a wooden chest at Cherbourg, thrown up by the sea. The chest had been thrown overboard from the "Alabama" during the encounter with the "Kearsarge." Maybe a superstitious sailor deemed it wise to rid his sinking ship of the "Devil." At Bordeaux, valuable engravings referring to Franklin's sojourn in France have come to light. Museums, galleries, are ransacking forgotten, dusty attics in hopes of discovering Franklin relics. Franklin is *à la mode*. A century hence, his memory will shine with still more lustre.

That Franklin's reputation as a model-citizen, a great philosopher, a great statesman, a genius, is universally recognized may be gauged by the fact that many of his works have been translated into French, German, Dutch, Finnish, Swedish, Italian, Spanish, Portuguese, Turkish, Greek, Gaelic, Brittany dialect, Japanese, Chinese, etc. I have failed to trace a Russian translation.

Where Franklin's works have found their way, there Civilization has progressed.

Many of his most important works remain to be translated. The Bibliothèque Nationale of Paris counts over 250 works by the illustrious American, yet it is far from satisfied. Many important works are missing. Maybe, the actual Franklin book-hunt will result in important discoveries. It is to be sincerely hoped. Meantime the following list of authors of works referring to Benjamin Franklin to be found at the French Library is not without interest:

- Jared Sparks.
- John Bigelow.
- P. Fievet.
- A. Liégaux-Wood.
- Em. Fenard.
- G. Munro.
- F. Lancelot.
- Edouard Laboulaye.
- P. A. Changeur.
- Dr. Stuber.
- J. G. Rosengarten.
- Barbeu-Dubourg.
- De La Mardelle.
- C. Farine.
- F. Lock.
- L. Bessière (Tabarly).
- G. D'Adda.
- L. François.
- Stevens.
- W. T. Franklin.
- E. du Chatenet.
- Emile Deschanel.
- A. Belot.
- Francis N. Thorpe.
- Ed. Robins.
- Ernest Choullier.
- Albert H. Smyth.
- A. C. Renouard.
- Le Compagnon de Simon de Nantua.
- Hannedouche.
- G. Piegnot.
- Francisco Ladislao A. D'Andrada.
- O. Hubner.
- D'Alibard.
- Karl Berdelli.
- E. Montroulez.
- Bernardo Maria Calura.
- Reschâd.
- Belval.
- Condorcet.
- H. de Triquetti.
- Paul de Lascaux.
- Mignet.
- Dr. Viger.
- Alph. Levray.
- James Parton.
- M. Seitte.
- Mme. Gustave Desmoulin.
- A. Genevray.
- E. Hale.
- W. A. Wetzel.
- John W. Jordan.
- Sydney G. Fischer.

and the list is not complete.

The Franklin book-lover casts a covetous eye on the many rare editions by his countryman. Many, priceless. Their place should be in America, but, a great consolation lies in the fact that they lie in the world's greatest library.

Paris, March, 1906.

The Social and Domestic Life of Franklin.
By Agnes Irwin, LL. D.
Dean of Radcliffe College, Cambridge, Mass.

[Read by title at the stated meeting of the Institute, held Monday, January 21, on the occasion of the 200th anniversary of Franklin's birth.]

It is not easy to say anything about Benjamin Franklin, and it is almost, if not quite, impossible to say anything that has not been said before, and said over and over, for more than a hundred years. He is as well known to us as one of our own contemporaries, opportunity and material for study are both ample, and he himself is not elusive nor inaccessible; his character holds no great surprises, no perplexing depths, no subtle incongruities; he has his own secrets and keeps them very well, but they are not dark secrets of personality. He is human, and he is many-sided, but he is not complex. It is because he has so many sides and so much power, that he makes us wonder at him. "Strange that Ulysses does a thousand things so well!"

What made him what he became? What made him what he was? Where did he come from? He was in no sense a typical New Englander; the New England conscience was not in him; there was something in his temper alien to the true New-England spirit. He was born in Boston, and his mother was a Folger from Nantucket; but Nantucket is a place by itself and only geographically a part of New England, and Franklin's father was an Englishman, born, bred and married in England, who emigrated with his wife and their three children and settled in Boston about twenty-five years before Benjamin Franklin was born. It has sometimes been said that he was like the Folgers, his mother's people; he was certainly like his uncle, Thomas Franklin, who died just four years before his nephew, Benjamin, was born, on the same day of the same month. "Had he died," said William Franklin to his father, "just four years later, one might have supposed a transmigration of souls." Franklin's rise in life remains one of the endless questions of heredity, environment, opportunity. An obscure family of freeholders in an inland county in England keeps on the even tenor of its way for more than two hundred recorded years; a member of the family emigrates to Massachusetts and works unremittingly at his humble trade of soap-boiler and tallow-chandler in order to maintain his large family of seventeen children. Of these seventeen, one, the fifteenth child and youngest son, writes his name high in the brief list of the unforgotten dead. Of the thirteen children who sat round their father's table at one time, and who all grew up and married, this youngest son was the only one who rose to distinction. The others did well enough, or stayed where they were. He must have been a clever little lad, for he learned to read so young that he "could not remember to have been ever without this acquirement," and his father destined him for the church, encouraged by his friends, who assured him that the boy would certainly become a man of letters. The father's brother, Benjamin, promised to give his namesake all his volumes of sermons. What a waste that would have been! He went to the grammar-school at the age of eight, and did very well, going up from the middle to the head of his class, from one class to another, but his father renounced his projects for the child on account of the expense of a collegiate education, and at the age of ten Benjamin Franklin left school and began to work for his father. We know very little of his early years, and all we know he has told us. He was a boy, very much like all other boys, and he lived the life of a boy on the water side, learning early to swim well and to manage boats, and he had a hankering for the sea, which may have been the desire to escape from home, so common in boys, which nearly made a privateer of Franklin's own son, or it may have been that appeal to the imagination of every coast-born New Englander, made by the wind from the sea. "A boy's will is the wind's will and the thoughts of youth are long, long thoughts." He was "generally a leader among the boys, and sometimes led them into scrapes," as in the case of the wharf, built with other people's property; and in a boat or canoe, he was "commonly allowed to govern, especially in any case of difficulty." His dislike for his father's trade, his passionate love of reading, his differences with his brother, the flight from home, the new life in a strange land—we know it all by heart, for he has told us.

The Autobiography is the one great source of information

about Franklin during the early years of his life. There are few letters, fewer traditions, no legends. His tie with his own family was never very strong, and he writes to his sister Davenport (about the year 1727), "Your kind and affectionate letter was extremely agreeable to me, and the more so, because I had not for two years before received a line from any relation, my father and mother only excepted. * * I should be mighty glad of a line from [Sister Mecom] and from Sister Holmes, who need be under no apprehensions of not writing polite enough to such an unpolite reader as I am. I think if politeness is necessary to make letters between brothers and sisters agreeable, there must be very little love among them. * * * Dear sister, I love you tenderly." Of this same sister Davenport, he writes to sister Mecom (19th June, 1730), "Yours of May 26th I received, with the melancholy news of the death of sister Davenport, a loss, without doubt, regretted by all that knew her, for she was a good woman. Her friends ought, however, to be comforted that they have enjoyed her so long, and that she has passed through the world happily, having never had any extraordinary misfortune or notable affliction, and that she is now secure in rest, in the place provided for the virtuous." He hears "the affecting news of his dear good mother's death" with the like decent composure, and, if he needs consolation, he finds it in the thought that "She has lived a good life, as well as a long one, and is happy." His letters to his mother are filial and kindly, with an occasional evidence of affectionate care for her comfort—for example, "the moidore enclosed towards chaise hire, that you may ride warm to meetings tbis winter;" and he takes pains to tell her about her grandchildren just the kind of news a grandmother wishes to hear. (16 Oct, 1749) "Your granddaughter is the greatest lover of her book and school of any child I ever knew, and is very dutiful to her mistress as well as to us." (Date uncertain) "As to your grandchildren, Will is now nineteen years of age, a tall, proper youth, and much of a beau. He acquired a habit of idleness on the Expedition, but begins of late to apply himself to business, and I hope will become an industrious man. * * * Sally grows a fine girl, and is extremely industrious with her needle, and delights in her work. She is of a most affectionate temper, and perfectly dutiful and obliging to her parents, and to all. Perhaps I flatter myself too much, but I have hopes that she will prove an ingenious, sensible, notable and worthy woman, like her Aunt Jenny. She now goes to dancing school." To his sister Jenny, indeed, he was truly a "loving brother." When he is a young man of twenty, he writes to her, "I always judged by your behaviour when a child, that you would make a good, agreeable woman, and you know you were ever my peculiar favorite.* * * I hear you are grown a celebrated beauty." When he is nearly forty years old, and has made his way in a strange land, she seems to feel "some uneasiness" about his attendance at worship, and his reliance on the efficacy of good works; he takes her admonition very kindly and is far from being offended with her for it. A year or two later, she entrusts her son, Benjamin, to his care, and his letters about Benny Mecom show a perfect comprehension of "the nature of boys" and of the ways of parents. "I have not shown any backwardness to assist Benny," he says to his sister (May, 1757), and indeed he had done for his nephew everything that he asked, except "embrace his quarrels and gratify his resentments." That there were family quarrels and that Franklin took no part in them, is clear: "I think our family were always subject to being a little miffy," he says in a letter written during the last year of his long life, and he tells a story of his kinsmen in Nantucket, which goes to show that if they kept away from him in his youth, they were quite as ready to keep away in his old age.

"By the way, is our relationship in Nantucket quite worn out? I have met with none from thence of late years, who were disposed to be acquainted with me, except Captain Timothy Folger. They are wonderfully shy. But I admire their honest plainness of speech. About a year ago I invited two of them to dine with me. Their answer was, they would, if they could not do better. I suppose they did better, for I never saw them afterwards, and so had no opportunity of showing my miff, if I had one."

His domestic life, when he had a home of his own, a wife and children, was the life of a busy tradesman, with many interests, some small, some great and growing greater. His letters to his wife are affectionate, but not lover-like; they show how implicitly he trusted her, and how faithfully she looked after his affairs for him; they show, too, how little she shared in his intellectual life, and how content he was that it should be so. In 1757, in a letter written from New York on the eve of his sailing for England, he writes to her to send him "the Indian Sealskin Hussiff with all the things that were in it. It will be an acceptable present to a Gimcrack great Man in London, that is my Friend." And he asks for the "two or three little Pieces on the Game of Chess, among my Books on the Shelves. One in French bound in Leather, 8vo. If you can find them yourself, send them. But do not set anybody else to look for them. You may know the French one by the work Echecs in the Titlepage.' From these descriptions we infer that she had very little curiosity and no French. The letters from England are longer and fuller, and leave nothing to be desired. He sends her "a crimson satin cloak, the newest fashion." Did she go to Christ Church in a crimson cloak, and her daughter Sally with her, in the scarlet feather, muff and tippet, that Billy Franklin sent her in the same invoice? Mrs. Franklin went to church, and her husband writes from London: "I had ordered two large print Common Prayer Books to be bound on purpose for you and Goodey Smith; and that the largeness of the print may not make them too bulky, the Christenings, Matrimonies, and everything else that you and she have not immediate and constant occasion for, are to be omitted. So you will both of you be reprieved from the use of spectacles in Church a little longer." At one time he sends a large case and a small box, containing carpets, gowns, silver, china—"to show the difference of workmanship, something from all the China Works in England." For Sally there were two sets of books, The World and the Connoisseur. Sally was very dear to him. He had, in 1757, a long intermitting fever, in which his landlady, Mrs. Stevenson, was very obliging and very diligent, but he writes to his wife, "I have a thousand times wished you with me, and my little Sally with her ready Hands and Feet to do, and go, and come, and get what I wanted." "My dear good Sally whose little hands you say eased your headache." he says in another letter. He hopes that "Sally applies herself closely to her French and Musick and that she continues to love going to church." "Sally's last letter to her brother is the best wrote that of late I have seen of hers. I only wish she was a little more careful of her spelling."

These five years in England must have been the happiest of Franklin's happy life. He says on his return, "I have had a most agreeable time of it in Europe. I have, in company with my son, been in most parts of England, Scotland, Flanders and Holland; and generally have enjoyed a good share of health." He was employed in business of importance; he was well off for his station; he was not unduly homesick, except when he was ill, and even then he was frank enough to write, "The agreeable conversation I meet with among men of learning and the notice taken of me by persons of distinction, are the principal things that soothe me for the present, under this painful absence from my family and friends." For now Franklin was famous. He had made his discovery; he had received the Copley Medal and was a Fellow of the Royal Society; when he went to the Commencement at Cambridge, his "vanity was not a little gratified by the particular regard shown [him] by the chancellor and vice-chancellor of the University, and the heads of colleges."

When he was in France with Sir John Pringle, in 1767, he went to court, and was presented to the King (Louis XV), who did Franklin the honor of taking some notice of him and spoke "very graciously and very cheerfully." In his account of the interview, Franklin makes no reference to the communication from the King to the Royal Society in which the King sends his express thanks to Mr. Franklin of Pennsylvania. He had suddenly become famous, but his head was not turned, though he showed his pleasure in his success, and he made friends everywhere. No wonder that he looked back on the poverty and obscurity in which he was born and bred, and counted himself a happy man. It was the late afternoon of his long day, and the twenty years that followed his return from England, in 1762, were to see the high-water mark of his public career; but his "vogue" had already begun, and the taking and making his likeness, which raged like an epidemic, was just beginning. The political situation made the sky very dark, and Franklin had more than his share of the public burden. His son William espoused the English side, and was estranged from his father; Mrs. Franklin died, and his little Sally married during his absence and not wholly to his mind; old age crept on apace and brought with it infirmities; but he kept on, vigorous, alert, undaunted, cheerful, full of interest, perfectly himself to the end. "The time of his life" was the time in Paris, the memory of which has lately been freshened by John Hay's article in the Century Magazine for January, 1906. He became the fashion, the idol of the populace and the darling of the court; his likeness was taken over and over again, in clay, in marble, in bronze, in oil, in pastel, in prints "(of which copies upon copies are sold everywhere)," until his face was "as well known as that of the moon;" "perhaps few strangers in France have had the good fortune to be so universally popular;" the society he gathered round him in his own house at Passy was absolutely to his liking; it was "the best in France—the true élite," at the time before the French Revolution of which Wordsworth wrote,

"Bliss was it in those days to be alive,"

a society of "princes all by intellect and many of them by birth." To be sure it did not satisfy Mr. Ralph Izard, who asked one day, "Why couldn't we have some of the gentlemen of France?" but Mr. Izard, although "untinctured with asperity upon every subject but one," was like the Nantucket cousins, and thought he could do better than dine with Dr. Franklin.

In this society Franklin played his part to perfection, thanks to "the amenity of wit and good humor" for which even Andrew Graydon gives him credit, and perhaps, thanks to the extreme wariness of his character, which was not more congenial to the feelings of Graydon than to those of Mr. Izard. He achieved the end for which he had been sent, and came back in 1785 to take his share in framing the Constitution. His two grandsons came home with him, Temple, who was his secretary, and his grandson Bache, his daughter's oldest son. The boy had gone abroad with him at the age of "half past six," to use his own phrase after his English had suffered for want of use, during four years at school in Geneva. He was "docile and of gentle manners, ready to receive and follow good advice" and his grandfather "loved him very much." He was bred to the trade of a printer while he was in France. Didot, to please Dr. Franklin, teaching the young Bache at the same time with his own son Firmin. There is extant, in the possession of one of Benjamin's grandchildren, a diary kept by the boy for his far-away mother. I have not seen it for many years, but I remember it as a perfect example of a boy's diary, irregularly kept, incomplete, wholly personal, and full of such obvious truths as "I went to M. Didot's to take a lesson. M. Didot went out, and left us at work. Firmin and I played."

Franklin had written from France, on his way to embark for America:

"I have continued to work till late in the day; 'tis time I should go home, and go to bed." But there was public work yet to do; and he did it until the end was near. He had public business enough to preserve him "from ennui, and private amusement besides in conversation, books, the garden, and *cribbage*." His daughter lived with him and was the comfort of his declining years. His grandchildren were constantly with him, and he took "great pleasure in their cheerful prattle." The younger ones are said to have been spoilt and unruly, and on one occasion, he was irritated to harshness by one of the little girls, but they stood in no fear of him, and he thought them promising. One must think, with Mrs. Logan (Deborah Norris) that "Dr. Franklin must have sensibly felt the difference between the éclat which he enjoyed at the Court of France, and the reception which he met with upon his final return to his native country. The mass of the population of Pennsylvania was, as it has been ever since (and may I not say ever was?) decidedly democratic; but there was a contrary spirit then dominant and thinly diffused over the surface of society which rejected the philosopher because they thought he was too much of that popular stamp. The first Constitution of our State after the Revolution, which was his work, was disliked; and I well remember the remark of a fool, though a fashionable party-man at that time, that it was by no means "fashionable" to visit Dr. Franklin. * * * Foreigners of the first distinction thought themselves happy in obtaining such a privilege, and a few of his old tried friends yet remained to cheer the evening of his eventful life."

Truly an eventful life; truly a remarkable person. "But so it sometimes is, a True Genius will not content itself without entering more or less into almost everything, and mastering many things more, in spite of Fate itself."

(Stated Meeting, held Wednesday, March 21, 1906.)
Improvements in the Diffraction Process of Color Photography.
BY HERBERT E. IVES.

The diffraction process of color photography, invented by Prof. R. W. Wood, of Johns Hopkins University, in 1899, is an application of the well-known three-color method of reproducing colors by photography. This method depends primarily upon the observations of Young, Helmholz and Clerk Maxwell, that all the colors of the solar spectrum may be counterfeited to the eye by mixtures of three narrow bands of color from the spectrum, these colors are *red*, near the Fraunhofer line C; *green*, near E, and *blue*, near F. For instance, red and green mix to give the eye a sensation of yellow indistinguishable from the true yellow of the spectrum; red and blue mix to give purple; and the three colors acting together produce a white whose difference from ordinary white light can be detected only by analysis with a spectroscope. What applies to spectrum colors applies equally well to the varied hues of nature. The coloring of such an object as a basket of fruit can also be duplicated to the eye by mixtures of the three primary colors. The tint of an apple, by a large proportion of red, less of green and blue; of a lemon, by nearly equal parts of red and green; of grapes, by a large proportion of blue.

The three-color process can be reduced to two problems; first, the production of three photographic negatives, each of which shall be an exact record of the amount of one of the primary colors requisite to mix with the others and counterfeit to the eye the color of the object photographed; second, some means of furnishing each record with its appropriate color and combining it with the others.

The solution of the first problem has been arrived at from experimental quantitative determinations of the mixing proportions of the primaries to produce the other colors. From these determinations three-color screens can be prepared, which, when used with suitable photographic plates, will yield three (black and white) negatives, each having the desired destitution of light and shade to form a record of one primary color. The negatives thus obtained are the basis of all three-color reproduction methods.

Numerous means have been suggested and tried for combining the three-color records with their corresponding colors. They may be placed in a triple lantern, each illuminated with its proper colored light and projected, superposed, upon a screen. The superposition may be effected by a system of mirrors, as in the Kromskop; by the use of three thin transparent films properly colored; by triple printing on paper, after the manner of much of the present-day magazine illustration.

A process which must be noted somewhat in detail because of its direct bearing on the recent development in diffraction color photography is the so-called Joly process*. Combination of the colors is effected in this by breaking up the three-color records into narrow lines, arranged in succession, a line of the red record, a line of the green, a line of the blue, and so on repeating across the picture. This triple record, whose lines should be close enough together to be indistinguishable by the eye, is mounted over a triple ruled color-screen,--a line of red pigment, a line of green, a line of blue, similarly spaced to the lines of the picture. The result, if the lines are fine enough—a condition never yet attained in the actual working of the process,—is that the eye blends the lines to form a structureless color picture in the form of a transparency.

The diffraction process, which is the subject of this paper, departs widely from the other methods. Its distinguishing feature is that for the production of the primany colors to view the records use is made of the diffraction grating, that is, of a transparent polished surface, usually of glass, ruled with fine parallel straight lines, several thousand to the inch. It is the property of a diffraction grating that if a bright line or point of light is viewed through it, not only will the light source be seen, but spread out to either side will be a series of spectra, those nearest the source being called spectra of the first order, the next, of the second order, etc. If the number of lines to the

---

* First published, as a matter of fact, by Louis Ducos du Hauron in 1869.

inch on the grating be increased the spectra are thrown farther from the central image, and vice versa.

The power of a grating to produce color is taken advantage of in the following way: Suppose we have a convex lens forming an image on a screen of a bright source of light, such as a gas flame. If the eye is placed where the image is formed the lens is seen uniformly and brilliantly illuminated. Suppose now a diffraction grating is placed over the lens. In addition to the image formed as before there will be produced a series of spectra. If the eye is placed in one of these the lens will, as before, appear illuminated, not, however, by light of the color of the source, but by the color of light striking the eye.

If now we can make one of our color records in the form of a diffraction grating of varying strength to correspond to the desired differences in the amount of the primary color, and

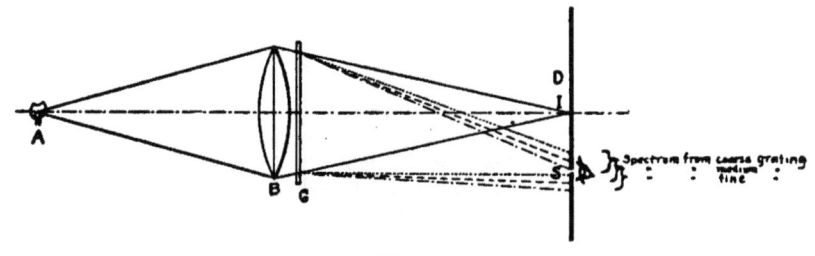

Fig. 1

place it over a lens, points can be found in the lateral spectra in which the lens (and the grating in coincidence with it) will appear as a colored picture. Further, since, as we have seen above, the distance of the spectra from the central image depends upon the fineness of the grating spacing, it is a simple matter to choose three gratings, one of which will send red to a chosen point, the second green, the third blue. Hence if we can make the three primary color records in the form of three diffraction gratings of three properly chosen spacings, each may be seen in its proper color by placing the eye in one of the diffraction spectra formed as above described.

In Fig. 1 we have represented the conditions for viewing diffraction color pictures. A is a source of light, B a convex lens, in front of which are three gratings G. On the screen D fall the central image I, and three spectra (only the first order spectra on one side are represented) so placed that the red of one, the blue of another, and the green of the third are superposed on the slit S, at which the eye is placed.

In Fig. 2 we have represented diagramatically a diffraction color picture of a red flower with green leaves on a blue ground. The coarse spacing of the lines in the flower represents a grating to send red light to the eye, say 2400 lines to the inch, the medium spacing of the leaf one to send green to the eye, say 3000 lines to the inch, the fine spacing of the background one to send blue, say 3600 lines to the inch. Mixed colors would be given by two or three gratings acting together.

To produce such gratings by photographic means the bichromated gelatine process, which lends itself well to the copying of minute structures, was used. In contact with a surface sensitized in this way was placed a glass grating; the image of the corresponding color record was then projected upon the surface

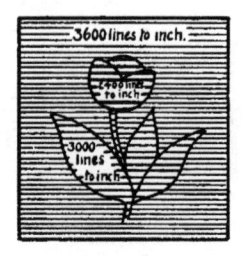

Fig. 2

for a sufficient time to give a full exposure. The grating was removed, another substituted and exposed under its corresponding color record, and so with the third. In this way all three grating pictures were printed, one on top of another,* forming a picture which by diffused light was transparent and quite invisible, showing its color only when viewed with the proper combination of lens and bright source of light. From the pictures made in this way copies could be made by simple contact printing on bichromated gelatine. Since a direct copy of a grating is still a grating, *i. e.*, a series of lines, the process is a positive one and copies are not reversed in light and shade as in making copies of ordinary photographs.

It is obvious that quite apart from its scientific interest the

---

*In practice it was found impossible to get three impressions on one gelatine surface and so two were made on one surface and the third on another, the two surfaces being afterwards placed in contact.

diffraction process promises very real advantages. For instance, the colors used are beyond question pure spectrum colors, and so there is no need to depend on dyes or colored glasses; also the ease and cheapness with which copies can be made places it in a class by itself among three-color processes. So perfect indeed did the process seem theoretically when first published that there was every reason to expect results fully comparable with the best of other methods.

This early promise was not fulfilled. A few pictures were obtained, interesting as scientific curiosities only. No dependdence could be placed in the results; some colors reproduced well, others did not; occasionally a good picture would be made, but the same procedure applied to another subject brought no success. Six years after its publication the process had made no progress and seemed fated to rank as a failure.

Last summer, through the courtesy of Prof. Wood, the writer was loaned a number of diffraction gratings, ruled on the Rowland dividing engines at Johns Hopkins University. Experiments with these revealed a fundamental defect on the above-described mode of making diffraction pictures. By finding means to overcome this defect results have been obtained of a remarkable degree of perfection.

The defect referred to is that the three gratings, in order to get their joint effect, were *superposed*, being, as we have seen, printed one on the other. In so doing the assumption was made that the effect of superposing gratings was to add their separate effects. As a matter of fact, additional, disturbing effects are introduced, partly due to the inability of the gelatine surface to take several grating impressions without mutual blotting out, and partly—chiefly, in fact—to the forming of a new compound grating. That is, if two gratings of different spacings are superposed, the two spacing periodically get in and out of step with each other, and this new periodic structure forms itself a diffraction grating. The new grating then forms its own series of spectra, which subtract light from the original ones. Therefore when the two gratings are superposed, the eye, instead of receiving a double quantity of light receives much less than the double quantity. Even more serious than this loss of light is the fact that the new spectra due to the two gratings together frequently fall in such a position as to introduce *false colors.* This is well illustrated by taking two gratings of different spacing and placing them on one another at right angles. Two sets of spectra will be formed, one by each grating, and parallel to it, and, in addition, a number of diagonally disposed spectra. As the gratings are turned into the same straight line all the spectra turn, and the additional diagonally-placed spectra take up positions between the spectra formed by the original gratings. Consequently, while the eye may receive red from one grating and blue from another, one of the spectra due to the two together may send some other color, such as green. This case actually occurred frequently, a pink rose reproducing as green, and red and blue color discs superposing to give green instead of purple.

These observations made clear the necessity for some method of obtaining the effects of the three gratings other than by superposition. It was at once seen that this could be accomplished by a procedure similar to the Joly process, namely, by having the grating elements in narrow juxtaposed strips. Some experiments had already been made by Prof. Wood with Joly pictures, not, however, with the specific purpose above mentioned, but rather to illustrate to possibility of making such pictures with very much finer color lines than it is possible to do by ruling alternating colored pigment lines for the observing screen. The mode of procedure involved laboriously ruling a special grating consisting of several lines of one spacing, followed by several of another, and then several of the third, repeating all the way across the plate. The width of each strip of lines was made to correspond to the width of an element of the Joly picture. From this grating a print was made on the special-line picture, which had been previously flowed with gelatine. This in turn was used to print gelatine copies.

A practical disadvantage of this plan, aside from the use of the special grating, is that one is restricted to the use of original Joly pictures of a certain definite spacing of line, determined by the limitations of the process employed in their production. A much more serious defect arises, however, in this way: The "Joly lines" if made, as they should be made, several hundred to the inch, themselves form a diffraction grating, which, as it is parallel to the three principal gratings, forms spectra superposed on those depended on to reproduce the colors of the object. This is quite as serious a defect as that arising from superposed gratings, and is sufficient to condemn the procedure.

From a consideration of these various difficulties it followed that some means of breaking the picture up into lines was imperative, and that that means should not involve the use of a special grating, nor of special Joly original pictures, difficult to obtain, and, most important of all, the narrow color strips or Joly lines must be arranged in some way so as not to give disturbing grating effects.

All of these ends were achieved by the following procedure:

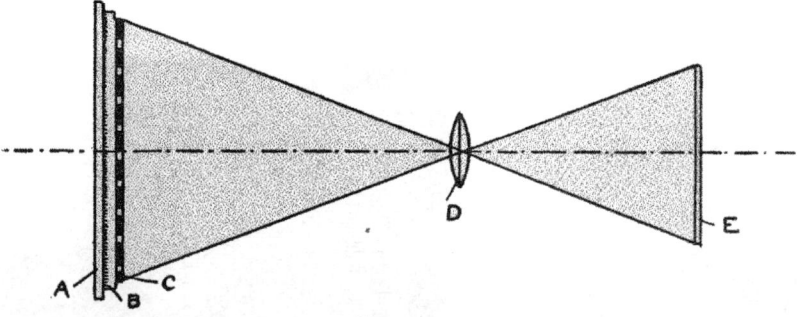

Fig. 3

In Fig. 3, which represents the method of making the improved diffraction pictures, A is the bichromated gelatine plate, rigidly fixed in position; B is a glass diffraction grating; C is a line screen, ruled with at least two hundred lines to the inch, with the opaque lines twice the width of the transparent;[*] D a lens, and E a positive color-record to be copied. The latter is an ordinary three-color positive containing no lines or structure,[†] and the grating is an ordinary continuously-ruled one. With say the red record at E and the corresponding grating at B, an exposure is made, resulting in a series of narrow strips. A second positive is then placed at E, the corresponding grating at B and the ruling C moved the midth of a transparent portion. A second exposure is then made, the opaque

---

[*] The opaque line screens were ruled by Mr. Max Levy, to whom, for his interest and generous assistance, the writer is greatly indebted.

[†] Positives from negatives made for the Kromskop were used.

lines shielding the previously exposed surface, and a similar treatment given to the third positive. There results finally a picture made up of alternating strips of three different gratings.

To eliminate the grating effects of the narrow strips of gratings considered as lines, the device is used of making the strips (Joly lines) run at right angles to the diffraction grating lines, so that the spectra produced by them are thrown off in another direction and do not enter the eye. Although the device is simple it is of extreme importance, and its adoption is rendered possible only by the plan described for making the pictures. The difficulties in the way of ruling a special grating with the three gratings disposed in a similar manner are practically in-

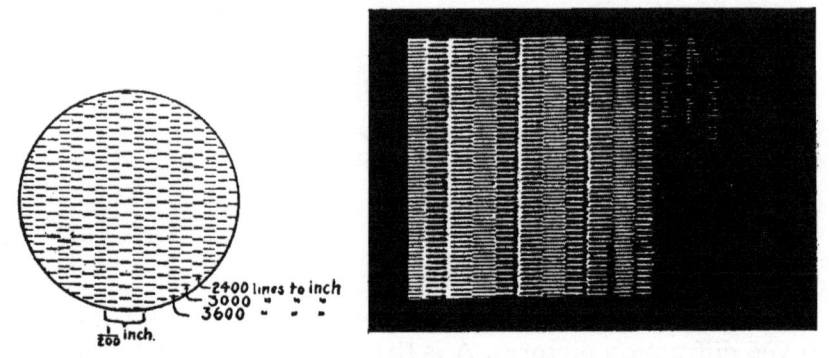

Fig. 4

superable. It is obvious that the strips of grating can be made as narrow as desired, easily narrow enough to be indistinguishable as such by the eye.

Fig. 4 gives an idea of the appearance of the finished picture under the microscope. The short, fine lines are the diffraction grating lines furnishing the three primary colors; 2400 to the inch for the red, 3000 for the green, and 3600 for the blue. The broad strips at right angles to the grating lines constitute the "Joly lines," of which there should be at least 200 groups of three to the inch.

When viewed with a lens and bright source of light the pictures made in this way are entirely free from the formerly-obtained defects. The colors are pure and brilliant, and, unlike ordinary Joly pictures, the color lines are too fine to be visible.

The results indeed approach those obtained with the Kromskop.

As a further modification of the original method the writer has found it possible to dispense with three gratings and obtain the colors with a single grating spacing properly used. To do this the source of light must be a rather long slit. Viewed through a grating the slit of course gives long spectra parallel to its length. If now the grating be rotated about, the perpendicular dropped from it to the slit, the spectra move in toward the slit. [The accompanying shift parallel to the length of the slit is compensated for by the slit being long.] So, by suitable rotation any desired spectrum color may be obtained at a chosen point. Starting with a grating of 3600 lines to the inch to give the blue when parallel to the slit, a rota-

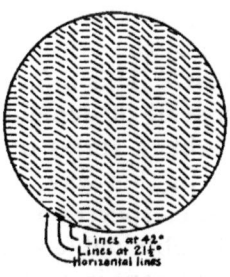

Fig. 5

tion of about $21\frac{1}{2}$ degrees will give the green, of 42 degrees the red. In the absence of suitable dividing engines to rule three properly-proportioned gratings this affords an exact and easy method of securing the three colors. It has the fourth advantage, that in printing copies such difficulties as securing perfect printing contact will affect all three colors alike, which is not the case with gratings of different degrees of fineness.

Fig. 5 shows a portion of a picture made in this way with one grating spacing.*

---

*After working out this idea the writer learned that some years ago Mr. Thorp, of Manchester, suggested the use of a single grating spacing to secure all three colors. Mr. Thorp's plan, however, was to use three sources of light and merely rotate the gratings until they "found" the source and each cleared the source belonging to the other two. He found a rotation of ten degrees convenient. As far as the writer knows this is the first publication of a plan to secure any desired color by rotation through a definite angle to be calculated from the wave length.

With these improvements probably the last word has been said on the diffraction pictures themselves. A very important improvement in the means for observing them, due to the writer's father, Mr. Frederic E. Ives, must be described.

The lens and bright light used by Prof. Wood do not form at all a conveinent arrangement, nor is it desirable to use artificial light. A convenient apparatus, easily set up, not liable to get out of order, and suitable for daylight use became desirable as soon as the pictures are perfected. The instrument about to be described was devised a few hours after the first pictures were obtained, and admirably fulfills its purpose. The greatest difficulty attending the use of daylight is that of getting sufficient light,—the illumination of the sky, toward which an instrument would naturally be pointed, is far from intense enough. This will be appreciated when it is remembered that only a very small portion of the original light is diffracted, perhaps ten per cent. at most. This difficulty has been overcome in a novel manner. Instead of depending on a single slit, as the narrow source of light, a series of slits is used, each furnishing one spectrum. In this way, with four slits, two first order and two second order spectra are utilized, yielding probably three times the light obtainable from a single slit.

Fig. 6 gives the instrument in section. A, B, C, D are the four slits; M a mirror; L_1 and L_2 lenses; P the diffraction picture; and S the slit through which the picture is observed. The lenses of course form an image of each slit at A^1, B^1, C^1, D^1; from each of these images, however, a certain amount of light is diffracted by the picture P; from B and C first order spectra fall on S, from A and D, second order. The use of second as well as first order spectra is a distinct advantage in that, as gratings never give a perfectly uniform distribution of light and color, certain desirable qualities of the picture are found in one order and not in the other, while if both orders are used the resultant evening up of qualities produces particularly satisfactory results.

By disposing the grating lines in a horizontal direction and using horizontal slits as sources the pictures may be viewed by both eyes, a desirable condition for convenience and comfort.

As an instrument the "Diffraction Chromoscope" is simplicity itself. It is, in fact, used much as the old stereoscope.

There are no adjustments; to use, it is merely placed before a window or Welsbach light and the pictures dropped to place. On looking into the eye slit before the introduction of the picture nothing is seen, the inside being perfectly black. The pictures themselves are transparent, colorless, and appear as plain pieces of glass under ordinary conditions of illumination. On placing them in the instrument the colors immediately flash out, a transformation which seems almost magical, affording a scientific demonstration of rare beauty.

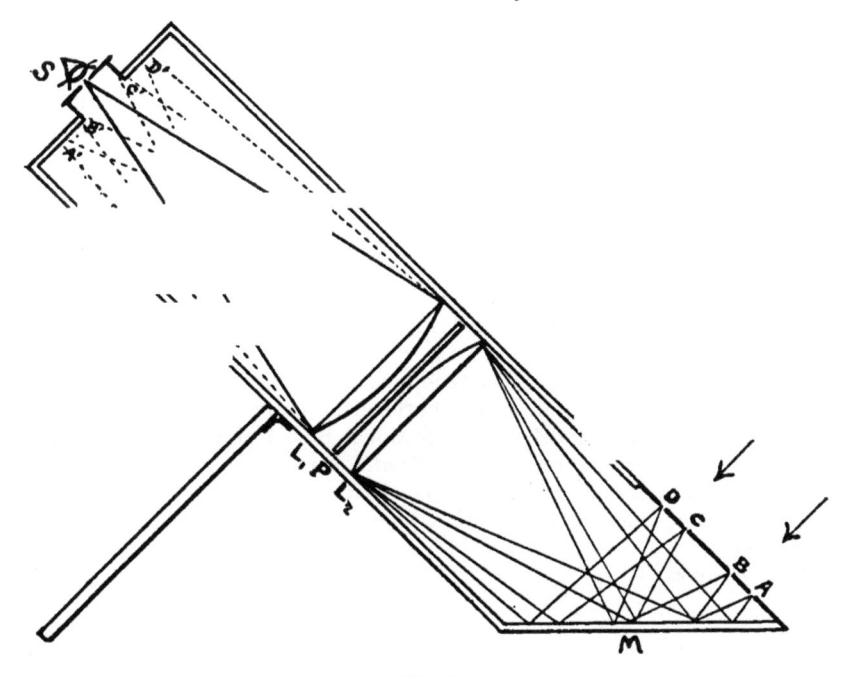

Fig. 6

Aside from the obvious use of the apparatus for scientific purposes it is expected that its simplicity and the perfection of the results will ultimately lead to many important uses. Now that the long standing obstacles in the way of success have been removed the process should develop rapidly. Such further steps as application to lantern projection and means for making the pictures directly in the camera are under consideration.

IMPROVED METHOD OF FIRE PROTECTION.

At the recent annual meeting of the National Fire Protection Association, held in Chicago, the following resolutions were adopted:

WHEREAS, The National Fire Protection Association was formed ten years ago "to promote the science and improve the methods of fire protection; to obtain and circulate information on this subject and to secure the coöperation of its members in establishing proper safeguards against loss of life and property by fire," and,

WHEREAS, In spite of all efforts up to the present time the terrible fire waste of this country has continued uninterruptedly, involving a loss per capita several times greater than in other countries, and,

WHEREAS, Public protection has not kept pace with the growth and increase of valuation in congested centers, and as the business conditions and prosperity of the country are liable to interruption if this increasing loss cannot be checked.

Be it Resolved. By the National Fire Protection Association in convention assembled, that an urgent appeal be and is hereby made to all interested to coöperate in bringing about better conditions by adopting improved methods of construction, by safeguarding hazards of occupancy and by introducing automatic sprinklers and other private protection with private water supplies quite in addition to the Public Fire Service, and,

Be it further Resolved, That we recommend that municipalities pass ordinances involving the adoption of an approved Building Code along the lines of the building recommended by the National Board of Fire Underwriters, and requiring the introduction of Automatic Sprinklers with private water supplies, in buildings of special occupancy and in so-called congested districts, to the end that the danger of sweeping conflagrations may be largely eliminated.

May 23, 1906.

A HUGE TOWAGE UNDERTAKING.

The immense floating dry dock *Dewey*, which has been sent out to the Philippines by the United States Government, has recently passed through the Suez Canal in tow of the 7,000-ton supply ship *Glacier*, assisted by two colliers, one of 5,000 and the other of 6,000 tons displacement. A powerful ocean-going tug, the *Potomack*, is standing by on the voyage as supplementary relief. Before the dry dock quitted America it was tested by placing the battleship *Iowa* and the cruiser *Colorado* in it; this test and others of a different character were carried out successfully. The total distance to be covered is close upon 12,000 miles, and it was estimated that a rate of 100 miles per day would be kept up.

The Plateau Country of the Southwest and La Mesa Encantada (the Enchanted Mesa).

By Prof. Oscar C. S. Carter.

There are many wonders and strange corners in the great Southwest. Arizona, New Mexico and the Southwest generally are fascinating, not only to the geologist, but the archæologist and ethnologist as well. There is an air of mystery and weirdness which anyone, even the least imaginative, is bound to feel. The cowboy in his free life is also under the spell. The average tourist is in a hurry to get to California, the land of flowers, and he speeds rapidly through the arid tract, much of which looks uninviting from the car window. It is the land of cliff dwellings, as well as the ruins of villages in the open country where one sees strange drawings and hieroglyphics on the rocks of bright red sandstone. It is the land of isolated volcanic peaks, enormous lava flows, painted deserts, high mesas and, last but not least, the greatest and grandest scenic wonder of the world, the grand cañon of the Colorado, in which all other cañons would be lost and whose side cañons even, vie with the Yellowstone cañon in depth. There dwell the Hopis and Acomas who build stone houses on bare rocky mesas several hundred feet above the desert, where not a blade of grass grows, and they make the mesas their permanent abode. These peaceful Pueblo Indians till the soil of the valleys under this harsh and trying environment and deficient rainfall. This southwestern region is also the home of their fiercer, restless and migratory brothers, the Navajos and Apaches. The cliff dwellings and other ruins are scattered over the greater part of the Southwest, and the Bureau of Ethnology under the Government has done excellent work in describing many of them as well as the Indian life of the present and past; but the West is a country of magnificent distances, and many places have never been visited. Many dwellings are in an excellent state of preservation, even the mud mortar being preserved. This is especially the case when they are in a sheltered spot, such as under overhanging cliffs where they are protected from the elements. Slowness of decay is characteristic of the arid region, where the air is dry, the altitude high, and the rainfall but eight to twelve inches a year. Eroson in that country produces weird and fantastic effects which we do not see in the East. There are places in Arizona and New Mexico where erosion has carved out some curious and bizarre forms from the sandstones. One of these localities is the valley in which is situated the Enchanted Mesa, in New Mexico. This far-famed mesa is in the southeastern part of the plateau region, very near the edge or boundary line. The plateau region or province contains so many natural and ethnological wonders that a description of it is necessary. It is sometimes called the Colorado plateau, because the Colorado River has cut its way through it. It is an elevated plateau from 4000 to 8000 feet above tide. It includes the southern half of Utah, the northern half of Arizona, Southeastern Colorado and Northwestern New Mexico, embracing an area of over 200,000 square miles. The Plateau Region begins in Utah, just south of the Uintah Mts. This interesting range of mountains is the only one in the United States having an east and west trend; they extend from the Wasatch Mountains to the Rockies. They are unlike other mountains. They have no peaks, just one enormous dome of sedimentary strata (sandstone), and not a piece of igneous rock found in the entire range. The Wasatch Mountains form the western boundary of the plateau region in Utah for a distance of 100 miles. These mountains have a north and south trend. In Arizona there is a low desert area in the south and the high plateau region in the north, a high wall or escarpment separates them. This escarpment enters Arizona from Utah on the north and trends from northwest to southeact. It is really the southern border or end of the Colorado plateau, which extends from Utah southward, often as a series of gigantic steps. The plateau country is the land of the mesa and cañon. The green of the upland and meadow so familiar in the East is wanting here. Instead, we find long lines of plateaus and mesas with steep precipitous slopes, mesas of great length and altitude. The name plateau might signify a level area easy of access and travel, but on the contrary in many parts it is extremely rugged and traveling is difficult. These glaring red and yellow tablelands can only be crossed when you know the country. The strata of the plateau consist of layers of sandstone shale and limestone of Palæozoic and Mesozoic ages, and lie nearly horizontal, with but a slight dip. Great fractures divide the country in the northwest part of the region into blocks of sandstone from ten to twenty miles wide. The region is full of volcanoes and enormous lava flows. A traveler might pitch his tent every night on a lava flow. The San Francisco peaks near Flagstaff, Arizona, are volcanoes nearly 13,000 feet high, snow-capped in August. At Ash Fork there is a bed of volcanic ash over 100 feet thick. These inland volcanoes, 400 or 500 miles from the ocean, are striking exceptions to the rule, that volcanoes are found along the shores of islands and continents. These flows are recent in a geological sense, and took place in Tertiary times. But some of the lavas are so fresh and little weathered that they are undoubtedly very recent and were perhaps seen by the Indians. At one place the lava entered the dwelling of a so-called Cliff Dweller, and it is a well known fact that these people were merely the ancestors of the present Pueblo Indians, and are not so ancient as is generally supposed. They made their homes often on the cliffs and mesas as do their descendants the Hopis and Acomas to-day. It is a well-known fact that the rainfall is slight on the plateau region, not sufficient for agriculture without irrigation, and irrigation is impossible there in some places on account of the enormous depths of the cañons. It is simply impossible to get water from their great depths to the rim. In some places, particularly near rivers like the Rio Puerco, artesian wells reach an underground supply. This is probably a young river. It has not a wide open valley which is a sign of age. It has not even cut a deep cañon as yet, but farther to the north, between the Santa Fé Railroad and the Grand cañon, artesian wells are out of the question, as the underground strata of limestone, sandstone and shale seem to be practically dry in many places. Part of the plateau along the western edge in Arizona, particularly the country that surrounds the San Francisco peaks and Mogollon Mountains, and the Cocanini plateau farther north near the Grand cañon, supports enormous forests of stately yellow pine (*pinus ponderosa*) which are far apart enough for a stage to drive through the forest. There is no undergrowth, but during the rainy season, in July, they are carpeted with wild flowers, such as the Indian paintbrush, pink and violet lupins, yellow primroses, iris, yellow marigolds, butter cups and wild roses, and in the ravines grow the pueriposa lily. This seems a brilliant patch of foliage for an arid region, but it must be remembered that the elevation of the plateau here (7000 to 8000 feet) is sufficient for a greater rainfall and the flowers spring up during the rainy season. It seems an apparent contradiction, but a journey of ten miles to the east of these pine forests, found at the base of the San Francisco peaks, brings us to the depressed area known as the Painted Desert, a veritable Sahara, which no one would think of crossing unless well supplied with water. Enough has been said to make the reader acquainted with the physiography and climate of the country in which the Enchanted Mesa is located. This is necessary because the traditions, habits, and modes of life of the Indians near by are profoundly influenced by their environment. Powell states (in 19th Annual Report of Bureau of Ethnology) "that a harsh environment begets profound faith. This is illustrated by the history of many cults. The Pueblo region was a gathering ground of many faiths; thirst and famine were ever threatening. Ceremonies with growth of corn and for rain and the reverence of beast gods are common." The start for the Enchanted Mesa is made from Laguna, a Pueblo Indian village, situated along the Santa Fé Railroad in New Mexico, sixty-six miles west of Albuquerque. We arrived at Laguna about midnight and were naturally anxious about accommodations.

The operator told us to walk across the track and we would find a half-breed asleep on the porch. It is the only frame house there, he said, all the others are adobe. We followed his directions and obtained first-rate accommodations with Mr. Marmon, who had lived among them for more than twenty years and married a Pueblo woman. They had an interesting family of boys and girls, some of whom were educated at Carlisle. Laguna is not a very ancient Pueblo. It was founded, according to Lummis, about 1699. Although Mr. George Dorsey, the well-known authority of the Bureau of Ethnology, says (Indians of the Southwest) that it was founded at the close of the Eighteenth Century by deserters from Acoma,

On the Road to Enchanted Mesa and Acoma

Enchanted Mesa—Dense Growth of Gramma in Foreground

Zuni and Sia. There are several villages in all that constitute Laguna, but the summer villages are not visible from the station. There is no doubt at all that the Indians were disturbed by the building of the railroad. Many families left and established summer houses and ranches and finally made them their permanent abode. Mrs. Marmon told me that many families had left. The population of the villages is 1100. They make excellent pottery, which they sell to tourists when the day trains stop at the station. Even at night you will find them there. Ten years ago the writer bought some ollas there at night. Although artistically decorated and of graceful shape, and superior to some other potteries, it has not the compactness, coherence and ring of the Acoma pottery, where they have the knack of firing their ovens more successfully. The town is really built of two-storied adobe houses that have flat roofs. It is built on the top of a slight eminence of light-colored sandstone, and the constant tread for many years of moccasined feet has worn a pathway or depression up the slight slope leading to the village. The dwellings are mud houses and not stone. The dwellings of Pueblo Indians differ according to location. In the cañons and on the steep mesas the houses are generally built of stone, which was abundant everywhere, but in the open country the houses were built of mud and pebbles, known as grout or cajon, a puddled mass, which was sun-dried before another layer was put on. Sometimes the structure was rendered firmer by horizontal or upright beams or poles. The women of Laguna are neatly dressed. They wear buckskin moccasins and blue dresses of ancient pattern. The men of Laguna wear a more modern costume and have no use for the more ancient dress. In the journey from Laguna to the Enchanted Mesa and Acoma we were accompanied by Mr. Marmon's son, who was educated at Carlisle. The journey takes from three to four hours, depending upon the kind of conveyance. Acoma, the cliff city, is but three miles beyond the Enchanted Meca, and they both lie in what is called a very wide cañon-like valley, with a more or less level floor in a plateau, in the young state of dissection, as the physiographers say. In other words, it is simply a grassy trough covered with gramma, and it has been carved from the plateau by erosion. It is more than ten miles long and from two to four miles wide, and is surrounded by vertical walls. The few writers who have written about this valley have merely called it a trough. What is needed is an exact topographic term, the mere mention of which will explain it exactly. This may seem a small matter to the layman, but the use of exact physiographic terms is important. Captain C. E. Dutton, in his article on Mount Taylor and the Zuni Plateau, (6th Annual Report U. S. Geological Survey) says: "Did it ever occur to the reader how poverty-stricken the (I will not say English exactly, but) Anglo-American language is in sharp, crisp, definite topographic terms? English writers seem to have gathered up a moderate number of them, but they got most of them from Scotland within the past thirty or forty years. They are not a part of our legitimate inheritance from the mother country. In truth, we have in this country some three or four words which are available for duty in expressing several sources of topographic characteristics. Anything that is hollow we call a valley; anything that stands up above the surrounding land we call a hill or mountain; but the Spanish or Mexican, if you prefer, is rich in topographic terms which are delightfully expressive and definite. There is scarcely a feature of the land which repeats itself with similar characteristics that has not a pat name." This may seem a rather radical statement to make, but Dutton wrote in 1885, and modern physiography was in its infancy then. There were practically no text books on physiography in use in our colleges then. Now there are excellent text books on the subject and our modern geologies have a chapter on physiographic geology. Physiography is practically a new science. It must be remembered also that England has a moist climate and is not at all arid, hence topographic terms which apply to an arid region were unknown there. The Spanish settled the Southwest and gave expressive names to the topographic features of an arid region. The excellent publications of the U. S. Geological Survey, which are a credit to the country and are praised as much in Europe as they are at home, treat physiography from a modern standpoint. This may seem a digression, but it is more important than scaling the mesa. Dutton was undoubtedly right, and Robert F. Hill, of the U. S. Geological Survey, has emphasized this want of physiographic terms in an article in National

Geographic Magazine, Vol. vii, page 291. He was formerly geologist for Texas, where Spanish names predominate, and he soon recognized their efficiency as topographic terms. Among one hundred or more some have become incorporated permanently in our physiographic literature. Among the most common are mesa—a dissected plateau or tableland; cuesta—an inclined plain or declivity, a tilted plain; arroyo—a dry stream course; sierra—a mountain range with serrated crest; cordillera—used in a collective sense for a number of mountain ranges; rio—a flowing river; rio chiquito—little river, for example, Colorado Chiquito, a branch of the Colorado known as Little Colorado; canyon—precipitous gorge cut by a river and other erosive agents; cajon—a box cañyon; plaza—not the popular meaning, an enclosed square in a city. The Mexicans use this term for a valley resembling a canyon in that it is bordered by sub-vertical walls, but differs in that its bottom instead of being narrow is of great breadth. This latter term would seem to describe the valley in which the Enchanted Mesa is located, but the author's description of it as a wide canyon-like valley with a more or less level floor, is sufficient. Throughout the canyon floor we find many curious and bizarre forms of erosion which are noticed in an arid region where we have horizontal layers of sandstone and shale. Towers, peaks, minarets, and castle-like forms, due to differential weathering, are common and remind one of the Garden of the Gods. It is truly an enchanted valley. This has lead Mr. Chas. Lummis, who has done much to popularize the Southwest and its inhabitants in his interesting books to say in rather vigorous language, "That it is the Garden of the Gods multiplied by ten, with ten equal but other wonders thrown in, plus human interest plus archæological value and an atmosphere of romance and mysticism." Mr. Hodge, of the Bureau of Ethnology, quotes this and endorses the above description and says the comparison is not overdrawn. (See National Geographic Magazine, Vol. viii.) The latter part of the description is correct, but in the Garden of the Gods some of the strata have a very steep dip, in fact almost stand on edge, and hence some of the forms of erosion are more bizarre and fantastic. The Acoma Indians who live on the high mesa with precipitous walls, three miles beyond the Enchanted Mesa, have a tradition that the Enchanted Mesa was formerly inhabited by their ancestors. Many years ago a storm arose and an enormous rock near the top was loosened and came crashing down, thus destroying the trail and all access to the mesa, leaving some old women of the tribe to starve. The mesa was accursed Katzimo on account of the storm, and the Indians say it must not be ascended. The credit of discovering this tradition is generally given to Bandelier, whose work among the Indians is celebrated. Mr. Lummis claims priority (in Science, October 25, 1897), and states that he first published a skeleton of the legend in 1885. It was years later, he states, before he could round out the story. In the Catholic World (for January, 1873) is a very entertaining sketch of a journey to Acoma. Whether the writer learned of this tradition from the Indians he does not state. Prof. Libbey, of Princeton, was the first man to scale the mesa, which is nearly 500 feet high. This was done by shooting a rope over with a cannon. The descent was then made by a breeches buoy, such as the life-saving guard use along the coast. He stated that he found no indication of its being inhabited. Mr. *Hodge*, of the Bureau of Ethnology, then ascended the mesa, using spliced ladders. He found sherds of pottery, stone axes, a shell bracelet and a pile of stones, called a cairn. The discussion was taken up by the newspapers and magazines *pro* and *con*, and was participated in by a number of authorities. The author wants to do justice to all of these men, who have done excellent work in their respective fields. So that in order to be fair he will quote their views *verbatim*. Prof. Wm. Libbey, the well-known geographer and traveler, states (in *Harpers' Weekly*, Vol. xxxxi, page 862), "While the men at Acoma were away in July, 1897, piously engaged in cutting timber to renew the crumbling roof of their church, another party of men, headed by two vandals, one was a scientist, Libbey, the other a journalist named Bridgeman, ascended the mesa. We set the cannon up on the eastern side of the southern point of the mesa. The large shot was provided with a projecting shaft and ring. To this was attached a cord, which led to a box where a long line was carefully coiled in such a way that the line left the box without being snarled up. The shot passed up in the air high above the mesa and fell on the western side. We spent two days in drawing ropes over the top of the rock until finally, by the aid of a team, the largest rope, about one inch in diameter, was pulled in place. This rope had a pulley attached to it, through which a smaller rope passed, and this pulley was hauled up exactly to the edge of the cliff. A curious seat was constructed with a board, which was suspended from a block which ran along the top of a large rope. One end of the smaller rope was attached to this block and the other fastened behind the team. The explorer seated himself in a chair and was drawn up to the top of the rock, and then another member, Pearce, was drawn up. We found ourselves upon the smaller end of the great rock and were cut off from the main portion of the mesa. We called for the ladder and climbed the mesa and wandered around for a long time, and when we came back to where we left our companions we called out to those below that the summit had never been inhabited; that there was no trace of houses ever having been constructed up there, and there were no remnants of pottery or fragments of household utensils or implements of any kind, no water-tanks for the storage of rain-water. One object alone looked as though it might have been built by human hands, and that was a small cairn like mass of stones. He spoke of stunted pine and cedar trees growing in scanty masses of earth here and there, but described the rest of the surface as perfectly bare and swept clean by the wind. A few grasses and plants, common on the plains below, were also found, and these, with one gray rat and a number of ordinary lizards, were all the evidences of life to be seen. All of which, he says, merely goes to show that he was in reality blinded by the spirits, and he remarked when he reached the plain once more that the mesa was disenchanted. It was an illusion of his mind that had been unduly excited by his trip up into the air only fit for spirits to breathe."

The following description and argument were written by Mr. *H*odge (see Nature, Vol. lvii, page 450).

"Acoma is the oldest settlement within the limits of the United States. Many of the walls still standing were seen by Coronado in 1540. Even then they were centuries old. The valley of Acoma is described as the Garden of the Gods. It is a level plain, bounded by mesas, and is clothed and carpeted by gramma. The mesas are variegated sandstone rising precipitously 300 to 400 feet. None are so precipitous as Katzimo or Enchanted Mesa, which rises 430 feet from the middle of the plain. This was one of the many wonderful home sites of the Acomas during their wanderings from the Mystic Shipapu in the far north to their present lofty dwelling place. Native tradition is distinguished from myth when uninfluenced by Caucasian contact, and may usually be relied on, even to the extent of disproving or verifying that which purports to be historical testimony. The Acoma Indians have handed down from Shaman to Novitiate, from father to son, in true prescriptorial fashion for many generations the story that Katzimo was once the homes of their ancestors, but during a great convulsion of nature at a time when most of the inhabitants were at work in the fields, an immense rocky mass became freed from the friable wall of the cliff, destroying the trail and leaving a few old women to perish. This tradition in its native purity was recorded twelve years ago (from 1898) by Mr. Chas. F. Lummis, and the same story was repeated by Acoma lips to Hodge while conducting a reconnaissance of the Pueblos in 1895. Hodge examined the talus especially where piled up high at the foot of the southwest cleft, up which the ancient pathway wound. It is covered with numerous fragments of pottery of a very ancient type, some of which was decorated by a vitreous glaze, an art, he states, lost to Pueblo potters. The talus at this point rises to a height of 224 feet above the plain, more than half way up the mesa could be formed only by earth washed from the summit during many centuries. His examination of the trail to a point within sixty feet of the top exhibited traces of what were evidently hand and foot holes that once aided ascent of the trail. He visited Katzimo a second time with Major George H. Pradt, Mr. A. C. Vroman and Mr. H. C. Hoyt. The ascent was made by ladders, and ropes and cameras and surveying instruments were carried with some effort, as the elevation above tide is 6000 feet, and they climbed in the hot sun. They fastened the rope to small pinions, where once was soil now found in the talus below. They soon found a sherd of pottery of an ancient type, plain gray ware, quite coarse in texture with a degraissant of white sand. The top of Katzimo is 2500 feet long. Great cedars grow gaunt and bare or lie prone and decaying. He was forced to the

Indian Pueblo of Laguna, New Mexico

Weatered Castellated Rocks on Road to Acoma

(Carter)

View of Enchanted Mesa from Edge of Cliff at Acoma. Showing Other Mesas in the Distance

Enchanted Mesa—End View

conclusion that the house walls, stone or adobe, were washed from the summit, there being no possibility that a trace remained. Nevertheless the pottery in the talus and on top and the distinct remains of a ladder trail all testify to former habitation. To the Acomas, Katzimo is still enchanted and is a subject in the study of mysticism."

The following forceful argument by Mr. Hodge is found in Harpers' Weekly, Vol. xli, page 1027:

"The fact that we found on the summit two fragmentary stone axes and a portion of a shell bracelet, several greatly worn potsherds of extremely ancient ware and a projectile point would not in itself be regarded as conclusive evidence that the mesa had actually been occupied as a village site; but when we know that the summit is, and evidently long has been, inaccessible to the Indians, that it has been washed by rains and swept by winds for centuries until scarcely any soil is left on the crest, that thousands of tons of soft sandstone have so recently fallen from the cliff, that their edges have not had time to become rounded by erosion; that the typography of the summit is such that not a cupfull of water now remains on the surface, save in a few eroded potsholes in the sandstone, but that it rushes over the precipice on every side in a hundred cataracts; that the rude stone monuments that Prof. Libbey believed might be due to erosion in a spot well protected from surface wash is artificial beyond preadventure, as anyone may judge from my photograph. That well-defined traces of an ancient ladder way may still be seen pecked in the rocky wall at the very cleft through which the traditional pathway passed, and above all the large numbers of ancient pottery fragments and other artefacts in the earthy talus around the base of the mesa, which must have washed from the summit, for they could have occurred there in any other way, we have incontrovertible evidence that the summit of Katzimo or Enchanted Mesa was inhabited prior to 1540, when the present Acoma was discovered by Francisco Vasquez de Coronado, and that the last vestige of the village has long been washed or blown over the cliff. After having seen hundreds of ruins and engaged in excavating dozens of others in the Southwest, I am firmly convinced that there is ample archæological evidence of former occupancy of the Enchanted Mesa by the Acoma Indians, and that their tradition to that effect is substantially true, and this opinion is entertained by every competent authority to whose attention these facts have been drawn."

George Wharton James, who has done much to popularize the Southwest in his interesting books, writes as follows (see Scientific American Supplement, Vov. xlvii, page 19,488):

"The evidence of human presence was found here all agree, but there is a vast difference between evidence of human presence and evidence of a large village or city.

(1). Had Acomas lived on the mesa they would have built houses not of adobe but of pieces of rock.

(2). Had such a city existed the ruined walls would have remained. Large blocks and pieces of sandstone would not have been eroded or washed away. The sloping conditions of the mesa not unusual. He makes a point that it would take a number of houses to store them and these houses would not blow away. Half a score or more of ruined cities exist on mesas storm swept and as exposed as the Enchanted Mesa, and yet the ruins are not swept away.

JAMES'S CONCLUSIONS.

(1). The mesa was undoubtedly the scene many times of human presence.

(2). Worn trail and other evidences clearly demonstrate that Indians have often visited it. These facts ought not to be accepted as conclusive evidence of the truth of the Acoma tradition that their ancient city was located there.

(3). Both Indian and white man are at fault in regard to the exact location of Katzimo, and that further researches will discover it and show far more positive and occular demonstration of its having been the occupied site of a larger city than the so-called Katzimo and Mesa Enchanted of the present discussion have done.

My reasons for advancing this last idea are:

(1). My firm belief in the general truth and reliability of the tradition.

(2). Unsatisfactory evidence adduced in favor of the village occupancy of the mesa hitherto known as Mesa Enchanted.

(3). My knowledge of the possibility of error both by Indian and white man owing to the lapse of centuries in determining the location.

(4). My actual conversation with Indians at Acoma who definitely assert that the scaled mesa is not their Katzimo, and that may be so. Some day they will conduct me to the real, genuine, sole and only Katzimo or Mesa Enchanted, where many ruins are to be found."

From the above statements it is evident that opinions vary, but all must agree that the pottery, ax-heads and other remains prove beyond all doubt that the mesa was visited by Acoma or other Indians before the trail was destroyed. The author has visited the Southwest on different occasions and examined many ruins and pictographs or inscriptions in different localities, is prepared to believe the mesa was inhabited. In fact it would be very strange if such an impregnable Gibralter or fortress isolated in the desert would not be used as a habitation. It was an ideal site, just such a one as the Indians would select. More secure from the migratory Apaches and Navajos than the mesa at Acoma. We are prepared to believe that it was something more than a mere look-out station to warn then of the approach of their enemies as they came up the valley. We believe, however, that the houses were adobe or mud houses, and entirely swept away by the cloud bursts which we have experienced in the plateau region. If the houses were built of stone the walls undoubtedly would be standing, as they do on other mesas and in the open country. Walls of very ancient ruins, such as we have noticed even on the very edge of the Grand Cañon more exposed to erosion than on the mesa.

The statement has been made, "had the Acomas lived on the mesa they would have built not of adobe but of rock."

This does not, necessarily, follow because the houses at Acoma are built of rock and adobe. The old Spanish church is built of adobe, the walls being several feet in thickness. The enormous grave-yard; which took them forty-five years to build, is made of earth brought from the valley below. In regard to the statement that the glazing of pottery is a lost art with the Pueblo Indians, there is room for difference of opinion. The author has examined considerable pottery of the so-called ancient or Cliff Dwellers' type, both fragmentary and whole, and has yet to see a piece which he considers has a true glaze. Briefly stated, the so-called fat clay is essentially silicate of alumina. A lean clay is silicate of alumina plus impurities, such as lime and the alkalies, potash and soda. These impurities come from the feldspars which produce the clay in nature by their decomposition. The tendency of these impurities is to make a clay slightly fusible or less refractory. If the temperature of a furnace or oven is too high, the clay is overburned and may vitrefy slightly on the surface, but we do not believe the temperature of these ovens with the kind of fuel the Indians use is ever sufficient to do this, and if it were possible it would not be a glaze.

Mr. Wm. H. Holmes, of the Bureau of Ethnology (4th Annual Report, page 268, Pottery of the Ancient Pueblos), says:

"A great deal has been said about the glaze of native American wares which exists if at all through accident. The surface of the white ware of nearly all sections received a high degree of mechanical polish, and the effect of firing was often to heighten this and give at times a slightly translucent effect, a result of the spreading or sinking of the coloring matter of the designs."

TOPOGRAPHIC MAPS OF THE GEOLOGICAL SURVEY.

How many persons would like a map of the district in which they live? Surely there are very many who would find convenient such a possession. To the people of many districts in the United States such a map—accurate, detailed, and of moderate size—is available for the small sum of five cents.

The United States Geological Survey is making a topographic map of the United States. This work has been in progress since 1882, and about three-tenths of the area of the country has been covered. The surveyed districts are widely scattered throughout the country. When the atlas is finally completed, every citizen of the United States may procure a federal map of his particular environment.

This great map is being published in atlas sheets of convenient size, which are bounded by parallels and meridians. The four-cornered division of land corresponding to an atlas sheet is called a quadrangle. The sheets are of approximately the same size, the paper dimensions being 20 by 16½ inches. Three scales, however, have been adopted. The largest scale of nearly 1 mile to 1 inch is used for thickly settled or industrially important parts of the country. For the greater part of the country an intermediate scale of about 2 miles to 1 inch is employed. A third and still smaller scale of about 4 miles to 1 inch has been used in the desert regions of the Far

West. A few special maps on larger scales are made of limited areas in mining districts.

The features shown on this map may be classed in three groups: (1) water, including seas, lakes, ponds, rivers, and other streams, canals, swamps, etc.; (2) relief, including mountains, hills, valleys, cliffs, etc.; (3) culture, that is, works of man, such as towns, cities, roads, railroads, boundaries, etc. All water features are shown in blue, relief is in brown, and cultural features are in black.

The sheets composing the topographic atlas are designated by the name of a principal town or of some prominent natural feature within the district.

They are sold at 5 cents each when fewer than 100 copies are purchased, but when they are ordered in lots of 100 or more copies, whether of the same sheet or of different sheets, the price is 3 cents each. Applications should be accompanied by the cash of by post-office money order (not postage stamps) and should be addressed to the Director United States Geological Survey, Washington, D. C.

THE WORLD'S MERCHANT MARINE.

Marine Engineering, of New York, has been comparing the growth of the American merchant marine with those of other nations, and in the course of an interesting article on the subject it is stated that "in no case does the American tonnage exceed 28 per cent. of that credited to Germany. Both France and Italy have a larger tonnage under construction that the United States, the differences running from 13,000 tons in the case of each of the above powers, to 35,000 tons in the case of Italy at a slightly more recent date, at which time even Holland exceeded the American construction by 8,000 tons. Aside from Great Britain." the article goes on, " there is not a single nation in the world which has naturally so great an interest in seaborne commerce as has the United States; not only are the latter's exports and imports immeasurably superior in both bulk and value to those of any other power, but the American coast line is enormous in extent, has many splendid harbors for the reception of ships, offers every facility which nature can offer for the proper development of a mercantile marine. It is simply a question of a difference in wages and in other operative costs between American ships and those of foreign nations. So long as this difference exists it will be manifestly impossible to operate American ships in direct competition with those of other powers, and this condition will persist just so long as American standards of living remain superior to foreign standards of living, and no longer.

"In other words the rehabilitation of the American merchant marine has come to be a question of either providing govermental assistance to shipbuilders and ship owners, or reducing the American scale of living to standards prevailing elsewhere in the world."—Chas. S. Lake, in *Modern Engineer*.

In Memoriam.

JOHN VAUGHAN MERRICK.

J. V. Merrick was born in Philadelphia, August 30th, 1828, and died at his home in Roxborough, Philadelphia, March 28th, 1906.

His parents were Samuel Vaughan and Sarah (Thomas) Merrick. The father's name is indelibly connected with the origin of the Franklin Institute, in 1824, as is fully described in the commemorative exercises of its fiftieth anniversary in 1874. S. V. Merrick's life was one of great activity and usefulness to the community. In addition to creating the works of the Southwark Foundry, he was the first President of the Pennsylvania Railroad. The first gas works in Philadelphia were built under his charge as engineer, in 1835. He was President of the Franklin Institute from 1841 to 1853, and until his death he was deeply interested in the success of this institute.

His eldest son, J. V. Merrick, the subject of this sketch, was graduated from the Philadelphia High School in 1843, being the youngest pupil, with one exception, ever admitted to this school. The same year he entered the works of Merrick and Towne as an apprentice, and was trained in the manual arts of the various mechanical departments. In 1849, he became the head of the firm of Merrick & Son's, and soon was recognized as one of the leading mechanical engineers of the city.

For the following twenty years his energies were devoted to the development of the manufacturing industries of the Southwark foundry. The works were enlarged, and its products further diversified. During the strenuous period of the civil war Mr. Merrick devoted his workshops, as well as his own personal services, to the nation's cause. His experience as a designer and constructor of steam machinery for ships was recognized by his selection by the Navy Department, in 1862, as a member of the Board of Experts to report on naval machinery.

Among the more important of his productions and of his works, which were put into successful operation, was the motive machinery for the following steamships: North Carolina, Cardenas, Quaker City, North America, Phineas Sprague, Alphonso, Continental, and also a number of tug boats for river and harbor service.

Among the prominent war ships built for the U. S. Navy were the machinery for the San Jacinco, Wabash, Yantic, Wyoming, Yazoo, New Ironsides. The latter being the first of our iron clads. The machinery for several revenue cutters and light draught monitors were also produced at his works during the civil war. The works also acquired a high repute for their miscellaneous products, such as pumping and blowing engines, the machinery for gas works, and apparatus used in sugar refining, &c.

He designed and patented a novel form of steam hammer, in which the steam cylinder was embraced in the moving ram, and steam supplied through a hollow stationary piston rod.

In recognition of his experience and ability Mr. Merrick was appointed in 1883, a member of the Board of Experts to report on improvements in the water supply for Philadelphia.

In 1870, owing to impaired health, he retired from active business, and devoted his remaining years largely to philanthropic work.

His zeal and activity in this line of duty were incessant to the close of his career. Always of a deeply religious nature, and recognizing in the church a most potent instrument for elevating the morals of the community, he advocated its cause with his efforts and his means,—by the most practical and efficient methods.

In 1873, he was instrumental in founding St. Timothy's Working Men's Club and Institute, Roxborough, an organization designed to furnish recreation and instruction, for a small fee, to that large class, who most need it. He devoted much of his time and energy to the interests of this institution, which still exists and has proved most beneficial to the neighborhood. In 1890, he and his wife founded St. Timothy's Hospital and House of Mercy, Roxborough, giving the land and the original buildings, together with a substantial endowment for its support, and he labored constantly for its development until his death.

The following list of offices held by Mr. Merrick, and in most of which he was an active worker, are indicative of his active and useful life:

Director of the Zoological Society since its origin, and Vice-President since 1886.

Trustee of the University of Pennsylvania since 1870; was senior trustee at his death; member of Committee on University and chairman of the Standing Committee on "Department of College and Philosophy."

Incorporator of the "Wagner Free Institute of Science," in 1864. Trustee from 1885 to 1894.

Trustee of the Episcopal Academy from 1874 to 1898.

Member of Board of Managers of the Episcopal Hospital and Chairman of the Building Committee from 1876 to 1900.

Vice-President American Society of Mechanical Engineers 1883-1885.

Delegate to the convention of the P. E. Church diocese of Pennsylvania since 1861, and for twenty-five years one of the four lay deputies to the triennial convention of the P. E. Church.

Since 1870, Trustee of the "Society for the Advancement of Christianity," State of Pennsylvania.

For over thirty years a director of the "Society for the Relief of Widows and Orphans of Deceased Clergy of the P. E. Church."

One of the founders of the "Free and Open Church Association," and President of it since its origin, in 1873.

Mr. Merrick was elected a life member of the Franklin Institute in 1849. Was a member of the Board of Managers from 1864 to 1884. President of the Institute for the three years, 1867-68-69, and a member of the Board of Trustees from 1887 to 1895.

He contributed the following papers to the *Journal* of the Institute.

"Evaporative Efficiency of Martin Boilers".........................lxi, 357
"Friction of Marine Steam Engines"................................liii, 132
"History and Construction of Iron Lighthouses"....................lxi, 145
"Marine Propulsion"...liii, 270, 417
"Mode of Increasing Draft in Marine Boilers"....................lviii, 390
"Notes on Steamship 'State of Georgia'".........................liii, 407
"Particulars and Performance of Steamship 'North Carolina'"....lviii, 411
"Screw and Engine of U. S. Steamship 'San Jacinto'".............liii, 50
"Steamboats on Western Waters".................................liii, 344
"Trial Trip of the Steamship 'Quaker City'"....................lviii, 269
"Use of Cornish Engines for the Water Supply of Cities"......lxxxviii, 229
"City Avenue (Schuylkill)Bridge"..............................cxxxix, 241

At the University Day exercises, held on February 22, 1906, the Trustees of the University of Pennsylvania conferred the honorary degree of Doctor of Science upon Mr. Merrick. In presenting the candidate to the Provost for this degree, Dr. S. Weir Mitchell said in part:

"Distinguished as an inventor and expert in many forms of industrial machinery, the nation owes him a debt of gratitude for the skill, the speed and the perfection shown in the construction of the engines needed for our ships during our civil war and in a time of utmost emergency. The honor we offer is also meant to express our appreciation of Mr. Merrick's thirty-six years of varied services to this University and of the rare personal qualities of which I spare him the gracious enumeration." JAMES CHRISTIE.

HIGH KITE FLYING.

The highest kite ascent yet recorded was made at the æronautical observatory at Lindenberg (Prussia) on November 25th 1905, 21,100 feet being attained. Six kites were attached to oneanother, with a wire line of nearly 16,000 yards in length. The minimum temperature recorded was 13 deg. F.; at starting the reading was 41 deg. The wind velocity at the surface of the earth was eighteen miles an hour, and at the maximum altitude it reached fifty-six miles an hour. The previous highest record by a kite was nearly 1,100 feet lower, and was obtained from a Danish gunboat in the Baltic.—*Model Engineer.*

Book Notices.

Morton Memorial: A history of the Stevens Institute of Technology with biographies of the faculty, trustees and alumni, and a record of the achievements of the Stevens family of engineers. Edited by Franklin De Ronde Furman, M.E., Professor of Mechanical Drawing and Designing; with an introduction by Alexander Crombie Humphreys, M.E., Sc.D., LL. D., President of Stevens Institute of Technology. (4 to., pp. 641, with preface and introduction. Hoboken, N. J. Stevens Institute of Technology, 1905.

This superbly printed and illustrated volume was originally planned as a souvenir to commemorate the twenty-fifth anniversary of the Stevens Institute of Technology. The idea of the volume originated with the President, Henry Morton, who was deeply interested in its publication, but his lamented and untimely death occurred before the work was completed. It appeared later, designated most appropriately the "Morton Memorial."

To all who know the history of this flourisihing and progressive institution, the generous tribute paid to the memory of its first President will be cordially endorsed. President Morton devoted the best years of his life to the development of the institute, and the foremost rank which it now occupies among the technical schools of the country is largely due to his ability as an organizer and administrator, no less than to those personal qualities, which enabled him to attract and enlist the coöperation of men of the highest professional standing to second his efforts in developing the strength and influence of the Institute.

Stevens Institute, to-day, is a monument to his memory—in the words of a resolution adopted by its Trustees—"more enduring than 'marble, or the gilded monuments of princes.'"

The friends of Stevens will find the present handsome memorial volume a complete history of the Institute from it origin. Included therein is a series of chapters devoted to the history of the achievements of the famous Stevens family, several of whom attained high repute as engineers, and one of whom (Edwin A. Stevens) was the founder of the Institute which bears the family name.

Following, are given biographies of the trustees, the faculty and the alumni.

The work is profusely illustrated with admirably executed engravings.

Taken all together, it is a publication highly creditable to all who have contributed to its production. W.

Sections.

ELECTRICAL SECTION: *Stated meeting,* held Thursday, May 17, 8 P.M. Mr. James Christie in the chair.

Present, twelve members and visitors.

Dr. Frederic A. C. Perrine, of New York, read the paper of the evening on "The Value and Design of Water-Power Plants as Influenced by Load Factor."

The paper was freely discussed by Messrs. Hering, Christie, Foster and the author.

The thanks of the meeting were extended to the speaker of the evening, and his paper was referred for publication. Adjournd.

RICHARD L. BINDER, *Secretary*

Franklin Institute.

(Stated Meeting, held Wednesday, May 16th, 1906.)

HALL OF THE INSTITUTE,
PHILADELPHIA, May 16, 1906.

MR. JAMES CHRISTIE in the chair.

Present, seventy-six members and visitors. Additions to membership since last report, eight.

The resignations of Messrs. William H. Lambert and C. Hartman Kuhn from the Board of Managers were reported.

Messrs. Coleman Sellers, Jr., and Samuel M. Vauclain were thereupon chosen to fill the vacancies.

The chairman then introduced Dr. Henry Emerson Wetherill, of Philadelphia, who presented an interesting account of "The Panama Canal," in which he dwelt specially on the the engineering and sanitary features of the work. The speaker's remarks were freely illustrated with the aid of lantern pictures.

Through the courtesy of Messrs. Williams, Brown and Earle, of Philadelphia, a series of lantern views of San Francisco were shown, exhibiting the destruction wrought by the recent earthquake.

Adjourned.

WM. H. WAHL, *Secretary.*

T
1
F8
v.161

Franklin Institute,
Philadelphia
Journal

Engineering

PLEASE DO NOT REMOVE
CARDS OR SLIPS FROM THIS POCKET

UNIVERSITY OF TORONTO LIBRARY

Lightning Source UK Ltd.
Milton Keynes UK
UKHW010623051218
333419UK00010B/903/P